Lecture Notes in Networks and Systems

Volume 18

Series editor

Janusz Kacprzyk, Polish Academy of Sciences, Warsaw, Poland
e-mail: kacprzyk@ibspan.waw.pl

The series "Lecture Notes in Networks and Systems" publishes the latest developments in Networks and Systems—quickly, informally and with high quality. Original research reported in proceedings and post-proceedings represents the core of LNNS.

Volumes published in LNNS embrace all aspects and subfields of, as well as new challenges in, Networks and Systems.

The series contains proceedings and edited volumes in systems and networks, spanning the areas of Cyber-Physical Systems, Autonomous Systems, Sensor Networks, Control Systems, Energy Systems, Automotive Systems, Biological Systems, Vehicular Networking and Connected Vehicles, Aerospace Systems, Automation, Manufacturing, Smart Grids, Nonlinear Systems, Power Systems, Robotics, Social Systems, Economic Systems and other. Of particular value to both the contributors and the readership are the short publication timeframe and the world-wide distribution and exposure which enable both a wide and rapid dissemination of research output.

The series covers the theory, applications, and perspectives on the state of the art and future developments relevant to systems and networks, decision making, control, complex processes and related areas, as embedded in the fields of interdisciplinary and applied sciences, engineering, computer science, physics, economics, social, and life sciences, as well as the paradigms and methodologies behind them.

More information about this series at http://www.springer.com/series/15179

Xin-She Yang · Atulya K. Nagar
Amit Joshi
Editors

Smart Trends in Systems, Security and Sustainability

Proceedings of WS4 2017

Springer

Editors
Xin-She Yang
School of Science and Technology
Middlesex University
London
UK

Amit Joshi
Sabar Institute of Technology
Gujarat Technological University
Gujarat
India

Atulya K. Nagar
Department of Mathematics and Computer Science
Liverpool Hope University
Liverpool
UK

ISSN 2367-3370 ISSN 2367-3389 (electronic)
Lecture Notes in Networks and Systems
ISBN 978-981-10-6915-4 ISBN 978-981-10-6916-1 (eBook)
https://doi.org/10.1007/978-981-10-6916-1

Library of Congress Control Number: 2017955656

Printed on acid-free paper

This Springer imprint is published by Springer Nature
The registered company is Springer Nature Singapore Pte Ltd.
The registered company address is: 152 Beach Road, #21-01/04 Gateway East, Singapore 189721, Singapore

International Advisory Committee

Dr. Nobert Jere, Namibia University of Science and Technology, Namibia

Dr. Pawan Lingras, Professor, Saint Mary's University, Canada
Hoang Pham, Professor and Chairman, Department of Industrial and Systems Engineering, Rutgers University, Piscataway, NJ
Dr. Hemant Purohit, George Mason University, USA
Dr. Naeem Hannoon, Universiti Teknologi Mara, Malaysia
Dr. Nagaraj Balakrishnan, Professor Karpagam College of Engineering Myleripalayam, Coimbatore, India
Prashant Bansod, SGSITS, Indore
Prof. Hipollyte Muyingi, Namibia University of Science and Technology, Namibia

Technical Programme Committee

Prof. David Delahaye, Saint-Martin, Cedex, France
Prof. Andrew G. Dempster, The University of New South Wales, Australia
Prof. Alan Dennis, Kelley School of Business Indiana University, Bloomington, Indiana, USA
Prof. Jitender Singh Deogun, Department of Computer Science and Engineering, University of Nebraska—Lincoln, Nebraska, USA
Prof. (Dr.) Avdesh Sharma, Jodhpur, India
Er. Nilay Mathur, Director, NIIT Udaipur, India
Prof. Philip Yang, Price water house Coopers, Beijing, China
Mr. Jeril Kuriakose, Manipal University, Jaipur, India
Prof. R.K. Bayal, Rajasthan Technical University, Kota, Rajasthan, India
Prof. Martin Everett, University of Manchester, England
Prof. Feng Jiang, Harbin Institute of Technology, China
Prof. Prasun Sinha, Ohio State University Columbus, Columbus, OH, USA
Dr. Savita Gandhi, Professor, Gujarat University, Ahmedabad, India
Prof. Xiaoyi Yu, National Laboratory of Pattern Recognition, Institute of Automation, Chinese Academy of Sciences, Beijing, China
Prof. Gengshen Zhong, Jinan, Shandong, China
Prof. Abdul Rajak A.R., Department of Electronics and Communication Engineering, Birla Institute of Dr. Nitika Vats Doohan, Indore, India
Dr. Harshal Arolkar, Immd. Past Chairman, CSI Ahmedabad Chapter, India
Mr. Bhavesh Joshi, Advent College, Udaipur, India
Prof. K.C. Roy, Principal, Kautilya, Jaipur, India
Dr. Mukesh Shrimali, Pacific University, Udaipur, India
Mrs. Meenakshi Tripathi, MNIT, Jaipur, India
Prof. S.N. Tazi, Government Engineering College, Ajmer, Rajasthan, India
Shuhong Gao, Mathematical Sciences, Clemson University, Clemson, South Carolina
Sanjam Garg, University of California, Los Angeles, California
Garani Georgia, University of North London, UK

Prof. Jonathan Clark, STRIDe Laboratory Mechanical Engineering, Tallahassee, Florida
Prof. Thomas Cormen, Department of Computer Science Dartmouth College, Hanover, Germany
Prof. Dennis D. Cox, Rice University, Texas, USA
Prof. Marcos Roberto da Silva Borges, Federal University of Rio de Janeiro, Brazil
Soura Dasgupta, Iowa City, Iowa, US
Prof. Gholamhossein Dastghaibyfard, College of Electrical & Computer Engineering, Shiraz University, Shiraz, Iran
Prof. Doreen De Leon, California State University, USA
Bartel Van de Walle, University Tilburg, Tilburg, The Netherlands
A.K. Chaturvedi, Department of Electrical Engineering, IIT Kanpur, India
Prof. Qin Bo, Universitat Rovira i Virgili, Tarragona, Spain, Europe
Prof. Dan Boneh, Computer Science Department, Stanford, California
Prof. Fatima Boumahdi, Ouled Yaich Blida, Algeria, North Africa
Prof. Nikolaos G. Bourbakis, Department of Computer Science & Engineering, Dayton, Ohio, Montgomery
Dr. P.D.D. Dominic, Department of Computer and Information Science, Universiti Teknologi Petronas, Tronoh, Perak, Malaysia
Dr. Pinar Kirci, Istanbul University, Engineering Sciences Department, Turkey

World Conference Industry Chairs

Eur Ing Dr. Malaya Nayak M.S., MBA, Ph.D., CEng FBCS CITP, SMIEEE CEO, IT Buzz Ltd, London
Mr. Bharat Patel, C Engg, FIE, FIETE, COO, Yudiz Solutions Private Limited, India

World Conference Presidential Chairs

Dr. Aynur Unal, Stanford University, USA
Dr. Durgesh Kumar Mishra, SAIT, India
Dr. Somsak Choomchuay, President—ECTI Association, Thailand
Dr. Anica Peters, NUST, Namibia
Chandana Unnithan, Victoria University, Melbourne, Australia

World Conference Presidential Co-chairs

Dr. Hemant Purohit, M-IEEE, George Mason University, USA
Chih-Heng Ke, M-IEEE, NKIT, Taiwan
Dr. Dharm Singh, M-IEEE, NUST, Namibia
Dr. Sheng-Lung Peng, National Dong Hwa University, Hualien, Taiwan
Dr. Valentina E. Balas, University Aurel Vlaicu Arad, Romania
Mr. Mihir Chauhan, G R Foundation, India

Organizing Secretary

Amit Joshi, Member ACM, IEEE, CSI, IACSIT—Singapore, IDES, ACEEE, MSA and Community Member—NPA

Preface

This LNNS volume contains high-quality papers presented at the WS4 2017: World Conference on Smart Trends in Systems, Security and Sustainability. The conference was held during 15–16 February 2017, London, UK, and organized communally by G R Foundation, Computer Society of India Division IV—Communication and IT Buzz Limited. It targets state of the art as well as emerging topics pertaining to ICT, Systems, Security, Sustainability and effective strategies for its implementation for Engineering and Intelligent Applications. The objective of this International Conference is to attract a large number of high-quality submissions and stimulate the cutting-edge research discussions among many academic pioneering researchers, scientists, industrial engineers, students from all around the world and provide a forum to researcher; propose new technologies, share their experiences and discuss future solutions for design infrastructure; provide common platform for academic pioneering researchers, scientists, engineers and students to share their views and achievements; enrich technocrats and academicians by presenting their innovative and constructive ideas; and focus on innovative issues at international level by bringing together the experts from different countries. Research submissions in various advanced technology areas were received, and after a rigorous peer-review process with the help of programme committee members and external reviewer, 31 papers were accepted with an acceptance ratio of 0.198. The conference featured many distinguished personalities like Mr. Amarjit Dhillon, GROUP CIO, Turning Point, London, UK, Prof. Xin-She Yang, Middlesex University, London, UK, Prof. João Manuel R.S. Tavares, Faculdade de Engenharia da Universidade do Porto, Portugal, Mr. Aninda Bose, Senior Publishing Editor and part of Global Acquisition Team at Springer, Germany, and Prof. Wim J.C. Melis, University of Greenwich, London, UK. Separate invited talks were organized in industrial and academia tracks on both days. The conference also hosted few tutorials for the benefits of participants. We are indebted to G R Foundation, CSI Division IV and IT Buzz Ltd for their immense support to make this conference possible in such a grand scale. A total of four sessions were organized as a part of *WS4 2017* including two technical, one plenary and one inaugural session. A total of 24 papers were presented in two technical sessions with high discussion insights. The total number of accepted submissions was 31 with a focal point on ICT and Systems, Security and Sustainability. Our sincere thanks to all sponsors, press, print and electronic media for their excellent coverage of this conference.

London, UK
Liverpool, UK
Gujarat, India
February 2017

Prof. Xin-She Yang
Prof. Atulya K. Nagar
Mr. Amit Joshi

About the Book

The volume deals with sustainability transitions which are transformations of major socio-technical systems of provision and use in areas such as energy, water, mobility, and food, towards more sustainable ways of production and consumption. The book provides insights of World Conference on Smart Trends in Systems, Security and Sustainability (WS4 2017) which is divided into different sections such as Smart IT Infrastructure for Sustainable Society; Smart Management prospective for Sustainable Society; Smart Secure Systems for Next Generation Technologies; Smart Trends for Computational Graphics and Image Modelling; and Smart Trends for Biomedical and Health Informatics. The book volume contains 31 high-quality papers presented at WS4 2017.

Contents

About the Editors

Prof. Xin-She Yang obtained his D.Phil. in Applied Mathematics from the University of Oxford. He then worked at Cambridge University and National Physical Laboratory (UK) as a Senior Research Scientist. Now, he is Reader in Modelling and Optimization at Middlesex University London and Adjunct Professor at Reykjavik University (Iceland). He is also an elected Bye-Fellow at Cambridge University as well as the IEEE CIS Chair for the Task Force on Business Intelligence and Knowledge Management. He is also listed on "2016 Thomson Reuters Highly Cited Researchers" list.

Prof. Atulya K. Nagar is the Foundation Chair as Professor of Mathematical Sciences at Liverpool Hope University and the Dean of Faculty of Science. He has also been the Head of Department of Mathematics and Computer Science since December 2007. A mathematician by training, he possesses multidisciplinary expertise in Nonlinear Mathematics, Natural Computing, Bio-Mathematics and Computational Biology, Operations Research, and Control Systems Engineering. He has an extensive background and experience of working in universities in the UK and India. He has been an expert reviewer for the Biotechnology and Biological Sciences Research Council (BBSRC) grants peer-review committees for Bioinformatics Panel; Engineering and Physical Sciences Research Council (EPSRC) for High Performance Computing Panel; and served on the Peer-Review College of the Arts and Humanities Research Council (AHRC) as a scientific expert member.

Mr. Amit Joshi is a young entrepreneur and researcher who has completed his graduation (B.Tech.) in Information Technology and M.Tech. in Computer Science and Engineering and pursuing his research in the areas of Cloud Computing and Cryptography. He has experience of around 6 years in academic and industry in prestigious organizations of Udaipur and Ahmedabad. He is an active member of ACM, CSI, AMIE, IACSIT-Singapore, IDES, ACEEE, NPA and many other professional societies. He has presented and published more than 30 papers in National and International Journals/Conferences of IEEE and ACM. He has also edited three books on diversified subjects. He has also organized more than 15 National and International Conferences and Workshops. For his contribution towards the society, he has been given Appreciation award by The Institution of Engineers (India), ULC, on the celebration of Engineers, 15 September 2014 and by SIG-WNs Computer Society of India on the Occasion of ACCE—2012 on 11 February 2012.

MITIGATE: A Dynamic Supply Chain Cyber Risk Assessment Methodology

Spyridon Papastergiou and Nineta Polemi

Abstract Supply chain services and logistic chains of the modern era have become dependent on ICT assets establishing a complex, interrelated and interactive cyber ecosystem. Isolated vulnerabilities in any of the cyber assets may have catastrophic impact in the whole supply chain. In this context, the paper proposes an evidence-driven Supply Chain Risk Assessment methodology which relies on high quality scientific and experimental based proofs and findings, (including simulation results and indicators, e.g. CVE) to optimize the evaluation and mitigation of the supply chain related risks and cyber threats.

Keywords Attacks · Vulnerabilities · SCADA · Cyber risks
BPMN model

1 Introduction

Logistics and supply chain services have become globally distributed, interconnecting authorities, ministries, companies, industry, agencies, Critical Information Infrastructures (CIIs) (e.g. transport networks, energy networks, telco networks), people, processes, services, products, and other elements that rely upon an interconnected web of transportation infrastructure and pathways, information technology, cyber and energy networks. A plethora of distributed ICT assets from these various supply chain business partners interrelate and interact; their isolated threats

The paper presents results from the E.C. project MITIGATE that the authors are in the Technical Management Team.

S. Papastergiou (✉) · N. Polemi
Department of Informatics, University of Piraeus, 80, Karaoli & Dimitriou str., 18534 Piraeus, Greece
e-mail: paps@unipi.gr

N. Polemi
e-mail: dpolemi@unipi.gr

X.-S. Yang et al. (eds.), *Smart Trends in Systems, Security and Sustainability*, Lecture Notes in Networks and Systems 18,
https://doi.org/10.1007/978-981-10-6916-1_1

and vulnerabilities may propagate in the whole supply chain putting in danger the whole supply chain.

In this context, the lack of visibility and traceability in the often-opaque processes and practices used to develop and acquire ICT related products and services from each actor increases the risk of not being able to detect and remediate intentional and unintentional compromise that may be introduced through a variety of means, including counterfeit materials and malicious software. Since, in our interconnected, automated electronic age, the modern supply chain itself is ICT-enable; its processes should also be protected from disclosure and exploitation.

Overall, there is a clear need for targeting, sophisticated global risk assessment frameworks to deal with the cascading effects risks, threats and vulnerabilities, associated with the ICT-based logistics and supply chains.

In this context, the paper introduces a rigorous, rational approach to risk management, which will produce high quality scientific and experimental based proofs and findings, (including simulation results, indicators and recommendations) to assist supply chain operators to evaluate and mitigate their risks. The paper presents a novel integrated collaborative Supply Chain Risk Assessment methodology, which enables the involved supply chain business partners to collaborate in the identification and classification of the various risks, while at the same time facilitating them in risk resolution and the creation of related supply chain plans.

2 Supply Chain Risk Assessment

The traditional goal of risk management [1–4] is to protect the business assets of an organization and minimize costs in case of failures and thus it represents a core duty of successful company management. Hence, risk management describes a key tool for the security within organizations.

However, risk management [5–13] is essential in the provision of logistics and supply chain services where a plethora of assets from various organizations interact and need to be protected from those threats that have impact in the whole supply chain. The estimation of risks and the development of appropriate protective measures for all (cross-partner) assets involved in the supply chain is required in the generation of a supply chain security policy ensuring the secure operation of any logistic chain or the secure provision of any supply chain service.

There is a family of ISO standards addressing the security management of the supply chains (including ISO 28000:2007 [14], ISO 28001:2007 [15], ISO/IEC 27005:2008 [16], ISO/IEC 27001:2005 [17], 27002:2005 [18]). However, their objectives are not to constitute a risk management method but, rather, to fix a minimal framework and to describe requirements, for the risk assessment process itself, for the identification of the threats and vulnerabilities allowing to estimate the risks, their level and then to be in a position to define an effective treatment plan. The generic nature of the above standards does not include aspects that promote the collaboration among the users. Although efforts have been made to standardize

Supply Chain (SC) security risk assessment, there is a lack of targeted methodologies. Lately the Medusa project (http://medusa.cs.unipi.gr/), developed a novel SC risk assessment methodology, compliant with ISO28001, ISO27001 and ISPS. Medusa can be used in order to assess the risks of a SC service. The derived overall risk values are used in order to generate a baseline SC security policy, identifying the least necessary security controls for each participant in the SC. In addition, Medusa assesses the risk of cascading threat scenarios within a SC. This enables the SC participants to fine-tune their security policies according to their business role as well as their dependencies. The Medusa methodology [18] concentrated on the business needs and requirements of the maritime SCs and a user friendly tool has been developed (MEDUSA tool (accessible via this link medusascsra.cs.unipi.gr)) that implements the methodology enabling port operators and SC business partners to perform risk assessments of the SC services that are involved. The ongoing CORE project (http://www.coreproject.eu/) aims to develop an innovative approach to designing global supply chains resilient (in real-time) to major disturbances caused by high impact events. However, CORE does not plan to provide a concrete risk assessment methodology for the global SC. Finally the ongoing project MITIGATE (http://www.mitigateproject.eu/) aims to realize a radical shift in SC risk management methodologies towards a collaborative evidence-driven approach (MITIGATE methodology) that alleviates the limitations of state-of-the-art risk management frameworks. The project will also integrate, validate and commercially exploit an effective, collaborative risk management (RM) system that will be parametrized for ports' SC services; the SC partners will be able to analyze all threats arising from their SC cyber assets, including threats associated with port CIIs interdependencies and associated cascading effects. In this paper we will outline the main steps of the MITIGATE methodology [19] which is compliant with ISO28000, ISO28001 and ISO27005 and can be used by SC business partners in order to estimate the cyber risks of their SC services; it lays a special focus on the cascading effects in multi-sector environments. Special emphasis will be paid in the provision of support for security processes associated with the dynamic (ICT-based) SCs. Therefore, we will include technical, policy, techno-economic and usability perspectives into the risk assessment process thus taking the viewpoints of a variety of stakeholders into account.

3 MITIGATE Supply Chain Risk Assessment (SCRA) Methodology

3.1 Goals and Assumptions

MITIGATE is a Supply Chain Risk Assessment (SCRA) methodology which aims to estimate and forecast the cyber risks of any Supply Chain Service (SCS) that its provision/delivery requires the interaction of various cyber assets from various business partners (cross-partners' cyber assets). The multi-objectives of the MITIGATE methodology are to: (a) identify and measure all cyber threats within a Supply Chain (SC) service; (b) evaluate the individual, cumulative and propagated vulnerabilities; (c) predict all possible attacks/threats paths and patterns within the SC cyber system (which consists of cross-partners' cyber assets); based upon specific propagation rules which consider the various attackers' profiles, the existing vulnerabilities and the cyber assets' dependencies; (d) assess the possible impacts; (e) derive and prioritize the corresponding cyber risks of the SC cyber assets; and (f) formulate a proper mitigation strategy.

In order to achieve the abovementioned objective, we make the following assumptions: the business processes of the SCS are linear (not cyclic) and thus the cross-partners' cyber assets are interconnected in one-way directed, linear paths (thus cyclic attacks are not possible); the cross-partners' cyber assets are used only for the provision/delivery of the SCS and are isolated from the partners' individual ICT infrastructure.

3.2 General Structure of the Methodology

The MITIGATE Supply Chain Service Risk Assessment (SCRA) methodology is triggered by any specific SCS and follows the standard RA Six (6) main blocks (i.e. Boundary Setting, Threat Analysis, Vulnerability Analysis, Impact Analysis, Risk Estimation, Mitigation Strategy).

The realization of each block contains several Steps (Sj) and Sub-steps (Si.j) as presented in the above Table 1. The main Steps of the proposed methodology are described in details in the following Sections.

3.2.1 Step 1: Boundary Setting

In this step, the Supply Chain Service (SCS) under examination will be decomposed (*Step 1.1*) and all the business partners involved are identified (*Step 1.2*). The next step (*Step 1.3*) is the identification and modeling of the main cyber or/and physical (controlled/monitored by a cyber system) processes that comprise the examined SCS. In this context the, following activities are performed: (a) *Activity 1*: *Supply Chain Service's Business Processes (SCSBPs) Identification*: All cyber or/and physical processes are defined and recorded; (b) *Activity 2*: *Business partners Association*: The business partners are linked to the defined SCS Business Processes; (c) *Activity 3*: *SCS cyber Assets Identification*: All cyber assets required for the provision of the examined SCS and the corresponding SCSBPs are identified and reported. The assets can be categorized into various types like Application Server, Desktop Application, Enterprise/Web Application, Mobile Application, Web Server, Operating System, Web Browser etc.; and *Activity 4*: *Assets Interdependencies Modeling*: Specification and Illustration of the interconnections that

Table 1 Mapping between SCRA main blocks and Sub-steps

Boundary setting	SCS cyber threat analysis	SCS vulnerability analysis	SCS impact analysis	SCS Risk estimation	Mitigation
S1.1: Goal and Objectives *of the SCS*	S2.1: SCS cyber threats' identification	S3.1: Identification of Confirmed Individual Vulnerabilities	S4.1: Individual Asset Impact Assessment	S5.1: Individual Asset Risk Assessment	S6: Risk Mitigation
S1.2: SCS business partners and participants identification	S2.2: SCS threat assessment	S3.2: Identification of Potential/Unknown (zero-day) Vulnerability	S4.2: Cumulative Impact Assessment	S5.2: Commutative Risk Assessment	
S1.3: SCS Modelling		S3.3: Individual Vulnerability Assessment	S4.3: Propagated Impact Assessment	S5.3: Propagated Risk Assessment	
		S3.4: Cumulative Vulnerability Assessment			
		S3.5: Propagated Vulnerability Assessment			

exist between the entities and the assets comprising the SCSBPs by taking into consideration the defined types of dependencies (e.g. hosting, exchange data/information, storing, controlling, processing, accessing, installing, trusted).

3.2.2 Step 2: SCS Cyber Threat Analysis

All threats related with the SCS cyber Assets reported in the previous step will be identified (*Step 2.1*) and evaluated (*Step 2.2*) in terms of their likelihood of occurrence. The methodology uses a five-tier scale (Very Low = 20%, Low = 40%, Medium = 60%, High = 80%, Very High = 100%) and the evaluation is based upon the following criteria: (i) the expected frequency of appearance based on the history of previous incidents; (ii) the participants' intuition and knowledge; and (iii) information retrieved from social media and existing repositories will be used in order to gather peripheral information and draw conclusions. Assume that for the particular threat T, a participant believes, based upon his expertise that the probability of occurrence of the scenario T to asset A is about 50% (corresponding to the threat level "Medium". Taking into consideration the information retrieved from the social media, he concludes that the probability is about 70%. In addition, he/she replies that T has a history of one incident in the last 12 months (corresponding to the threat level "High"). Thus, based on the answers of the participant and on the used scales, the Threat level assigned is High = 80%.

3.2.3 Step 3: SCS Vulnerability Analysis

The purpose of a vulnerability analysis is to identify, quantify and prioritize the vulnerabilities that exist in the cyber assets of the SCS under examination. The vulnerabilities fall into two categories: (i) confirmed vulnerabilities (*Step 3.1*) that have been reported and can be retrieved from the online databases (e.g. https://web.nvd.nist.gov, http://www.cvedetails.com); and *potential/unknown (zero-day)* vulnerabilities (*Step 3.2*) that exist in the assets but have not been disclosed yet. The vulnerabilities of both categories will be accompanied with specific attributes (Access vector, Access complexity, Authentication, Threat and Vulnerability Categories, Exploitable) defined by the CVSS.

Then the severity of all (confirmed and zero-day) vulnerabilities is estimated (Step 3.3). For the calculation of the *Individual vulnerability levels*, we use the CVSS metrics (the Access Vector (AV), the Access Complexity (AC) and the Authentication (Auth)) retrieved from the online databases; in particular, it is used a qualitative mapping from the CVSS metrics onto one category of the five-tier scale (ranging from "Very Low" (VL) to "Very High" (VH)). For example, taking a vulnerability V with an Access vector "Adjacent", an Access Complexity "Medium" and an Authentication "Single", the resulting Individual Vulnerability Level coming from the mapping is "Medium".

In addition to the *Individual* vulnerability level, the methodology introduces the following metrics: (a) the *Cumulative Vulnerability Level (CVL)* (Step 3.4) that measures the probability that an attacker can successfully reach and exploit each of the (confirmed and zero-day) vulnerabilities in a given vulnerability chain (the chain of sequential vulnerabilities on different assets that arise from consequential multi-steps attacks); and (b) the *Propagated Vulnerability Level (PVL)* (Step 3.5) that measures the weakness of the vulnerability chains to be exploited due to exposure of a specific vulnerability.

3.2.4 Step 4: SCS Impact Analysis

Impact analysis refers to the assessment of the expected consequence if a malicious action is realized. Thus, this step estimates the impact of the successful exploitation of each individual vulnerability (confirmed and zero-day) to an asset. In order to calculate the *Individual Impact level* (*Step 4.1*) for each vulnerability on the asset, the CVSS metrics retrieved from the online databases are used; in particular, a qualitative mapping from the three-security criteria Confidentiality (C), Integrity (I) and Availability (A) covered in the CVSS Impact metric onto a five-tier scale (ranging from "Very Low" (VL) to "Very High" (VH)) has been defined.

Similarly to the vulnerability analysis, the methodology proposes/introduces: (a) the *Cumulative Impact Level (CIL)* (*Step 4.2*) defined as the impact that occurs after a specific asset/vulnerability combination has been exploited by an attacker (cf. the Cumulative Vulnerability Level in Step 3.4); and (b) the *Propagated Impact Level (PIL)* (*Step 4.3*) defined as the overall impact that occurs when an attacker exploits a specific asset/vulnerability combination and further moves on into the network (cf. the Propagated Vulnerability Assessment in Step 3.5).

3.2.5 Step 5: SCS Risk Estimation

In this Step, various risks estimations will be performed. After collecting all Threat levels (Step 2), Vulnerability levels (Step 3) and Impact values (Step 4) for a specific asset we calculate the following:

- The *Individual Risk level* (*Step 5.1)* that represents how dangerous all threats are to a specific asset.
- The *Commutative risk level* (*Step 5.2)* that refers to the risk of a threat s occurring due to a vulnerability z that exist in the asset given that all independent vulnerability chains starting from other assets have been exploited.
- The *Propagated risk level* (*Step 5.3)* that refers to the risk of a threat s occurring due to a vulnerability v that exist in an asset given that v may cause the exploitation of various vulnerability chains.

4 Conclusion

Although there is a plethora of cyber security and Supply Chain Security standards and conventions (e.g. ISO27001, 27005, NIST Framework, ISO28000, ISO28001, ISPS) there no specific user friendly Supply Chain Risk Assessment methodology implementing these standards that the business partners in a logistic chain can use to estimate and manage its risks and their cascading effects. MITIGATE targets to contribute to the effective protection of the ICT-based supply. In the literature, and its main characteristics are:

- *Holistic view*: provides a holistic view of the ICT infrastructure required for the provision of the supported SCS in order to identify and evaluate all SC cyber threats and risks within the SC.
- *Collaborative*: aims to promote collaboration between business partners.
- *Business-centric*: The importance of the business partners in the provision of the supply chain service is considered (providing more targeted risk assessment results).
- *Compliance with standards*: is compliant with a range of existing standards such as ISO/IEC 27005, 27001, 27002, ISO 28000, 28001 and IMO ISPS, as a means of increasing its adoption and longevity.
- *Implementable*: adopts a sequential step-by-step approach with clear inputs and outputs so it can be easy be implemented in an ICT tool.
- *Sector-independent*: could be applied in various sectors (e.g. maritime, business, health, transport) where supply chain services are offered.

- *Auditable*: The results produced may be easily compared with other risk assessment methodological approaches since it uses standardized notations.
- *Privacy aware*: The Signed Security Declaration legally binds the business partners to privacy aware SCSRA.

The MITIGATE methodology will be implemented in a collaborative tool during the MITIGATE project, enabling all business partners within a SCS to perform their SCRA. It will be localized in the maritime sector so maritime stakeholders can test its functionality and performance.

Acknowledgements This work has received funding from the European Union's Horizon 2020 research and innovation programme under grant agreement No 653212, project MITIGATE. Finally, special thanks to University of Piraeus, Research Center for its continuous support.

References

1. Inventory of Risk Management/ Risk Assessment Methods: European Union Agency for Network and Information Security (ENISA). https://www.enisa.europa.eu/topics/threat-risk-management/risk-management/current-risk/risk-management-inventory/rm-ra-methods
2. Giannopoulos, G., Filippini, R., Schimmer, M.: Risk assessment methodologies for Critical Infrastructure Protection Part I: A state of the art. Joint Research Center Publication (JRC) (2012)
3. Polemi, N., Ntouskas, T.: Open issues and proposals in the IT security management of commercial ports: the S-PORT national case. In: 27th IFIP International Information Security and Privacy Conference, pp. 567–572. Springer (IFIP AICT 376), Greece (2012)
4. Ntouskas, T., Polemi, N.: Collaborative security management services for Port Information Systems, International Conference on e-Business, (ICE-B 2012), SciTePress, pp. 305–308, Italy (2012)
5. Haimes, Y., Santos, J., Crowther, K., Henry, M., Lia, N.C., Yan, Z.: Risk analysis in interdependent infrastructures. In: IFIP International Federation for Information Processing, vol. 253, pp. 297–310. Springer (2008)
6. Aung, Z.Z., Watanabe, K.: A framework for modeling interdependencies in Japan's critical infrastructures. In: 3rd IFIP International Conference on Critical Infrastructure Protection (CIP-2009), pp. 243–257. Springer, USA (2009)
7. De Porcellinis, S., Oliva, G., Panzieri, S., Setola, R.: A holistic-reductionistic approach for modeling interdependencies. In: 3rd IFIP International Conference on Critical Infrastructure Protection (CIP-2009), pp. 215–227. Springer, USA (2009)
8. Hokstad, P., Utne, I.B., Vatn J.: Risk and Interdependencies in Critical Infrastructures. In: A Guideline for Analysis.Springer Series in Reliability Engineering (2013)
9. Kotzanikolaou, P., Theoharidou, M., Gritzalis, D.: Assessing n-order dependencies between critical infrastructures. Int. J. Crit. Infrastruct. **6**(1/2), 63–110 (2013)
10. Pederson, P., Dudenhoeffer, D., Hartley, S., Permann M.: Critical Infrastructure Interdependency Modeling: A Survey of U.S. and International Research. Idaho National Laboratory (2006)
11. Rinaldi, S.M., Peerenboom, J.P., Kelly, T.K.: Identifying, understanding and analyzing critical infrastructure interdependencies. IEEE Control Syst. 11–25 (2001)
12. Theoharidou, M., Kotzanikolaou, P., Gritzalis, D.: Risk assessment methodology for interdependent critical infrastructures. Int. J. Risk Assess. Manag. **15**(2/3), 128–148 (2011)

13. Zio, E., Sansavini, G.: Modeling interdependent network systems for identifying cascade-safe operating margins interdependency. IEEE Trans. Reliab. **60**(1) (2011)
14. Specification for security management systems for the supply chain (ISO 28000:2007): International Standardization Organization (ISO/IEC), Geneva, Switzerland (2007)
15. Security management systems for the supply chain — Best practices for implementing supply chain security, assessments and plans — Requirements and guidance (ISO/IEC 28001): International Standardization Organization, Geneva, Switzerland (2007)
16. Information security risk management (ISO/IEC 27005): International Standardization Organization, Geneva, Switzerland (2011)
17. Information technology—Security techniques—Information security management systems—Requirements (ISO/IEC 27001:2013): International Standardization Organization, Geneva, Switzerland (2013)
18. Information technology—Security techniques—Code of practice for information security controls 2nd ed. (ISO/IEC 27002:2013): International Standardization Organization, Geneva, Switzerland (2013)
19. Polemi, N., Kotzanikolaou, P.: Medusa: a supply chain risk assessment methodology. In: 4th Cyber Security & Privacy Innovation Forum, Belgium, pp. 79–90 (2015)

Small Polar Codes Concatenating to STBC Applied to MIMO Systems

Madiop Diouf, Idy Diop, Ibra Dioum, Birahime Diouf, Sidi Mohamed Farssi, Khaly Tall and Lamine Sane

Abstract This paper uses a concatenation of small Polar Codes length (N=32) and Space Time Block Code, this Polar-STBC is applied to no diversity (SISO), SIMO, MISO and MIMO systems. Minimum Mean Square Error using Successive Interference Cancellation (MMSE-SIC) is a soft output used to the receiver in order to improve Bit Error Rate (BER) and finally Successive Cancellation Decoder (SCD) is placed to the decoder in order to improve the BER and Frame Error Rate (FER). Comparison between several STBC without concatenation schemes and this small Polar-STBC shown that the proposed allows minimizing the BER and FER performances.

Keywords Polar codes · STBC · MIMO · MMSE-SIC · BER

The authors express their sincere thanks to, The Excellency African Center for Mathematics Computing Science Information Communication Technology (CEA-MITIC), for financial support in that paper.

M. Diouf (✉) · I. Diop · I. Dioum · B. Diouf · S. M. Farssi · K. Tall · L. Sane
Department of Computer Science, Polytechnic Institute (ESP)/Cheikh Anta DIOP University (UCAD), Dakar, Senegal
e-mail: madiop.diouf@esp.sn

I. Diop
e-mail: idy.diop@esp.sn

I. Dioum
e-mail: ibra.dioum@esp.sn

B. Diouf
e-mail: dioufbira11@yahoo.fr

S. M. Farssi
e-mail: farsism@yahoo.com

K. Tall
e-mail: khaly.tall@esp.sn

L. Sane
e-mail: lamine.sane@esp.sn

X.-S. Yang et al. (eds.), *Smart Trends in Systems, Security and Sustainability*, Lecture Notes in Networks and Systems 18,
https://doi.org/10.1007/978-981-10-6916-1_2

1 Introduction

The diversity spatial is one of methodologies using the capacity in MIMO systems to combat channel fading. The principle is to convey several replicas of the same information through each antenna. By doing this, the probably of losing the information decreases exponentially [1].

STBC provides full spatial diversity in the collocated MIMO systems, but it doesn't have the coding gain over fading channels. In the documentation many approach of concatenate STBC techniques to other codes have been proposed [2–4]. In [5, 6], a concatenation scheme of good encoding and decoding named Polar Codes in with STBC called Polar-STBC of long length have been discussed and achieved sufficient gain due to this concatenation. In [7] the authors propose the antennas detection and reduce the complexity of the receiver by offering Maximum Likelihood detection algorithm. In [8] we proposed propose a linear filter detection MMSE-SIC using a small Polar Code which allowed to reduce the complexity while maintaining the BER performance.

This paper proposes a small length Polar-STBC scheme presented in [6, 9]. We applied it with a soft output MMSE-SIC at the receiver at first time followed by a Successive Cancellation decoder (SCD) to compare the BER and FER performance between Polar-STBC and STBC only with small Polar length.

The rest of the paper is organized as follows. Section 2 gives a brief review of Polar codes. The system model Polar-STBC and the Soft Output detector are presented in Sect. 3. Section 4 gives firstly numerical simulations of the Polar-STBC and STBC only systems using MMSE-SIC at the receiver, and secondly the SCD algorithm results using Polar-STBC and several STBC are applied to the detector. While Sect. 5 conclude the document.

2 Polar Codes

Polar codes are defined as the first codes that achieve the channel capacity [5]. *PC* (*N, K*) denotes a polar code of block length $N = 2^n$ and dimension K, W is a Binary-input Discrete Memoryless Channel (B-DMC), I(W) is the symmetric capacity [5] and Z(W) is the reliability parameter. Let A and its complementary in $\{1, \ldots, N\}$ A^c respectively denote information and frozen bits sets. The construction of polar codes is based on channel polarization [5, 8].

Polar coding

The relations of polar coding are

$$x_1^N = u_1^N G_N \tag{1}$$

$$G_N = B_N \begin{bmatrix} G_N/2 & 0 \\ G_N/2 & G_N/2 \end{bmatrix} \tag{2}$$

B_N denotes a permutation matrix, $G_1 = [1]$, u_1^N an information set, x_1^N a code word, y_1^N the received word, G_N a generator matrix, and $u_1^N = \{u_1, u_2, \ldots, u_N\}$. In polar coding if u_1^N follows a uniform distribution then $W_N^{(i)}$ is the channel really seen by u_i [5]. The most reliable channels $W_N^{(i)}$ are used to carry the information bits and the least reliable contain frozen bits $Z\left(W_N^{(i)}\right) \leq Z\left(W_N^{(j)}\right)$, with $i \in A$ and $j \in A^c$[10].

3 System Model

In our system model, we propose double encoding, a small polar coding following to STBC, after their concatenation. The items are sent to the MIMO systems. To cooperative diversity system, Rayleigh channel and Additive White Gaussian Noise (AWGN) are also use. We used the same encoding offer to the first section. Alamouti MIMO system is shown in Fig. 1, where N_t represents transmit antennas and N_r receiver antennas, BPSK modulated data stream is used in this model.

After polar encoding, these polar code words are STBC encoded and fed to the N_t transmitting antennas by using

$$\begin{pmatrix} x_1 & -x_2^* \\ x_2 & x_1^* \end{pmatrix} \tag{3}$$

where $x_i\{i=1, \ldots, N_r\}$ are the ith transmit antenna and $\aleph_t^{N_r}$ represents AWGN which is modeled by independent and identically distributed (i.i.d) samples with variance $\sigma_0/2$. The received signal at the time t is $y_t^{N_r}$ such that:

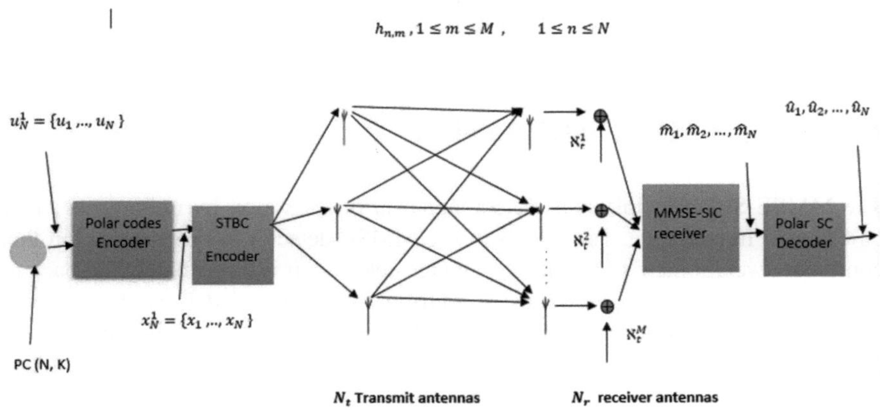

Fig. 1 The system detailed diagram block

$$y_t^{N_r} = \sum_{i=1}^{N_r} h_{j,i} x_i + \aleph_t^{N_r} \quad (4)$$

$h_{j,i}$ is the gain of frequency between the *i*th transmit and the *j*th receive antenna, and H is the MIMO matrix channel.

We adopt in this paper this strategy, the rows of each coding scheme represents a different time instant, the columns are the transmitted symbol. Assuming that each symbol has duration T, then at time $t+T$, the symbols $-x_2^*$ and x_1^* where $(.)^*$ denotes the complex conjugate, are transmitted from antenna 1 and antenna 2 respectively.

Suppose that we have two receive antennas, the receivers symbols are developed in [9, 10].

$$y_t^{N_r} = \begin{pmatrix} h_{1,1} & h_{1,2} \\ h_{2,1} & h_{2,2} \end{pmatrix} \begin{pmatrix} x_1 & -x_2^* \\ x_2 & x_1^* \end{pmatrix} + \aleph_t^{N_r} \quad (5)$$

The STBC decoder provides the followings signals:

$$\widehat{x}_1 = h_{1,1}^* y_1^1 + h_{2,1} y_2^{1^*} + h_{1,2}^* y_1^2 + h_{2,2} y_2^{2^*} \quad (6)$$

$$\widehat{x}_2 = h_{2,1}^* y_1^1 - h_{1,1} y_2^{1^*} - h_{1,2} y_2^{2^*} + h_{2,2}^{2^*} y_1^2 \quad (7)$$

which after changing in (14) and (15) we obtained:

$$\widehat{x}_1 = \left(|h_{1,1}|^2 + |h_{2,1}|^2 + |h_{1,2}|^2 + |h_{2,2}|^2 \right) x_1 + h_{1,1}^* \aleph_1^1 + h_{2,1} \aleph_2^{1^*} + h_{1,2}^* \aleph_1^2 + h_{2,2} \aleph_2^2 \quad (8)$$

$$\widehat{x}_2 = \left(|h_{1,1}|^2 + |h_{2,1}|^2 + |h_{1,2}|^2 + |h_{2,2}|^2 \right) x_2 - h_{1,1} \aleph_2^{1^*} + h_{2,1}^* \aleph_1^1 + h_{1,2} \aleph_2^{2^*} + h_{2,2}^* \aleph_1^2 \quad (9)$$

After these two encoding, a soft output as MMSE-SIC is placed to the output, the proposed algorithm is presented to the following section

4 Proposed MMSE-SIC Receiver

The MMSE-SIC principle is to detect signal in one iteration by nulling out other co-Channel Interference. The idea is to use MMSE detector in order to exploit this combining weight matrix [11]. If the signal is detected, it is immediately fed back to the linear combining process and its contribution is cancelled from the received signal in the next detection iteration and found the next the minimum MSE as shown [8].

In [8], the matrix W satisfies $WH = 1$. The MMSE detector is replaced by:

$$W^i = \left(H^{(i)H}H^{(i)} + \sigma_n^2 I\right)^{-1} H^{(i)H} \tag{10}$$

The final decision is given by:

$$MSE^{(l)} = k_l^2\left(1 - (H_l^{(i)H} w_l^{(i)}\right) \tag{11}$$

The decision for each antenna is given by

$$\widehat{x}_l^{(i)} = w_l^H y_l \tag{12}$$

$$\widehat{x}_l^{(i)} = \text{sign}\left\{\text{Real}\left(w_l^H y_l\right)\right\} \tag{13}$$

where $H_1^{(l)}$ represents the first column of the channel matrix $H^{(l)}, w_1^{(l)}$ denotes the weight matrix of first antenna, k_l^2 is the transmit signal of antenna l.

The MMSE-SIC algorithm during a STBC encoding (at time T and T+1) can be described as a recursive procedure as follows.

Algorithm 1: the recursive MMSE-Algorithm

Initialization: set i = L

for i = 1 to $L^{(l)}$

Step 1. compute the weight matrix using (10)

Step 2. Determine MSE for each antenna l using (12)

Step 3. Determine $l_i = argmin_{l_i \in L_i}\{MSE_{L_i}\}$

Step 4. Determine the estimate (12)

Step 5. Update the receiver by cancelling the contribution of estimate from received signal

Step 6. $y_{i+1} = y_i - H_l^{(i)H}\hat{x}_l^{(i)}$

Step 7. $i = i + 1$, return to Step1.

This algorithm allow us to determine the first BER performance of MMSE-SIC on SISO, SIMO, MISO and MIMO in Simulation 1.

Table 1 Polar-STBC and STBC only Parameter with SC Decoding

Parameters	Value
Decoding algorithm	Successive Cancellation
Channel characteristic	Rayleigh fading
Polar codes length	32
Polar codes rate (r = k/N)	0.5
fft size	64
Modulation	BPSK
Number of transmitting and receiving antennas of Polar-STBC	($N_t = 2$ and $N_r = 2$)
Number of transmitting and receiving antennas of STBC only	$N_t = 2$ and ($N_r = 2, N_r = 3$)

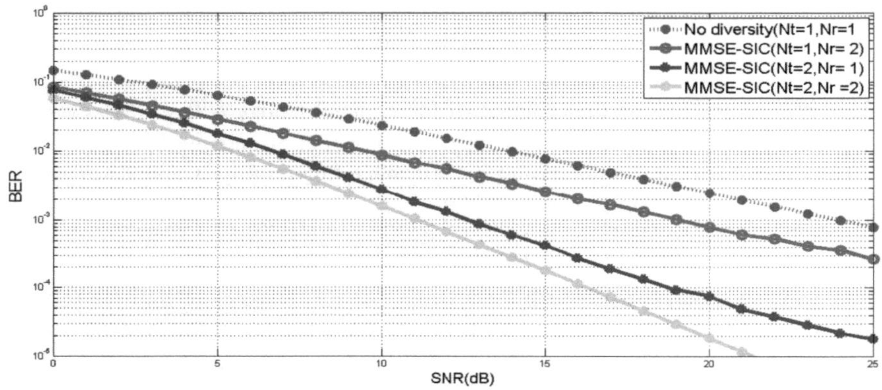

Fig. 2 Comparison between the different diversity schemes

5 Polar Successive Cancellation Decoding

In essence, the SC algorithm is an efficient method of calculation $N = 2^n$ Probability pairs, corresponding to n recursively defined channels. The SC decoding algorithm is wholly described in [5, 12]. We use this algorithm for computing the code word in this paper.

The parameters simulation are placed on the Table 1.

6 Simulation Results 1

In this section, we present the BER results for the proposed MMSE-SIC receiver using Polar-STBC in Fig. 2.

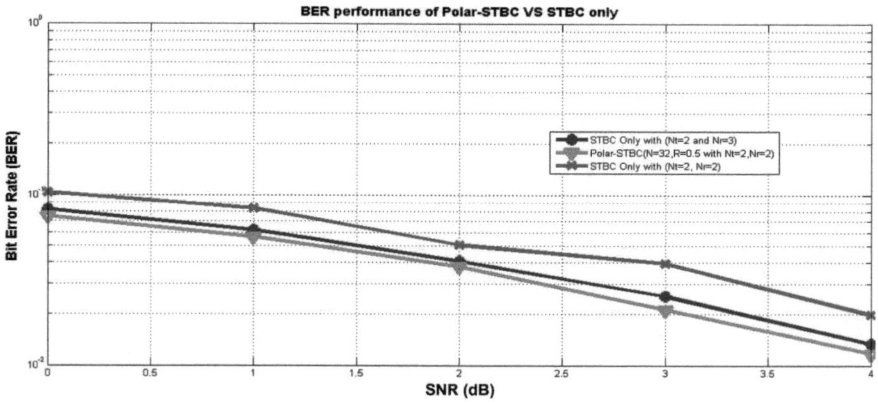

Fig. 3 BER performance between small polar-STBC and STBC only

For instance, at an SNR = 20 dB, the SISO system (1Tx, 1Rx) gives a BER = 5×10^{-3} while the SIMO (1Tx, 2Rx) gives a BER = $2 \times 5\ 10^{-3}$, the MISO system (2Tx, 1Rx) gives a BER = 8×10^{-5} while the MIMO (2Tx, 2Rx) gives BER = 3×10^{-6}. Further at 10^{-5} we noted an improvement over the SISO (1Tx, 1Rx), SIMO and MISO respectively.

7 Simulation Results 2

In Fig. 3, BER performances is analysis at 6×10^{-2} for Polar-STBC using $N_t = 2$ and $N_r = 2$ versus STBC only using $N_t = 2$ and $N_r = 2$ and $N_t = 2$ and $N_r = 3$.

We noted a big improvement when the BER is 6×10^{-2}, the SNR for Polar-STBC 2 * 2 MIMO antennas is about 1 dB improvement over the STBC only using $N_t = 2$ and $N_r = 2$ and better than STBC only using $N_t = 2$ and $N_r = 3$ around 0.3 dB improvement.

At the other hand we introduces the frame error rate performance versus SNR per receiver antenna (e.g. SNR = $2 * E_s/N_0$) in Fig. 4.

Its shown that the FER at 6×10^{-2}, a slight improvement of Polar-STBC using $N_t = 2$ and $N_r = 2$ versus OSTBC only using $N_t = 2$ and $N_r = 2$ about 1.7 dB, but also outperform STBC only using $N_t = 2$ and $N_r = 3$ MIMO around 0.4 dB. These results illustrate that the STBC used in MIMO provide transmit diversity communication over fading channel, but also the coding gain is improved by using polar channel coding.

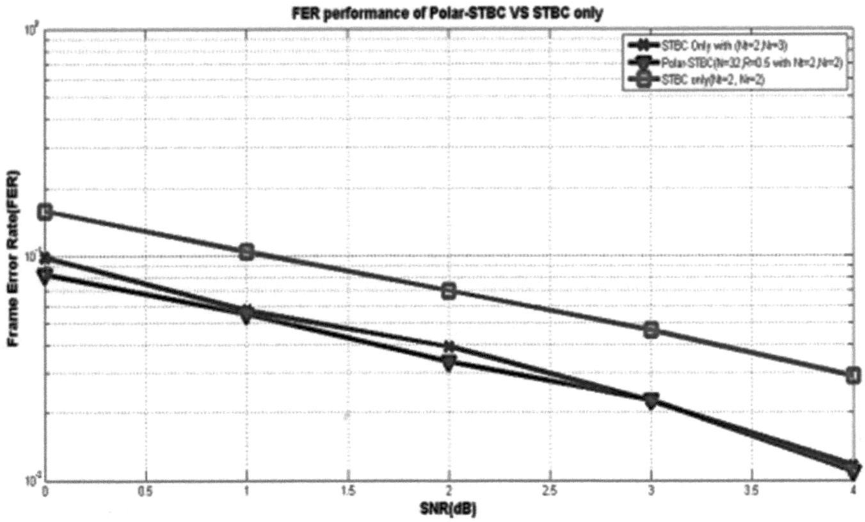

Fig. 4 FER performance between small polar-STBC and STBC only

8 Conclusion

In this paper, after concatenation of small polar codes and STBC codes in order to obtain maximum diversity gain and coding gain, MMSE-SIC receiver is used at the output to further exploit the soft values. In addition, applying successive cancellation decoding the BER is further reduced. The results obtained show that the Polar-STBC has better performance than the STBC even with more antennas. This proposed scheme inherits the advantages of Polar Codes that have both low encoding and decoding complexity but also the advantage of good receiver MMSE-SIC and very good decoding SCD. Indeed, this task is particularly challenging. This proposition open many perspectives such as architecture implementation for small Polar-STBC codes.

References

1. Gesbert, D., Shafi, M., Shiu, D., Smith, P., Naguib, A.: From theory to practice: an overview of MIMO space-time coded wireless systems. IEEE J. Sel. Areas Commun. **21**(3), 281–302 (2003)
2. Mehran, F., Maunder, R.G.: Wireless MIMO systems employing joint Turbo-like STBC codes with bit-level algebraically-interleaved URSCs. In: Proceedings of the IEEE International Wireless Symposium, Apr 2013

3. Futaki, H., Ohtsuki, T.: LDPC-based space-time transmit diversity schemes with multiple transmit antennas. In: 2003-Spring Proceedings of the 57th IEEE Semiannual Vehicular Technology Conference (VTC), vol. 4, pp. 2589–2593, Apr 2003
4. Hou, J., Siegel, P.H., Milstein, L.B.: Design of multi-input multioutput systems based on low-density parity-check codes. IEEE Trans. Commun. **53**(4), 601–611 (2005)
5. Arikan, E.: Channel Polarization: a method for constructing capacity-achieving codes for symmetry binary-input memoryless channel. IEEE. Inf. Theory **55**(7), 3051–3073 (2009)
6. Zhao, S., Sun, Q., Feng, M.-K., Zheng, B.: A concatenation scheme of polar codes and space-time block codes in multiple-input multiple-output channels. In: 6th International Congress on Image and Signal Processing (CISP 2013)
7. Jegadeeswari, E., Arunachalaperumal, C., Bavithra, K.B.: Design of low complexity polar coded spatial modulation. Middle-East J. Sci. Res. **23**(5), 981–986 (2015)
8. Diouf, M., Diop, I., Dioum, I., Farssi, S.M., Diouf, B., Tall, K.: Soft output detection for MIMO systems using binary polar codes. In: 5th International Conference on Multimedia Computing and Systems—IEEE Conference, 29 September–1 October 2016, Marrakech, Morocco. To appear in Decembre 2016
9. Zhao, S., Sun, Q., Feng, M.-K., Zheng, B.: A concatenation scheme of polar codes and space-time block codes in multiple-input multiple-output channels. In: 6th International Congress on Image and Signal Processing (CISP 2013)
10. Cortes-Pena, L.M.: MIMO space-time block coding (STBC) simulations and results design project. In: Personal And Mobile Communications, Georgia Tech (Ece6604), Apr 2009
11. Tran, X.N., Fujino, T.: Combined MMSE-SIC multiuser detection for STBC-OFDM. IEICE Trans.Commun. **E89-B**(5) (2006)
12. Yuan, B., Parhi, K.K.: Successive cancellation decoding of polar codes using stochastic computing. In: Proceedings of the IEEE International Symposium. On Circuits and Systems (ISCAS), May 2015
13. Diouf, M., Diop, I., Dioum, I., Farssi, S.M., Diouf, B., Tall, K.: Study of polar codes MIMO channels polarization using MIMO system. In: Computing Conference Science And Information, London, pp. 681–690 (2016)

Target Tracking in Wireless Sensor Networks Using NS2

Umair Shafiq Khan, Nazar A. Saqib and Muazzam A. Khan

Abstract Wireless sensor network has enormous number of sensors spread over a geographic zone. Each mote in WSN has the ability to communicate with other motes and can process the data in limited capacity. Localization is a technique in which we estimate the location of sensor nodes. In wireless sensor network, nodes locations are not fixed. So to track the target, many solutions come to the mind like GPS, RFID, Bluetooth, Multilateration and Trilateration. In this paper, we use a simple algorithm which reduces the energy usage by minimizing the communication overhead and track the target efficiently.

Keywords Object tracking · Wireless sensor network · NS2

1 Introduction

Day by day improvements in technology have made the deployment of efficient, small, less power and distributed devices. These independent devices are known as nodes. The cost and implementation of these nodes are not much expensive, so it is an active research area. These nodes are also known as motes. Each sensor node can transmit the data and has limited processing capability. When the multiple nodes coordinate with each other, they make a sensor network, which has the capability to process a large data in very detail. As sensor networks are deployed on Ad hoc basis and co-ordinate to carry their task.

In the past, sensor network contains a small number of sensor nodes, which used to directly connected to central location through a gateway. But as the requirements are increasing and the number of nodes are also increasing, it is not practically

U. S. Khan · N. A. Saqib · M. A. Khan (✉)
Department of Computer Engineering, College of Electrical & Mechanical Engineering (NUST), National University of Sciences & Technology, Islamabad, Pakistan
e-mail: muazzamak@ce.ceme.edu.pk

X.-S. Yang et al. (eds.), *Smart Trends in Systems, Security and Sustainability*, Lecture Notes in Networks and Systems 18,
https://doi.org/10.1007/978-981-10-6916-1_3

possible to transmit all the data to the central processing station, because every node has a battery and to transmit data for a long distance more and more power is required, that's why the more focus is on the distributed nodes processing, so the nodes can process the information and can co-ordinate with other nodes. It is possible only when each and every node is connected with the network.

In the Sensor networks, nodes are required to find and known the exact location of other sensor nodes to communicate and share the data with those nodes. So, normally nodes are designed to overcome different environmental conditions and line of sight constraints like reflection, refraction etc. Mostly we do not have fixed infrastructure of environment for the communication purposes of the sensor nodes. So, nodes should be efficient to overcome this problem (Fig. 1).

To find the approximate location of nodes, a technique is used, which is Multilateration. It is also known as Hyperbolic Navigation in radio navigation army systems. In multilateration each node broadcast its location, so the calculation is made on the basis of distance which requires the frequency of the waves received and after how much time it is received.

This paper comprises on four sections. Section 2 includes the literature review and previous research. In Sect. 3, Proposed methodology is described whereas Sect. 4 is related to simulation and Sect. 5 concludes the results and future work.

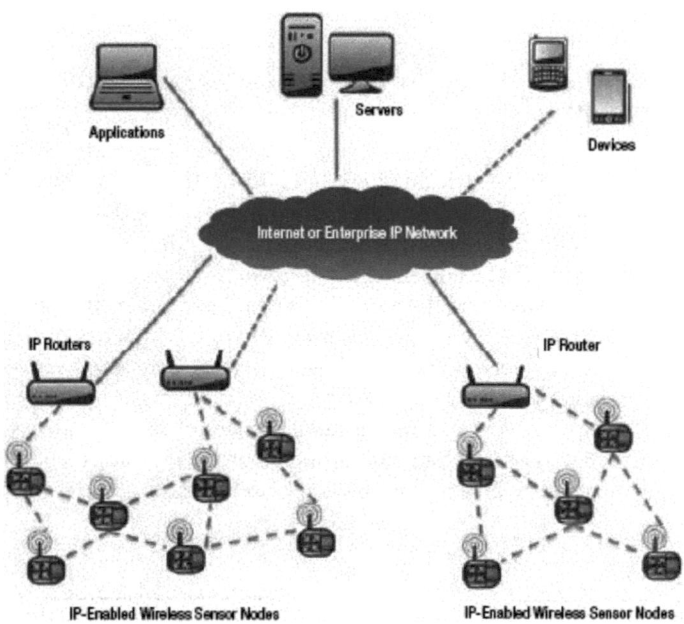

Fig. 1 Wireless sensor network [23]

2 Literature Review

Object tracking in WSN is a difficult task. People proposed many techniques to track the object or node in WSN. Braginsky and Estrin proposed the solution to find the object in Wireless Sensor network by using GPS [1]. But there are numerous solid reasons against the use of GPS in Sensor nodes. Firstly, GPS is an expensive module which increases the cost as well as the size of Sensor node. Secondly, GPS works on the phenomenon of Line of sight (LOS); it means GPS works only in outdoor environment. The topography of location and foliage also effects on the results, which a GPS device produces.

Some other tracking techniques like Infrared, Ultrasound and Radio proposed by Saikat Ray and Rachanee in the IEEE INFOCOM [2]. In Infrared, a unique RF identity is broadcasted by the nodes. Other nodes receive this RFID and track the location on the basis of distance. Other technique is by using Ultrasound, it is also a distance based technique, but gives the better results as compared to Infrared because it measures the time of flight of Ultrasound w.r.t the reference of Infrared. Another proposed solution is by using Radio wave; systems used the signal strength of radio waves to measure the location. It also gives a better approximation due to the ability of penetration through various materials (Fig. 2).

Recursive trilaterations/multilaterations are the most proposed techniques for localization [3]. In this, an upper hierarchy is organized for the network. Some nodes know the positions of other nodes and they act as a beacon for the other nodes who doesn't know the locations and send their location periodically. It is quite possible that beacons are not connected to all the nodes present in WSN. So,

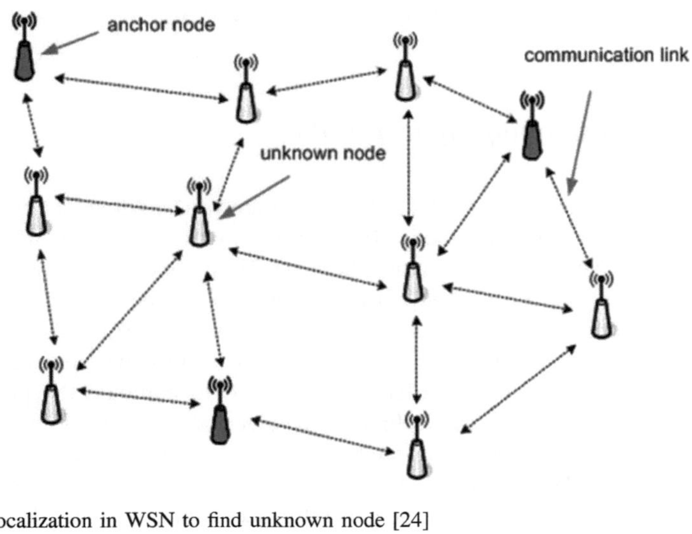

Fig. 2 Localization in WSN to find unknown node [24]

such nodes which will receive information from beacons, those nodes calculate its own position and act like as beacons for others. It is also known as iterative multilateration.

Trilateration is the most commonly used algorithm for localization. GPS also works on this principle [4].

Wensheng Zhang proposed a tree-based algorithm for WSN [5]. When node sense the target, a tree is constructed in which root collects the data and process it. When target moves, tree is reconstructed if necessary. Some other tree-based target tracking techniques like Dynamic Object Tracking (DOT) [6], Deviation Avoidance Tree (DAT) [7], and Optimized Communication and Organization (OCO) [8] have been proposed to overcome the time and energy consumption issues of centralized approaches. In [9] Nguyen suggested energy-efficient algorithm by adding some features to LEACH [10]. To avoid the node failure during target tracking, MZA Bhuiyan, suggested a monitor and backup scenario for WSN target tracking [11].

Elham Ahmadi proposed adaptive clustering method for target tracking which collects the data and select the cluster-head according to the structure [12]. There are some other methods used for dynamic clustering are IDSQ [13], DELTA [14], and RARE [15]. To track the high-speed targets Youngwon Kim introduced an algorithm which uses Doppler effect to calculate the target location and he also studied the signaling delays at different sampling time [16].

Some hybrid techniques like HPS, DCAT [17] and DPT [18] also exists which overcomes the issues presents in one technique and fulfils the requirements of more than one proposed techniques. Most of these hybrid techniques are based on voronoi diagrams/divisions and predicts the location of target using least square method.

B. Thiyagarajan et al. studied some centralized, distributed and existing techniques for target tracking in wireless sensor networks [19].

2.1 *Global Positioning System (GPS)*

In Global Positioning System (GPS), 24 satellites are working together and operate in orbit around earth. Each satellite is 20.200 km away from the each and complete two circles every day around the earth. These satellites work in such a manner that four satellites cover one location or region each and every time. The technique used to estimate the location of nodes is called Time of Arrival (TOA) in which GPS receiver is constantly receiving the information being sent by the four satellites and finally another technique called trilateration is used to compute the position of vehicle. Using Time of Arrival technique, receiver is also able to measure its latitude, altitude and longitude. This method is less costly, as generally all vehicles are outfitted with GPS beneficiary. Be that as it may, this system prompts some undesirable issues, for example, not generally accessible, or it works only for outdoor application (Table 1).

Table 1 Target tracking techniques comparison

Technique	Accuracy	Availability	Synchronize	Security
GPS	No	No	Yes	Yes
RFID	No	Yes	Yes	Yes
Bluetooth	No	Yes	Yes	No
Multi-lateration	Yes	Yes	Yes	No
Trilatera-tion	Yes	Yes	Yes	Yes

To ascertain the area of hub, for 2D situating GPS collector needs to access no less than three satellite signs and for 3D no less than four satellite signs are required. The signs created by satellite can without much of a stretch be irritated or blocked which thus cause the mistake in position of hubs or inaccessibility in urban thick situations. Additionally, not all GPS beneficiaries will give precise result since GPS recipient can have restriction blunder of ±10 to 30 m. So this position worth is not helpful in the application which requires precise restriction data. This sort of mistake can be minimized with the assistance of Differential GPS (DGPS) technique. In this strategy, GPS collector figures its position with reference to definitely known physical area. At that point, the telecom of this distinction is done in the system and every single adjacent GPS beneficiary right their registered position data taking into account the differential data communicated by any one GPS recipient. The impediment of this strategy is that it requires some altered ground-based reference station for data broadcasting [20].

2.2 *RFID*

In this technique, nodes communicate with each other with the help of RFID. Remote Sensor Node communicates bundle to its neighbor having separation of one hop utilizing radio channel. All the while, it trades the information with neighbor utilizing portable RFID tag/user set [21]. Eun-Kyu Lee at el. has assumed that only few nodes have GPS receiver and remaining have an RFID tag/reader set. There were different tags used in this technique.

2.3 *Multilateration*

This technique is used for the target tracking in Wireless Sensor Networks. In which, a node sends the signal to the other nodes at the defined time and measures the distance from the other nodes. Multilateration is the most common technique used in the army navigation system, where it is also called as hyperbolic navigation. These systems are not very secure, as these systems sends the signal at defined time, so the signal can be observed by using oscilloscope and can be hacked.

2.4 Trilateration

Trilateration is the technique is which location of nodes exactly or relatively measured by using the triangles, circles and spheres. This technique doesn't measure the angles and it has practical application like Global Positioning System.

We use different methods to measure the precise distance among those different nodes in WSN.

1. ***Time of Arrival (TOA)***

In Time of Arrival, we measured the time, speed and distance between two different nodes. The most important thing is that both nodes should be synchronized to use this technique.

2. ***Time Difference of Arrival (TDoA)***

In this technique, a wave is transmitted from different emitters, the difference of arrival time of those transmitted waves is calculated and on that basis, we measure the distance of nodes from other nodes.

3. ***Received Signal Strength Indicator (RSSI)***

In the method, we use the attenuation model. Signal attenuation is calculated during the transmission and then on the basis of that attenuation, distance is measured by using the formula of that specific environment attenuation model.

3 Proposed Methodology

Sensor nodes have limited capability of data processing and life time of network is a very important factor which cannot be ignored. So most of the techniques described above, rather they have much communication overhead or they consume excessive energy. There is need of a simple and efficient algorithm which track the target accurately and enhance the network life by minimizing communication and computational overhead. To make it simple some assumptions have to be made at this point.

Assumptions:

1. Every node knows its own location.
2. Every node knows the location of its neighbors.
3. All nodes are distributed uniformly.

Nodes communication range is double form their sensing range. When a node sense the target, it calculates the distance based on RSSI and sends the information to cluster-head. Cluster-head predicts the node movement direction based on previous node information which sensed the target. Then cluster-head calculates the co-ordinates of target in that direction by simply using distance formula.

Basic flow chart is shown in the Fig. 3.

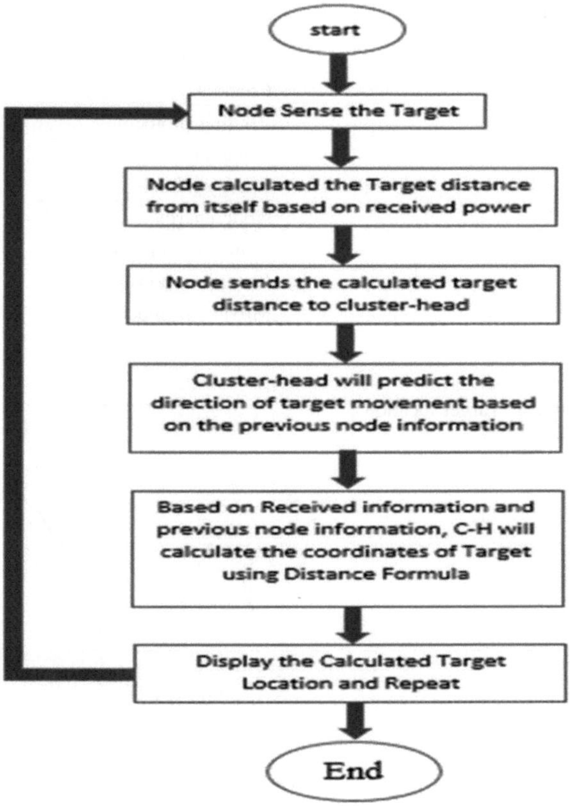

Fig. 3 Flow chart of proposed methodology

4 Simulation and Results

To evaluate the proposed algorithm, we simulate this algorithm in NS2 with the following parameters (Table 2).

Simulation study shows the proposed methodology gives better results as compared to the base [22]. Tracking error significantly reduced by using different patterns of clustering and adjusting the sensing range between nodes as shown in Figs. 4 and 5.

Cluster-head predicts the movement of target on the basis of previous node information. When the algorithm starts, initially cluster-head doesn't have any information regarding the previous node. So, the error reduces once cluster-head gets the previous node information.

Routing packets have to travel within the cluster, which reduces the routing overhead and it has direct impact on the life time of network.

Table 2 Simulation parameters

Parameters	Values
No. of nodes	52
Channel	Wireless
Routing protocol	AODV
Antenna	Omni
Energy model	Energy model
Node energy	3 J
Transmission power	0.175 W
Receiving power	0.175 W
Movement model	Random-Way point

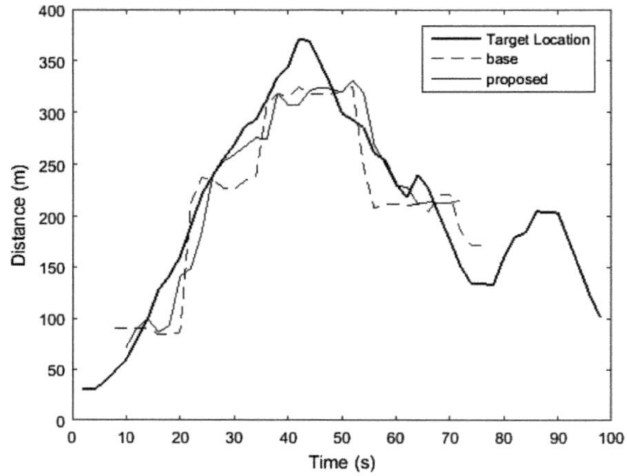

Fig. 4 Target location

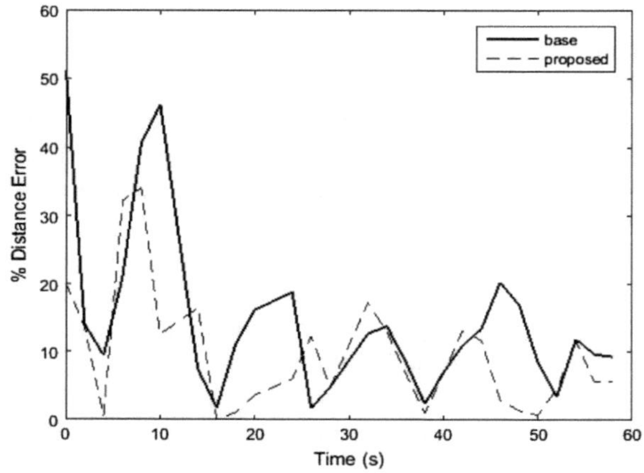

Fig. 5 Tracking error

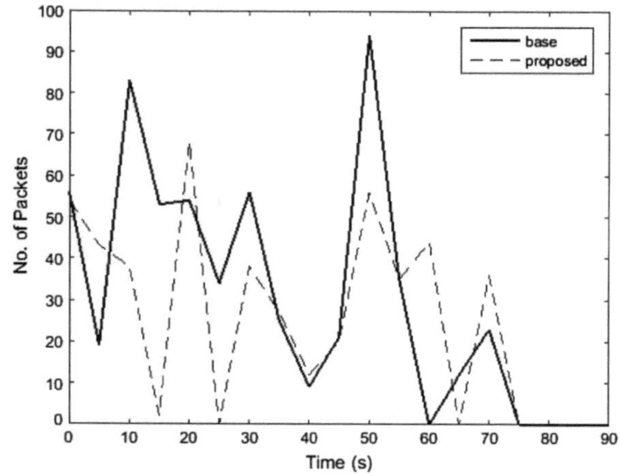

Fig. 6 Routing overhead

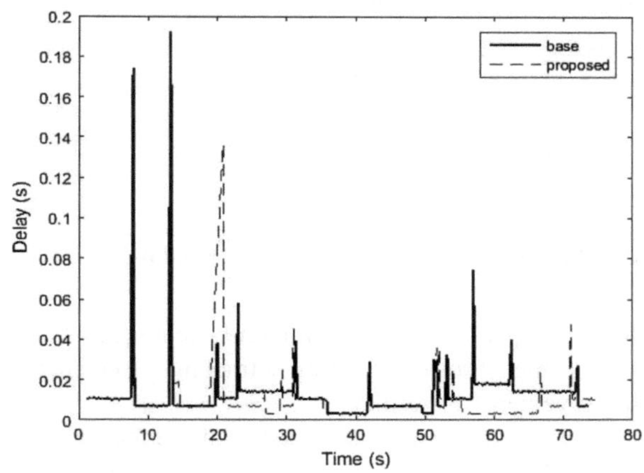

Fig. 7 End-to-End delay

Simplicity of algorithm reduces the end-to-end delay as it can be shown in Fig. 7. However, there are some spikes in the graph when the target moves from one cluster to another there is a short delay between switching.

Energy consumption of network reduces as it has to process less data rather than complex calculations. And processing is distributed between all nodes present in the network. Every node calculates the distance after sensing the target and only sends the calculated value which significantly reduces the communication overhead and saves the battery life of network nodes.

Figures 6 and 7 illustrates that proposed methodology minimize the routing overhead and end-to-end delay and increased the network life as shown in Fig. 8.

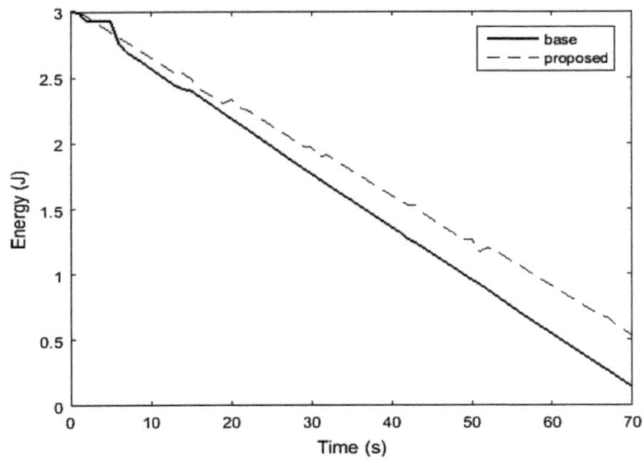

Fig. 8 Network life-time

Table 3 Simulation results

Topology	Tracking error (%)	Network life-time (s)	Normalized routing overhead	E-E delay (ms)
Base	7.468	73.09	1.31	35.29
Proposed	5.189	74.85	1.1	11.81

5 Conclusion and Future Work

In this paper, we studied different techniques of target tracking and proposed a simple methodology for target tracking. Evidences in Table 3 shows significant improvements in results.

Future work can be extended with the feasibility of this algorithm with different protocols. And somehow tracking error can be reduced by working on different patterns of sensor nodes distribution.

References

1. Braginsky, D., Estrin, D.: Rumor routing algorithm for sensor networks. In: Under Submission to International Conference on Distributed Computing Systems (ICDCS-22), (2001)
2. Ray, S., Ungrangsi, R., De Pellegrini, F., Ari, T., David, S.: Robust location detection in emergency sensor networks. In: IEEE INFOCOM (2003)
3. Bulusu, N., Estrin, D., Girod, L., Heidemann, J.: Scalable coordination for wireless sensor networks: self-configuring localization systems. In: Proceedings of the Sixth International Symposium on Communication Theory and Applications (ISCTA 2001), Ambleside, Lake District, UK, July 2001

4. Brain, M., Harris, T.: How GPS receivers work. http://www.howstuffworks.com/gps1.htm
5. Zhang, W., Cao, G.: DCTC: dynamic convoy tree-based collaboration for target tracking in sensor networks. IEEE Trans. Wirel. Commun. **3**(5) (2004)
6. Tsai, H.-W., Chu, C.-P., Chen, T.-S.: Mobile object tracking in wireless sensor networks. Comput. Commun. **30** (2007)
7. Lin, C.Y., Peng, W.C., Tseng, Y.C.: Efficient in-network moving object tracking in wireless sensor networks. Department of Computer Science and Information Engineering, National Chiao Tung University (2004)
8. Tran, S.P.M., Yang, T.A.: OCO: optimized communication & organization for target tracking in wireless sensor networks. In: Proceedings of the IEEE International Conference on Sensor Networks
9. Nguyen, L.T., Defago, X., Beuran, R., Shinoda, Y.: An energy efficient routing scheme for mobile wireless sensor networks. IEEE Int. Symp. Wirel. Commun. Syst. (2008)
10. Chan, H., Luk, M., Perrig, A.: Using clustering information for sensor network localization. In: Proceedings of the International Conference on Distributed Computing in Sensor Systems (DCOSS'05) (2005)
11. Bhuiyan, M.Z.A., Wang, G., Wu, J.: Target tracking with monitor and backup sensors in wireless sensor networks. In: Proceedings of 18th International Conference on Computer Communications and Networks (2009)
12. Ahmadi, E., Sabaei, M., Ahmadi, M.H.: A new adaptive method for target tracking in wireless sensor networks. Int. J. Comput. Appl. (0975 – 8887), **22**(9) (2011)
13. Zhao, F., Shin, J., et al.: Information-driven dynamic sensor collaboration for tracking applications. IEEE Signal Process. Mag. (2002)
14. Walchli, M., Skoczylas, P., Meer, M., Braun, T.: Distributed event localization and tracking with wireless sensors. In: Proceedings of the 5th International Conference on Wired/Wireless Internet Communications, 23–25 May 2007
15. Olule, E., Wang, G., Guo, M., Dong, M.: RARE: an energy efficient target tracking protocol for wireless sensor networks. In: 2007 International Conference on Parallel Processing Workshops (ICPPW 2007) (2007)
16. An, Y.K., Yoo, S.M., An, C., Wells, B.E.: Doppler effect on target tracking in wireless sensor networks. Comput. Commun. **36**, 834–848 (2013)
17. Chen, W.-P., Hou, J.C.: Dynamic Clustering for Acoustic Target Tracking in Wireless Sensor Networks (2003)
18. Yang, H., Sikdor, B.: A protocol for tracking mobile targets using sensor network, sensor network protocols and applications. In: First IEEE International Workshop on Sensor Network Protocols and Applications, pp. 71–81, Anchorage, Alaska (2003)
19. Thiyagarajan, B., et al.: Target tracking using wireless sensor networks: survey. Int. Conf. Adv. Res. Comput. Sci. Eng. Technol. Article No. 57 (2015)
20. Sunita, S.S., Yadahalli, R.M., Tamboli, A.S.: Vehicular Ad-Hoc network localization techniques: a review. Int. J. Electron. Commun. Technol. **3**(2) (2012)
21. Lee, E.-K., Yang, S., Oh, S.Y., Gerla, M.: RF-GPS: RFID assisted localization in VANETs, department of computer science. In: IEEE 6th International Conference, University of California, Los Angeles (2009)
22. Guo, M., Olule, E., Wang, G., Guo, S.: 2010, Designing energy efficient target tracking protocol with quality monitoring in wireless sensor networks. J. SuperComput. **51**, 141–148 (2010)
23. Security in Wireless Sensor Networks. http://rtcmagazine.com/articles/view/101228. Accessed 16 Aug
24. Massive MIMO. http://www.commsys.isy.liu.se/en/research (2016). Accessed 18 Aug 2016

Taking Away the Green Screen—A Brief Discussion of a Multidisciplinary Approach to IoT via, Smart Parking and Interaction Design

Muftah Fraifer, Helen Hasenfuss and Mikael Fernström

Abstract "See a need, Fill a need" Wedge et al. (Robots 20th Century Fox, [1]) though these words can aptly sum up an entrepreneurial spirit, they are also a potentially good attitude for finding relevant solutions to global environmental issues. The important question that arises is how the need is filled? Long term, sustainable and creative solutions are required to reduce the impact of the human footprint. A multidisciplinary approach has great potential, as it is capable of combining a variety of domains and enriching the knowledge base. This paper will discuss a smart parking project to address the issues of carbon emissions, time management and congestion in cities whilst the 2nd project portrays a new approach to interacting and designing computers inspired by efficient biological systems.

Keywords IoT · Self-organizing · Smart parking · Sustainability
Future computing · Video processing

1 Introduction

The Internet of Things (IoT) is a trending topic in current research, as are the means by which this concept can be successfully implemented. This paper describes two very diverse projects that attempt to demonstrate the considerations of designing technology for today and for tomorrow. Section 1 describes a smart parking system

M. Fraifer (✉) · H. Hasenfuss · M. Fernström
Interaction Design Centre, CSIS-University of Limerick, Limerick, Ireland
e-mail: Muftah.Fraifer@ul.ie

H. Hasenfuss
e-mail: Helen.Hasenfuss@ul.ie

M. Fernström
e-mail: Mikael.Fernstrom@ul.ie

X.-S. Yang et al. (eds.), *Smart Trends in Systems, Security and Sustainability*, Lecture Notes in Networks and Systems 18,
https://doi.org/10.1007/978-981-10-6916-1_4

based on the Internet of Things (IoT) whilst Sect. 2 illustrates a theoretical approach to 3D computing that aims to improve interactivity and efficiency through the application of self-assembly. The advantage of examining such projects is that the current considerations and designs can inform the development of intricate systems such as the IoT. For example, in the exploration of self-assembly, qualities such as the ability to self-repair or build new pathways can be useful in a continuously growing network. The remediation of existing technologies to carry out new functions is a trait explored in the smart parking project. To briefly clarify remediation refers to the trend in the cultural arts/media domain: to blend the old and new or to make the old new again. Both projects address concepts such as sustainability and conscientious energy usage and how the ability to affect these concepts must be accessible to all stratas of society. This approach requires planning, thought, effort, time and most of all energy. It is also clear that this approach is currently in conflict with the manner in which modernised society is expected to live. Life is fast-paced, with new gadgets, games, apps, products emerging every few months and technology has become disposable.

The smart parking example highlights the potential to re-use CCTV technology, by integrating the present infrastructure and combining it with a new approach of handling data—the Internet of Things. The example of a self-assembling computer predominantly attempts to illustrate that the investment of time and energy is translated into a more meaningful relationship with technology and its dispensability. Despite the diversity of each project mentioned, a common element is Time. It is an important consideration and how it can be used more efficiently.

2 Towards Internet of Things Framework

The International telecommunications Union (ITU) recommended four dimensions for IoT as it is explained in different applications, which are:

- Item idenification ("tagging").
- Sensor and wireless sensor networks ("feeling")
- Embedded system ("thinking")
- Nano-technology ("shrinking")

Most Internet connections worldwide are currently devices used directly by humans, such as laptops and mobile phones, which most people experience as a mainly human computer-human interaction [2, 3]. As the Internet of Things (IoT) model emerges there may be possibilities of connecting all the objects that surround us to the network, in one form or another. The modern IoT landscape can be broken up into seven key verticals of adoption as in Fig. 1.

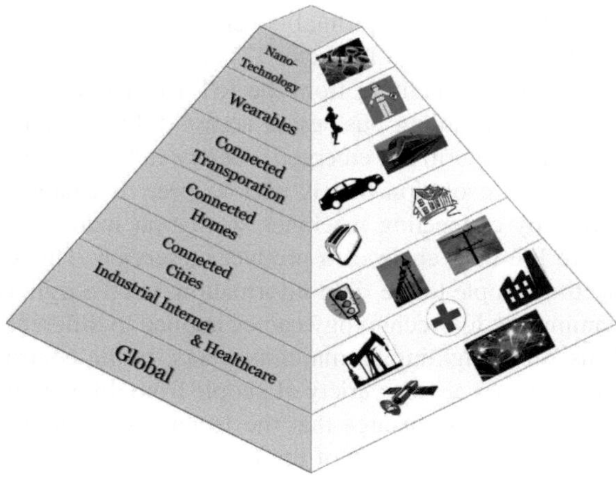

Fig. 1 The IoT 'landscape' and how it is expanding, i.e. from nano- to global domains

The IoT will have smart objects that exchange information between themselves and the number of the objects connected to the Internet will continue to grow exponentially. This idea of IoT is possible with the Internet protocol IPv6, which can handle a larger number of addresses (IPv4: 232, Ipv6: 2128) in order to support the new Internet-enabled devices. IPv6 also removes the need for Network Address Translation, provides for automatic configuration, better routing, real Quality of Service measurement and improved privacy and security due to built-in authentication [3, 4]. As we move into this new era of ubiquity, a future scenario of telecommunications will be dominated by machine to machine (M2M) communication instead of people accessing information or requesting a service through his/her mobile device. The IoT may facilitate control of physical things and new services may be developed that derive information by connecting between virtual and physical devices. The traditional paradigm was anytime, anywhere connectivity for anyone, however it will now include connectivity of anything as well [5, 6]. Such a technological transition will require continuous development of new applications and new business models to meet new needs and expectations.

Some key requirements that need to be considered for the infrastructure of the IoT systems. Firstly, IoT must facilitate people's needs, including open governance, security, and privacy, scalability and flexibility. The IoT should be an open worldwide infrastructure, similar to the present Internet. Secondly, the IoT is about interacting with the objects around us, including static objects, and potentially augmenting them [7]. Thirdly, designing an open, scalable and reliable infrastructure requires open standards. It will be a enormous network, considering that every object will have its own address and identity (IPv6). Moreover, the IoT will need to be flexible to adapt to changing requirements and technological development, and be sustainable to provide a basis for essential investments [7]. Fourthly, the IoT can also be a profitable venture for stakeholders, and various businesses not limited solely to the IT sector [7].

Any user benefiting from the IoT will incite interest and motivate other participants. By enabling businesses across different industries to develop valuable services, a new term has emerged recently called Industry 4.0, that indicates to a fourth industrial revolution that is peer to peer. It enables systems and smart manufacturing by connecting machines so that an industry can create networks along the entire life-time chain of a product or service. This can improve business by connecting people to the right information, via the right device at the right time. In the community IoT technology can be applied to different areas in the environment, e.g. Bins, street lighting, community parks, traffic control, art projects, etc. Since this encompasses a wide variety of people from different technologically knowledgeable backgrounds, it ensures that the technology maintains an approachable standard. A significant advantage of the IoT is its ability to recycle existing technology and also to be embedded into the environment. Embedded technology is not always seen and enables user augmentation rather than user de-skilling. It is important that technology does not render the user useless but rather aid them in achieving more or becoming more efficient in their tasks. For example, smart parking reduces the time spent looking for a space, reduces the emissions, contributes to a positive emotional state (driver does not become frustrated, angry, pressurized, etc.).

Involving communities in IoT projects also ensures that the technology and knowledge of it, is accessible to a larger number of people. It promotes a sense of "socio-technical responsibility" and emphasises the fact that the more care that is given in maintaining and understanding the IoT, the better the network works [8]. By working together knowledge is maintained and spread with respect to improving existing system, finding creative solutions when something goes wrong, or how to protect it.

3 Smart Parking System Prototype: A Design, Scenario, and Prototype as an Example

Parking in capitals is a significant challenge in particular in areas where the public transport facilities are not as extensively available. Carbon emissions, available spaces, waiting and driving time are just some of the factors that affect resources and would greatly benefit from the application of well-designed technology.

Fraifer and Fernström [9] designed a medium fidelity prototype where parking and CCTV-nodes are emulated in an office environment (using toy cars). The prototype has been implemented in out-door environment as well. They have acquired very promising results that confirm the feasibility of using this parking system using Galileo board as microcontroller to interact with all entities. Figure 3 shows out-door environment during implantation process.

3.1 System Prototype

The prototype contains 3 components [9]: an Android App that connects the user with a parking server (PS). And an IoT cam (web-camera) that is connected to a microcontroller (Intel Galileo) and Lastly, a Parking Server which hosting a database of the system.

The IoT cam uses a DHCP connection to connect the PS, which is on a fixed IP. A simple blob-detection algorithm (using OpenCV library) as computer vision monitors and reports changes in a defined grid of polygons [9].

IoT cams connect to the Parking Server (PS), which aggregates the data and updates the database of available car parking spaces in close to real-time. It should be possible to define and redefine parking areas by system operators. The system will operate in close to real-time, we can aim to improve the parking experience rather than forcing and regimenting it.

3.2 Technique of Sensing

ISODATA method is used for thresholding (Classification) to obtain a high grade of accuracy, and to consistently find the contours for each rectangle (car). Each camera (node) must be placed vertically to cover the desired locations as shown in Fig. 2. The camera, in a real and simulated environment (CCTV nodes), is fixed in position with a side view of the parking space as in Fig. 2. The proposed system depends on the OpenCV Algorithms, which has been used to detect vacant parking spaces [9]. The system can identify available vacant parking spots, the process of finding the vacant parking process explained below in Sect. 3.6 in Pseudo code. A virtual reality (VR) environment was created, as shown in Fig. 3, to test and evaluate the prototype in a laboratory setting (also, the University's Ethics Committee suggested a potential insurance issue while conducting experiments). During the experiment, the test subject drove the car and attempted to use the app to search for a parking space. The Moderator observed and listened (Think-Aloud Method) during these

Fig. 2 OpenCV algorithms design for each parking spot

activities. The analysis of all 7 experiments combined these observations and the data streaming live from each camera (node), using either the main PC or Galileo board.

3.3 Interaction with MySQL Database

Linux, Apache, MySQL, PHP (LAMP) server has been used as platform to interact with all entities. The system has real-time communication between the mobile app and the server, using a RESTful API. Figure 4 shows the detection process of in-door and out-door environments.

Fig. 3 VR environments using emulator during testing the protptype

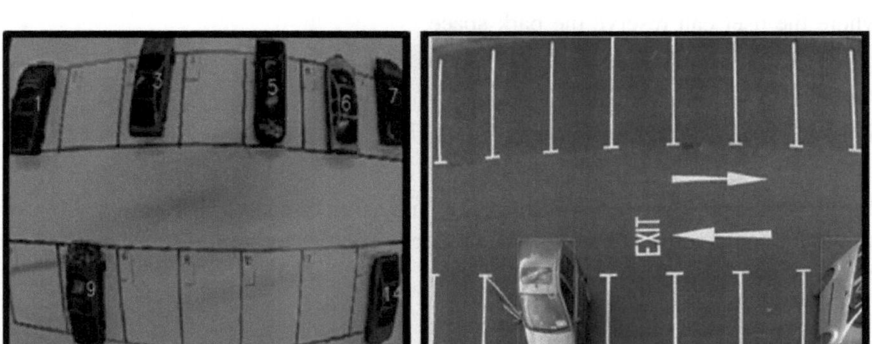

Fig. 4 The testing scenario of our prototype (admin GUI) during run the code

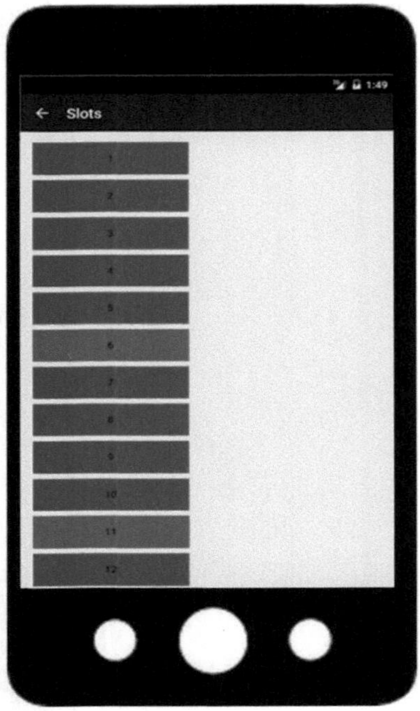

Fig. 5 An activity GUI of shows status of parking

3.4 Implementation

An IoT parking prototype based on CCTV-Camera (nodes) have been proposed and tested in indoor and out-door environments.

2 codes written in Java programming language, ***SystemApp*** (OpenCV) is used to detect the statue or changes of the car park and send it every 60 s to the database in almost real-time and the other ***iApp*** (GUI) is used to retrieves these changes to the user. Both codes are interacting via the MySQL database

Figure 5 shows the implementation of the system shows red and green bottoms where the user can reserve the park space.

3.5 Graphical User Interface (GUI)

Through the app, car parking information and services can be delivered to the user in a functional and user-friendly way. When users register with the system they are given unique IDs.

3.6 Program Code (Pseudo-Code for the Detection Algorithm)

```
//These variables are for checking the  car in park. If
there is a car on parking spot 1, "occupied" is true. If
there is no car "occupied" is false
boolean occupied1 = false, occupied2 = false,...
occupied20 = false;
// If a contour is detected in the parking spot the
contour represents a parked car (occupied parking)
if(contours.size() != 0 )
{
// checking each contour
for(int idx = 0; idx < contours.size(); idx++)
           {
//To get bounders of rectangular of contour.
Rect contourRect=Imgproc.boundingRect(contours.get(idx));

// If bounding rect is between the two aspect ratio x,y.
if((contourRect.width < min_x)||(contourRect.height <
min_y)||(contourRect.width > max_x)||(contourRect.height
> max_y))
                  continue;
// Get aspect ratio.
double ratio = [(double)contourRect.height
/(double)contourRect.width];
      if ((ratio < x)||(ratio > y))
                  continue;
If (The detected contour is a car)
      {
    // Draw rectangle of the car.
       R = rectangle of car,
    // Get the position of each car.
       Pi = center position of car(i),
    // Give each occupied parking spot an ID.
       ParkID(i) = occupied_parking_spot(i),
      }
}
```

4 Future Computing

What does efficiency mean? The dictionary definition states: (esp. of a system or machine) achieving maximum productivity with minimum wasted effort or expense [10]. Designing, working or living efficiently means therefore taking only what is needed and not living in excess. Many systems already exist which embody this concept. They are biological systems and provide a vast resource of inspiration for science and technology developments. For example, the Euplectella aspergillum (venus flower basket) is a silicone based lifeform that has created a tower structure with maximum strength through optimum use of materials. This lifeform has inspired structural engineering concepts as well as advances in fiber optics [11]. Similarly, when encountering technology that explores self-assembly or self-organization the core inspiration originates from ant, termite or bee colonies. These systems are efficient and represent many concepts that can be adapted to man-made systems. Another important consideration with respect to creating a more efficient computing network/system is contained in the technological potential. To demonstrate, even though the following comparison is not altogether fair as stated by the author it does highlight the vast quantity of computing power available to the average user in today's society: "The iPhone 6 uses an Apple-designed 64-bit Cortex A8 ARM architecture composed of approximately 1.6 billion transistors… Put simply, the iPhone 6's clock is 32,600 times faster than the best Apollo era computers and could perform instructions 120,000,000 times faster" [12].

Most of us have access to technology that is so powerful but we only access and use a small percentage of it. It can be considered very inefficient. Therefore, what about a type of computer that could adapt to the functions that the user requires, a computer that you could simply add more components to in order to increase computing capacity or vice versa? In 2002 Butera suggested a paintable computer that could do this [13]. It is a computer made up of many smaller mini-computers that work together to create something greater than the individual—very similar to our perceptual Gestalt that 'the whole is greater than the sum of its part' [14]. Ideally this type of multiagent computer could learn and adapt to a user's requirements. It would provide only the functions that the user needs and thereby not wasting energy, storage capacity or computing power on extraneous functions.

- How would it learn? Communication protocols between self-assembling agents are still current research topics [13, 15–17]. Similar to neural networks or an ant colony each individual agent has its programming (behavior, learning ability, reactions). When it comes in contact with other agents it can communicate and interact. In this interaction messages are passed which contain instructions (e.g. build a cube, build a bridge, fill a gap, build a keyboard, etc.) that allow the agents to decide whether to assemble or dis-assemble and which functionality to represent (act as a pressure sensor, pulse relay, etc.). Since these agents would adapt to the user, they would learn the most used functions the user requires and become more efficient in creating the required interface.
- What would it be made of? The scale at which these mini-computers are envisioned to exist are around the 2–6 mm. This means that there is vast scope for the use of smart polymers and bio-degradable materials [18–20].
- How would it be made? The matter of designing an agent that carries out the physical tasks suggested is still open for experimentation. Current agent designs have favoured geometric shapes such as the cube or rhombus dodecahedron

[17, 21, 22] due to their reduced complexity with respect to programming and their affordance to self-assemble. However other avenues of interest incorporate origami [21], biologically inspired features [23–27] or based on magnetics [28]. The advantage of a system that can to an extent behave autonomously is the reduced energy required as input into the system. Rather than instructing each individual agent: where to be, when to carry out a task and what it has to do, it is more efficient to let the agents interact with each other and figure out the most optimum method or path to take in order to complete the task. Concepts such as self-repair and working without a hierarchical chain of command are worthwhile elements that can be adapted into other domains such as IoT or cloud computing.

In the line of research of micro- and nanobots the question of power is also a significant obstacle. However, advances in this area will be of greater benefit for the area of green technology. Since on these scales it is not possible to rely on traditional powering solutions like batteries concepts like electrolytes, polymagnets, piezo- or triboelectric generators are being explored. Similar to using the energy from the surrounding (solar, wind, waves, etc.) exploiting the piezoelectricity found in cells, sound, friction or light outside the visual spectrum are possible avenues of research. The methods of extracting this energy may then also be applicable in other areas. This is the benefit of multidisciplinary exchange. The overarching concept for such an approach to computing is that the user invests time into teaching the agents and thereby is more emotionally invested in the technology. If time effort, energy and care are invested there is a greater likelihood that the system or object which results, will last longer. It is no longer a throwaway item. Since it is a new avenue of research it is possible not only to design for the function but to design for sustainability with respect to materials, recycle-ability, reusability. Since people have experienced a computer-wave it is possible to learn from its development and consider more greener approaches particularly in the manufacturing domain. In relation to the designing so that the user is encouraged to invest time and care, it is necessary to create an emotional connection. Since human is emotional creatures it is a key element of design. It can have many benefits, for example evoking positive effect in users will enable problems to be solved creatively and with an open mind but also will improve tolerance and coping strategies when technology does not work according to expectations [29].

5 Conclusion

Two projects have been described in order to highlight that addressing today's problem is as relevant as considering how to shape tomorrow's technology in order to avoid past mistakes. Work done in the area of self-assembling or self-organising networks may be of benefit for large-scale IoT projects that require flexibility with respect to adapting to expanding or reducing components. Similarly, to the above-mentioned IoT community projects, if there is shared interest and investment into a project it has a higher probability of success and being maintained. A significant trend of developing technology is that it should ideally cater for the beginner as well as the expert. Whilst this is not always possible, ideally for new systems to take hold they must be robust, stable and accessible. Defining a common protocol or approach that underlies the IoT will be useful in integrating global communities as well as creating a stable system. Lastly another important change in the development of the IoT is the environmental consideration. It is an important element that influences the types of project currently being considered (traffic, community areas, home, health, etc.) but also attempts to highlight the attitude towards technology. With finite resources, it is vital to adopt a more sustainable approach. Rather than accepting the attitude that technology is disposable, systems that incorporate physical and functional adaptability have the potential to shift this focus.

References

1. Wedge, C., Saldanha, C.: Robots 20th Century Fox (2005)
2. Ashton, K.: That 'internet of things' thing. RFiD J. **22**, 97–114 (2016)
3. Atzori, L., Iera, A., Morabito, G.: The internet of things: a survey. Comput. Netw. **54**, 2787–2805 (2016)
4. Baoyun, W.: Review on IoT. J. Electron. Meas. Instrum. **23**, 1–7 (2009)
5. Gershenfeld, N., Krikorian, R., Cohen, D.: The IoT. Sci. Am. **291**, 76 (2004)
6. Haller, S., Karnouskos, S., Schroth, C.: The internet of things in an enterprise context. In: Future Internet Symposium. pp. 14–28. Springer, Berlin, Heidelberg (2016)
7. Uckelmann, D., Harrison, M., Michahelles, F.: Architecting the internet of things (2016)
8. Dulong de Rosney, M.: Alternative internet(s): the benefits and challenges of distributed services http://blogs.lse.ac.uk/mediapolicyproject/2014/11/26/alternative-internets-the-benefits-and-challenges-of-distributed-services/
9. Fraifer, M., Fernström, M.: Smart car parking system prototype utilizing CCTV nodes: a proof of concept prototype of a novel approach towards IoT-concept based smart parking. In: 2016 IEEE 3rd World Forum on Internet of Things (WF-IoT), pp. 649–654 (2016)
10. Apple Dictionary Aa.: Apple Inc, Mobile app (2009)
11. Ray, M.: Biological Fibre Optic Nanotech—Venus Flower Basket Youtube (2013)
12. Puiu, T.: Your smartphone is millions of times more powerful that all of NASA's combined computing in 1969. http://www.zmescience.com/research/technology/smartphone-power-compared-to-apollo-432/
13. Butera, W.: Painting the Computer, Painting the Computer, unpublished thesis (Doctorate of Philosophy in Media Arts and Sciences), Massachusetts Institute of Technology (2002)
14. Dewey, R.: A Psychology: An Introduction (2007)
15. Lifton, J.: Pushpin computing: a platform for distributed sensor networks (2002)
16. Uhrmacher, A., Weyns, D.: Multi-Agent Systems. CRC Press/Taylor & Francis, Boca Raton (2009)
17. Bojinov, H., Casal, A., Hogg, T.: Multiagent control of self-reconfigurable robots. Artif. Intell. **142**, 99–120 (2002)
18. Office, L.: Centimeter-long origami robot. http://news.mit.edu/2015/centimeter-long-origami-robot-0612

19. Breger, J., Yoon, C., Xiao, R., Kwag, H., Wang, M., Fisher, J., Nguyen, T., Gracias, D.: Self-folding thermo-magnetically responsive soft microgrippers. ACS Appl. Mater. Interfaces **7**, 3398–3405 (2015)
20. Kim, J., Li, W., Choi, Y., Lewin, S., Verbeke, C., Dranoff, G., Mooney, D.: Injectable, spontaneously assembling, inorganic scaffolds modulate immune cells in vivo and increase vaccine efficacy. Nat. Biotechnol. **33**, 64–72 (2014)
21. D'Andrea, R., Muehiebach, M.: Cubli. http://www.idsc.ethz.ch/research-dandrea/research-projects/cubli.html
22. Larry Hardesty, M.: Self-sculpting sand. http://news.mit.edu/2012/smart-robotic-sand-0402
23. Christensen, D., Hawkes, E.: MicroTugs | Biomimetics and Dextrous Manipulation Lab. http://bdml.stanford.edu/Main/MicroTugs
24. J. P. L., J.: JPL Robotics: Research Tasks, Gripping foot mechanisms for anchoring and mobility in microgravity and extreme terrain. http://www.robotics.jpl.nasa.gov/tasks/showTask.cfm?FuseAction=ShowTask&TaskID=206&tdaID=700015 (2012). Accessed 12 June 2016
25. Dynamics, B.: Boston Dynamics: Dedicated to the Science and Art of How Things Move. http://www.bostondynamics.com/robot_cheetah.html
26. Dynamics, B.: Boston Dynamics: Dedicated to the Science and Art of How Things Move. http://www.bostondynamics.com/robot_bigdog.html
27. Petersen, K., Nagpal, R., Werfel, J.: Self-organizing Systems Research Group. http://www.eecs.harvard.edu/ssr/projects/cons/termes.html
28. International, S.: Magnetically actuated micro-robots for advanced manipulation applications (2014). https://www.youtube.com/watch?v=uL6e3co4Qqc. Accessed 12 June 2016
29. Parsons, A.: Emotional Design: Why We Love (or Hate) Everyday Things20063Donald A. Norman. Emotional Design: Why We Love (or Hate) Everyday Things. Basic Books, A member of the Perseus Books Group, 257 pp. $15.95, NY, New York (2004) (paperback). J. Consum. Mark. **23**, 115–116 (2006)

Modeling SCADA Attacks

Eleni-Maria Kalogeraki, Nineta Polemi, Spyridon Papastergiou and Themis Panayiotopoulos

Abstract SCADA systems play a vital role in the efficient operations of the ports' Critical Infrastructures (CIs) and their Maritime Logistics and Supply Chain Services (MLoSC). In this paper we provide a process-centric modeling approach using BPMN 2.0 specification in order to visualize an attack likely to be detected on SCADA systems. The SCADA model serves as a study on how security concepts (e.g. security paths, vulnerabilities, propagation of attacks) can be represented with modeling notations.

Keywords Attacks · Vulnerabilities · SCADA · Cyber risks
BPMN model

1 Introduction

Industrial Control Systems (ICS) are critical components of industrial automation systems designed to collect and store data, to delineate and control industrial processes [1]. ICS systems are mainly distinguished into Distributed Control Systems (DCS) and Supervisory Control and Data Acquisition (SCADA) systems. SCADA are systems that monitor and control dispersed assets of industry plant via centralized data acquisition and supervisory control techniques [2].

SCADA infrastructures, used in several industrial sectors such as oil refineries, Liquefied Natural Gas, (LNG), power plants, manufacturing, transportation,

E.-M. Kalogeraki (✉) · N. Polemi · S. Papastergiou · T. Panayiotopoulos
Department of Informatics, University of Piraeus, 80, Karaoli & Dimitriou str., 18534 Piraeus, Greece
e-mail: elmaklg@unipi.gr

N. Polemi
e-mail: dpolemi@unipi.gr

S. Papastergiou
e-mail: paps@unipi.gr

T. Panayiotopoulos
e-mail: themisp@unipi.gr

X.-S. Yang et al. (eds.), *Smart Trends in Systems, Security and Sustainability*, Lecture Notes in Networks and Systems 18,
https://doi.org/10.1007/978-981-10-6916-1_5

pharmaceutical pulp etc. [2–4], play a vital role in the maritime industry [5]. Most physical processes within a MLoSC service (e.g. vehicles and cargo loading/unloading, LNG distribution and storage) are executed with autonomous or semi-autonomous mechanical physical systems and machineries (e.g. ships, trucks, cranes, electronic gates/fences) under the control of sophisticated SCADA systems.

The importance of SCADA systems in the operations of ports' Critical Infrastructures (CIs) and the high impact of a SCADA security breach, attracts the attention of adversaries to perform malicious activities including damage, corruption, piracy, terrorism, robbery and physical/cyber-attacks, causing human casualties, economic loss, political disruption, environmental harm [1, 6].

Attacks in the SCADA systems hosted in ports or maritime transport companies may cause disruption or damage of critical mechanical devices (e.g. container cranes, safety and mechanical systems that operate locks and dams) and even worse they may cause loss of life, stealing of cargo, destruction of ship. The effects (in terms of thermal radiation, overpressure blast wave and flying shrapnel) of the explosion of an LNG tanker or in the ports' LNG storage facilities or terminals due to a hacked SCADA system, could lead to lack of energy stock, which could be critical during cold waves, affecting the environment (degradation, fragmentation or loss of ecosystems), the economy and more important the citizenship wellness and health integrity.

In this paper, we model security concepts of an attack scenario occurring in SCADA systems by structuring a BPMN model of a credible attack scenario on SCADA assets in the Maritime Industry. The critical MLoSC Vehicle Transport Service (VTS) serves as the demonstration scenario that the attack is performed in this paper; in particular we focus on the vehicles and transportation processes of the VTS that are facilitated by SCADA systems. Following a process-centric approach, we visualize the operations of these SCADA systems using Business Process Management Notation (BPMN), a de facto ISO standard for business process modeling. We then decompose the involved SCADA systems and identify their individual components/assets, their interrelations, interactions and interdependent sub-processes. Finally we use this BPMN-model (SCADA model) in order to reveal the propagation effects of a specific attack. The aim of the paper is to reveal the usability of a process-centric BPMN—modeling in attack simulations.

The remaining of the paper is structured as follows: Sect. 2 refers to work related on SCADA security issues and risk assessment process models. Section 3, describes the business model of SCADA systems architecture. Section 4, includes the description of the attack scenario and the development of the SCADA BPMN model and its viewpoints. In Sect. 5, we draw conclusions and directions for further research.

2 Related Work

Until the previous decade, SCADA systems were operating as stand-alone systems via private networks. In recent years, their network topology has changed. SCADA systems are performing through large and long-distant networks communicating with a majority of heterogeneous systems [7]; this interconnection increases the security threats and their cascading effects requiring a rigorous protection of their security [1].

Currently, supply chain security risk assessment methodologies contribute towards the estimation of physical and cyber risks of the supply chains and of the individual business partners involved [8]. However, the individual risks of all interdependent ICS assets involved in a supply chain is a complex problem [9] requiring advanced risk assessment methodologies at asset level based upon graph and modeling theories as proposed by the new European Commission project MITIGATE.[1]

Estimation of security risks on SCADA systems, assumes deep analysis and comprehension of parameters such as the attacker's profile, the causes of vulnerabilities and the prominent cyber-attack vectors. The attacker's profile in SCADA networks appears to have one or a combination of the following characteristics: terrorist, disgruntled employee and insider, state or non-state hacker, malware for Confidentiality, Integrity and Availability (CIA) purposes, hobbyist and script kiddie chasing challenge, hacktivist intruding for political reasons [4, 7].

The underlying causes of vulnerabilities in current SCADA system architecture are the following: (i) the misconfiguration of the wireless devices, (ii) the high level of interdependency among transportation infrastructure systems, (iii) deficiencies in security controls as lack of cryptography policies used in SCADA networks [3] or unskilled, naive employees revealing passwords to colleagues ignoring the potential risk [7], (iv) the accessibility of the ports' transportation systems via networks, devices and software components either directly (wired) or remotely (wireless) for scheduled or corrective maintenance purposes.

Some remarkable cyber attack vectors against SCADA systems are database attacks; backdoors and holes in the network perimeter; Cinderella attack on time provision and synchronization, communications hijacking and man-in the middle attacks. These attacks fall in five categories: (i) on the Communication stack, (ii) on the UDP port (attacks appear on the transport layer), (iii) at application layer (lack of security control to many of the SCADA protocols, (iv) on the hardware, (v) on the software.

BPMN global standard is undoubtedly a good practice for modeling such security attacks; there is a considerable work on modeling attack scenarios using the Business Process Modeling Notation (BPMN) specification. A BPMN representation aims on cyber mission impact assessment (CMIA) [10]; a Model-Based Systems Engineering (MBSE) human-centered approach is proposed for designing

[1]http://www.mitigateproject.eu/.

interventions to improve the delivery of maritime security services and to understand the use and generation of security information in support of cargo unloading operations at a marine terminal [11]. It presents unloading port operations and sets security measures but it does not include SCADA architecture operations and attack scenarios models. The domain model for information systems security risk management (ISSRM) is used to present how possible attack scenarios and countermeasures are defined with BPMN [12]. It analyzes the security requirements and represents risk treatments in BPMN, although, some concepts are not presented arguing that extensions in BPMN are required. The process-centric and technical centric approach proposing an architectural methodology by establishing collaboration between the service providers and potential users of the collaboration tool [13]. The methodology is established in the Norwegian SW national initiative MIS project. An analysis and evaluation of security aspects in BPMN is described in [13], initiating security-related extensions in BPMN.

Finally, a process-centric threat model in BPMN 2.0 is developed to maintain a security level in an ever-changing Internet of Services avoiding downtime [14]. Our work is based upon this process-centric approach applied to MLoSCs threats and it concentrates on modeling the process-threats in SCADA internal system operations according to the SCADA topology. During our attack analysis scenario, we model security threats using current elements of BPMN 2.0. We also consider visualizing security concepts at lower level such as representing types of SCADA vulnerabilities with the BPMN 2.0 specification.

3 SCADA Technical Model for Maritime Industry

SCADA systems, as described in Sect. 1, facilitate maritime and shipping operations, as they provide services referring to vessel control and monitoring, stevedoring and cargo warehousing. Thus, our SCADA model provides procedures within the MLoSCS to support the VTS, namely, to unload vehicles from the vessel and transport them to car terminal. The SCADA model infrastructure is divided into four successive levels: the bottom level is known as field or technological level including sensors, amplifiers etc.; the distribution level consists of remote control stations; the supervision level, is the core of SCADA system, comprises of the Master Terminal Unit (MTU) and the Human Machine Interface (HMI); finally the top level is the enterprise level communicating with the Port Community System (PCS) of Port Authority.

A SCADA system in a port's CI can serve as a common process automation system, used to gather data from sensors and instruments located at remote points and to transmit them at a central point for either control or monitoring purposes. The system is, thus, able to analyze and display information improving the performance of vehicles shipping operations, reducing waste of time and providing cost saving. In order for a SCADA system to meet its objectives it incorporates a bundle of components (Fig. 1): (i) Human Machine Interface (HMI) regarding

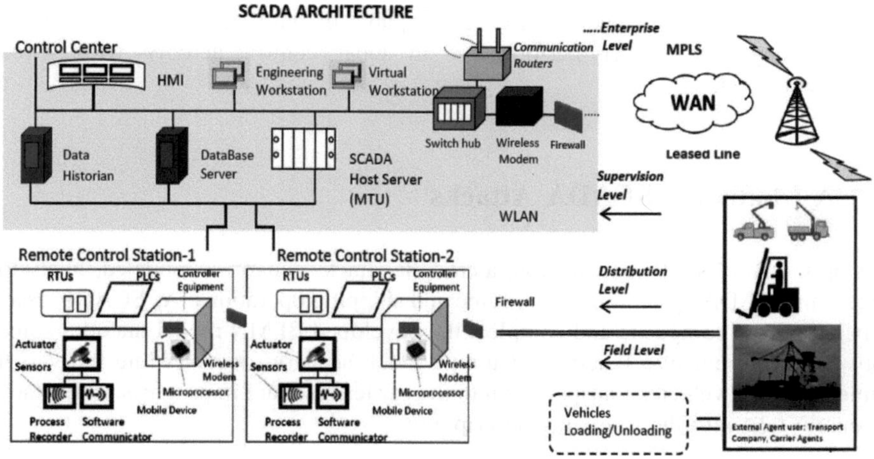

Fig. 1 SCADA system overall infrastructure used for stevedoring; loading/unloading vehicles from vessels and their transportation to car terminals

input-output devices that act as central point of human monitoring and control of the vehicles' unloading processes, storing databases, and display of statistical control charts, and reports; (ii) Programming Logic Controllers (PLCs), solid-state computers reading signals from devices such as sensors, switches, keyboards capable of monitoring the motion of objects and automating harsh industrial processes by providing electromechanical operations during vehicles' loading and unloading of the vessels; (iii) Remote Terminal Units (RTUs) devices are converting electrical to digital signal, microprocessors electronic devices connecting physical equipment with the supervisory stations via WLAN and receive messages from the master station to control the interconnected objects; (iv) a SCADA Master Terminal Unit (MTU), which is, typically, a real-time data repository; (v) a data historian Database for storing data commands; (vi) Supervisory stations which are responsible for communicating between SCADA equipment such as RTUs or PLCs and HMI feeding telemetry data to the central monitoring station of a SCADA system; RTUs are considered microprocessors electronic devices connecting physical equipment with the supervisory stations via WLAN and receive messages from the master station to control the interconnected objects; (iv) a SCADA Master Terminal Unit (MTU), which is, typically, a real-time data repository; (v) a data historian Database for storing data commands; (vi) Supervisory stations which are responsible for communicating between SCADA equipment such as RTUs or PLCs and HMI software for data acquisition and (vii) Trailer Cranes and Forklifts for vehicles unloading from the vessel which are also monitored and controlled via sensors of the SCADA system. All communications between the aforementioned SCADA components are performed using wired and/or wireless network connections. RTUs perform better in large geographical areas than PLCs, owing to their great wireless communication. PLCs are capable of multiple inputs and outputs; hence, they have better performance in local control activity with wired communication.

4 A Model for SCADA Attacks

Using the BPMN 2.0 specification, a credible attack scenario is modeled, likely to occur in SCADA systems over maritime and shipping operations. As SCADA asset operations are composite and complex, the developed BPMN model includes some abstract representative functions. Current attack scenario occurs while the vessel unloading and vehicles transportation to the car terminal are in progress. The attack scenario falls into the following description:

> *"The criminals manage to compromise some critical elements of the SCADA system by exploiting various vulnerabilities, one of which is especially interesting as the MTU user is a domain administrator with elevated privileges to most SCADA assets at port. The criminals use the credentials of the MTU administrator to gain remote access to the SCADA system and by exploiting the vulnerability that allow them to execute code, they penetrate into the HMI system. As a consequence, they identify the precise location of the relevant vehicles within the port and some especially valuable truck-based transports. Then, the hackers send their location and delivery times to their own drivers and change the relevant data to collect the vehicles."*

The BPMN model is structured along the four successive levels of SCADA architecture, described in Sect. 3, assumed as the discrete parts of different SCADA users operations within the process, depicted with the swimlane element of BPMN. The process analysis refers to the operations taken via SCADA to move the vehicles initiating from sensors sending the geolocations to RTUs/PLCs; then remote control devices transfer the message to supervisory station assets which they record, process and send additional directions to remote control stations. The latter converts the signal in a way that it is readable from actuators, which they finally transmit the information to sensor mechanisms of trailer cranes and move the vehicles.

In the current attack scenario, criminals because of the limited security controls, penetrate into the HMI system, and alter the geo-location of vehicles according to their mean purposes. To achieve this, they exploit two types of vulnerabilities, found in HMI hardware; the gain privileges and code execution types; Fig. 2 shows the BPMN model of the attack scenario. The attacker's performance over the process is depicted with the event BPMN elements. More specifically, gain the MTU administrator privileges and execute code to penetrate into the HMI system shown with the escalation (circle) event element of BPMN 2.0. The potential of such attack generates alternative paths in the process, initialized from the exclusive gateway diamond element of BPMN 2.0 after the vehicles delivery event message (Fig. 2): (i) the normal path, where criminals have not performed the attack on SCADA assets, ends up with the vehicles transport to car terminal (shown with the terminate end event of BPMN 2.0 in Fig. 2.) (ii) the improper path where criminals

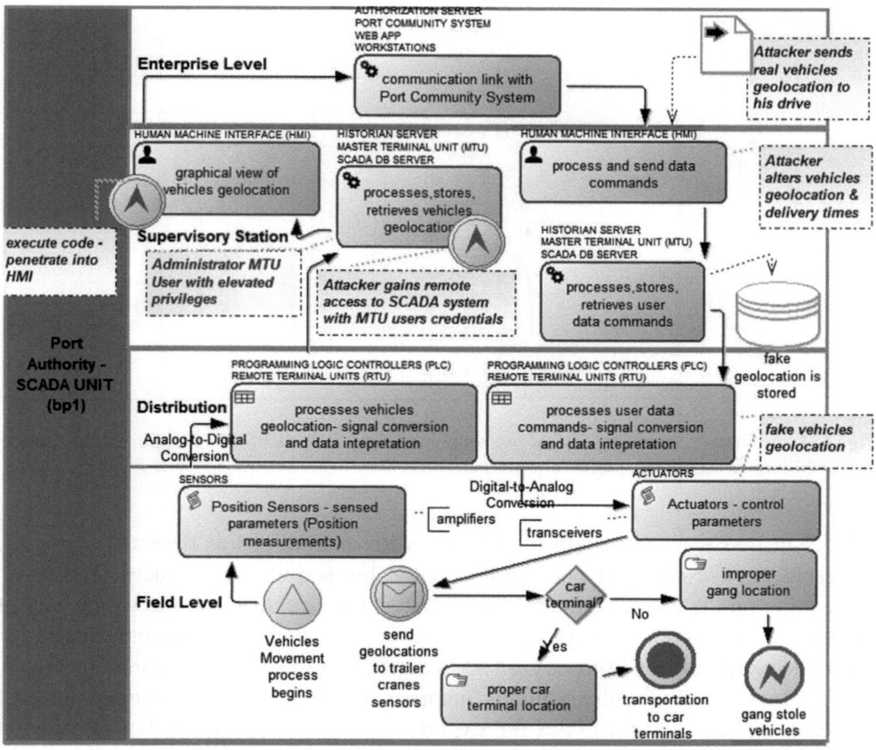

Fig. 2 Execution steps of the "vehicles robbery" attack during vessels unloading

have performed the attack and process closes with the gang committing the vehicles robbery (shown with the error end event of BPMN 2.0 also depicted in Fig. 2).

In the described attack scenario, attackers exploit two types of vulnerabilities; gain privileges, execute code, which are both presented in the BPMN model with the throwing escalation event of BPMN 2.0. This is because both vulnerabilities reach an escalation in order to be exploited. Likewise, we propose to express some important types of vulnerabilities that are possible to appear in SCADA assets with the boundary event elements of BPMN 2.0, namely, events that appear in boundaries of BPMN task elements, as shown in Fig. 3. For example, the vulnerability type "directory traversal" is expressed with the interrupting conditional intermediate event, as it declares the condition that a part of the application is vulnerable. Alike, a "Denial of Service (DoS)" is characterized with the interrupting cancel event as in this case the user is deprived of the services. Vulnerability types such as "memory corruption" or "SQL injection" are presented with the interrupting error event as they generate errors. Additionally, we could estimate the profile of the attacker which can be one or a combination of the ones described in Sect. 2 (terrorist, disgruntled employee etc.); here could be a terrorist. Attacker can be presented in

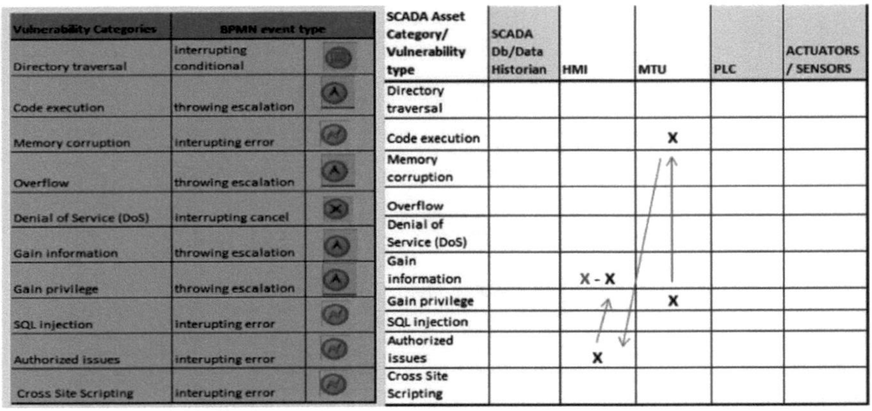

Vulnerability Categories	BPMN event type
Directory traversal	interrupting conditional
Code execution	throwing escalation
Memory corruption	interupting error
Overflow	throwing escalation
Denial of Service (DoS)	interrupting cancel
Gain information	throwing escalation
Gain privilege	throwing escalation
SQL injection	interupting error
Authorized issues	interupting error
Cross Site Scripting	interupting error

SCADA Asset Category/ Vulnerability type	SCADA Db/Data Historian	HMI	MTU	PLC	ACTUATORS / SENSORS
Directory traversal					
Code execution			X		
Memory corruption					
Overflow					
Denial of Service (DoS)					
Gain information		X - X			
Gain privilege			X		
SQL injection					
Authorized issues		X			
Cross Site Scripting					

Fig. 3 Left image: mapping SCADA vulnerabilities with BPMN events. Right image: vulnerabilities of SCADA assets

BPMN in two ways; showing his activity with annotations as depicted in Fig. 2; represented by a different pool and recorded his activities into successive tasks. Depending on the use case, the business analyst can decide each time which way of model to adopt. In our SCADA model, we followed the annotation representation to show the exact impact of the adversary's attacks on the SCADA assets. In Fig. 3, the left image shows vulnerabilities mapping with the BPMN boundary events, while the right image illustrates the exploitation of vulnerabilities that impacts on SCADA assets; HMI, MTU are affected explicitly, though SCADA Db/Data Historian, PLC, actuators/sensors are affected implicitly as a cascading effect. The alteration of the geolocation in the HMI platform is shown in the right image of Fig. 3 with red "X".

5 Conclusions

SCADA systems are critical components of current industrial automation systems underlying highly security risks. SCADA systems are mainly used in Ports' Critical Infrastructures (CI) and their Maritime Logistics and Supply Chain Services (MLoSC) providing cost reduction and saving time of their operations and processes. As cyber-attacks rapidly increase in the maritime ecosystem, SCADA security needs to be highly considered. In this paper, we identify and model SCADA system architecture to examine the consequences and propagation of an attack on its components. Following a business-centric approach, we develop a BPMN model for SCADA threats, to visualize cyber criminal activity on SCADA assets and identify ways to model security risks using BPMN specification; we indicate how to represent vulnerability types using the BPMN event elements. The future work of the authors includes the development of business and ontology models for risk assessment purposes.

Acknowledgements This work has received funding from the European Union's Horizon 2020 research and innovation programme under grant agreement No 653212 project MITIGATE and the Erasmus + project OMI-KA2 project. The authors also thank all partners of these projects. Finally, special thanks to the University of Piraeus, Research Center for its continuous support.

References

1. Mattioli, R., Moulinos, K.: Analysis of ICS-SCADA cyber security maturity levels in critical sectors, European Union Agency for Network and Information Security (ENISA) (2015)
2. Stouffer, K., Pillitteri, V., Lightman, S., Abrams, M., Hahn, A.: Guide to industrial control systems (ICS) security, NIST, Special Publication 800-82 Rev.2 (2014)
3. Igure, V.M., Laughter, S.A., Williams, R.D.: Security issues in SCADA networks. Comput. Secur. **25**, 498–506 (2006)
4. Nicholson, A., Webber, S., Dyer, S., Patel, T., Janicke, H.: SCADA security in the light of Cyber-Warfare. Comput. Secur. **31**, 418–436 (2012)
5. The U.S. Department of Homeland Security/Office of Cyber and Infrastructure Analysis (DHS/OCIA): Consequences to seaport operations from malicious cyber activity, Critical Infrastructure Security and Resilience Note, Operational Analysis Division (2016)
6. Parfomak, P.W., Frittelli, J.: Maritime Security: Potential Terrorist Attacks and Protection Priorities, CRS report for Congress, Congressional Research Service (2007)
7. Daryabar, F., Dehghantanha, A., Udzir, N.I., Mohd Sani, N.F.B., Bin Shamsuddin, S.: Towards Secure Model for SCADA Systems. In: Proceedings 2012 International Conference on Cyber Security, Cyber Warfare and Digital Forensic, CyberSec 2012, pp. 60–64 (2012)
8. Papastergiou, S., Polemi, N., Papagiannopoulos, I.: Business and threat analysis of ports' supply chain services. In: Tryfonas T., Askoxylakis I. (eds.) HAS 2015. LNCS, vol. 9190, pp. 642–653 Springer Cham (2015)
9. Ma, Z., Smith, P.: Determining risks from advanced multi-step attacks to critical information infrastructures. In: Luiijf, E., Hartel, P. (eds.) CRITIS 2013. LNCS, vol. 8328, pp. 142–154. Springer Cham (2013)
10. Musman, S., Temin, A., Tanner, M., Fox, D., Pridemore, B.: Evaluating the impact of cyber attacks on missions. In: 5th European Conference on Information Management and Evaluation, ECIME 2011, pp. 446–456 (2011)
11. Braxton, M.O., Rowell, M.K., Butler, K.A., Haselkorn, M.: A Human-Centered Approach to Modeling Security Information and its Impact on Cargo Operations for Guidance in the Design of Interventions in the Information Sharing Environment, MODSIM World (2016). http://www.modsimworld.org
12. Altuhhova, O., Matulevičius, R., Ahmed, N.: Towards Definition of Secure Business Processes. In: Bajec, M., Eder, J. (eds.) Advanced Information Systems Engineering Workshops, CAiSE 2012, LNBIP, vol. 112, pp. 1–15. Springer, Berlin, Heidelberg (2012)
13. Kosuri, K., Fjørtoft, K.E.: E-Collaboration for maritime operations using business process modeling. Int. J. Adv. Sci. Technol. **34**, 101–112 SERSC (2011)
14. Meland, P.H., Gjære, E.A.: Representing threats in BPMN 2.0. In: 7th IEEE International Conference on Availability, Reliability & Security (ARES2012), pp. 524–550. IEEE (2012)

Automatic Lung Field Segmentation Based on Non Negative Matrix Factorization and Fuzzy Clustering

Ganesh Singadkar, Shubham Talbar, Parang Sanghavi, Bhavin Jankharia and Sanjay Talbar

Abstract Obtaining accurate and automated lung field segmentation is a challenging step in the development of Computer-Aided Diagnosis (CAD) system. In this paper fully automatic lung field segmentation is proposed. Initially, a visual appearance model is constructed by considering spatial interaction of the neighbouring pixels. Then constrained non-negative matrix factorization (CNMF) factorized the data matrix obtained from the visual appearance model into basis and coefficient matrices. Initial lung segmentation is achieved by applying fuzzy c-means clustering on the obtained coefficient matrix. Trachea and bronchi appearing in the initial lung segmentation are removed by 2-D region growing operation. Finally, the lung contour is smooth by using boundary smoothing step. The experimental results on different database shows that the proposed method produces significant DSC 0.987 as compared to the existing lung segmentation algorithms.

Keywords Constrained non-negative matrix factorization · Fuzzy c-means clustering · Lung segmentation

---

G. Singadkar (✉) · Sanjay Talbar
Department of Electronics & Telecommunication Engineering, Shri Guru Gobind Singhji Institute of Engineering and Technology, Nanded, Maharashtra, India
e-mail: singadkar.ganesh@gmail.com

Sanjay Talbar
e-mail: sntalbar@sggs.ac.in

Shubham Talbar
Department of Biologically Inspired System Science, Indian Institute of Technology, Jodhpur, Rajasthan, India

P. Sanghavi · B. Jankharia
Dr. Jankharia's Imaging Centre, Mumbai, Maharashtra, India

X.-S. Yang et al. (eds.), *Smart Trends in Systems, Security and Sustainability*, Lecture Notes in Networks and Systems 18,
https://doi.org/10.1007/978-981-10-6916-1_6

1 Introduction

Computer-Aided Diagnosis (CAD) is widely used for faster and efficient diagnosis of lung diseases such as lung cancer, Interstitial Lung Disease (ILD), etc. These abnormalities are predominantly located inside the lung parenchyma. Majority of the existing methods on lung segmentation work on the basis of contrast between the lung parenchyma and the chest region. Most of the times these intensity based lung segmentation fail to differentiate between lung tissues and chest regions. Therefore in order to overcome these limitations it is very necessary to consider the spatial relationship between the neighboring voxels to achieve accurate lung segmentation results.

From available literature, lung segmentation algorithms can be classified into five different categories: Intensity-based, shape based, statistical model based, neighboring anatomy guided and machine learning based methods. Hu et al. [1] proposed an intensity based automatic lung segmentation method using morphological operations for border refinement. Several different intensity based lung segmentation algorithms such as region growing, watershed [2], graph search [3] and fuzzy connectedness are also proposed in the literature. Although intensity based segmentation algorithms are easy and computationally efficient, they may fail because of the attenuation difference between the normal and abnormal lung regions. Shape-based methods use the anatomical information of the peripheral organ. An atlas of the lung is constructed from the anatomical knowledge which guides the segmentation process. However creating an ideal atlas for the lung that is suitable for all patients is a challenging task. Other algorithms such as level sets, graph cut and active contour are also considered as shape based method. In these methods a boundary curve is initialized inside the image. Defined internal and external forces will adopt the desired boundary of the object of interest. However, because of the pathological condition, it will be very easy for such methods to adopt incorrect boundary of the lung. Itai et al. [4] suggested 2-D parametric deformable model based lung segmentation, which utilizes the lung boundary as an external force for the deformable model. Silveira et al. [5, 6] proposed robust level set based segmentation algorithm, where externally connected edge points are used for stopping the boundary curve. The boundary curve is initialized into the chest region to achieve the segmentation of left and right lung. Sluimer et al. showed that shape model of the normal lung could be used to guide the segmentation of the diseased lung. Statistical model based methods are employed to obtain precise boundaries of the lung. El-Baz et al. [7, 8] proposed the iterative Markov random field model based lung segmentation which uses the unique interaction of the neighboring pixels. Initial lung segmentation is corrected by applying iterative refinement step. This method works well for the abnormal tissues than the intensity based segmentation method. Anatomical model based method uses the atomical knowledge of peripheral organs (e.g., rib cage, heart, spine) to guide the segmentation process. Brown et al. [9] proposed knowledge-based lung segmentation. Algorithm matches the extracted object with the anatomical model. Prasad et al. [10] developed the pathological lung segmentation algorithm in which the curvature of rib was used to help the lung segmentation but a clear disadvantage of this method is that it is hard to define the curvature information of the lung apex and lung lobe. Many times it leads to over segmentation near mediastinum. Machine learning based methods are used for the segmentation of the abnormal regions because it utilizes intensity, texture, shape and anatomical features of the image. The extracted features are given to classifiers such as support vector machine, random forest, neural network, etc. For extracting features of the diseased lung, many methods have been proposed such as 3-D adaptive multiple feature method (AMFM), texton-based approach, intensity based approach, local binary pattern, a histogram of gradient. The major challenge of the machine learning based approach is the selection of the most discriminant feature. If relevant features are not selected, then it may lead to misclassification. Mansoor et al. [11] applied the machine learning based method in combination with the region growing and neighboring anatomy guided segmentation to segment the abnormal lung.

Very few times Nonnegative matrix factorization (NMF) has been used for image segmentation. NMF is one of the most powerful machine learning tool used for dimensionality reduction, data mining and many a time it can also be used for feature extraction due to its capability to discriminate amongst the various objects or modalities. Recently Hosseini-Asl et al. [12] used NMF for the lung segmentation. In their method, NMF is applied to each axial 2-D CT image by converting it into classical context image [13]. A context vector was created from the original voxel by considering the spatial interaction of neighboring voxel. Finally, K-means clustering was applied on the H matrix to discriminate between the lung and chest voxels. A major limitation of the Hosseini-Asl et al. [12] method is that the number of clusters in the lung CT image should be known in advance to the segmentation.

To overcome the limitations of the methods proposed in literature, we proposed CNMF and Fuzzy C-means based method for the lung segmentation. In the proposed approach, visual appearance model of the image is obtained by considering the spatial interaction of the neighboring voxel. The visual model is converted into voxel in new feature H by applying CNMF algorithm. FCM classifies each voxel in H matrix based on its membership value of tissue classes. We have used FCM to cluster the voxel into the lung and chest regions.

The paper is organized as follows Sect. 2 present the description of the method. Experimental results and discussion are given the Sects. 3 and 4 provides the conclusion.

2 Description of the Method

The framework shown in Fig. 1 illustrate the basic step of our proposed lung segmentation approach. The image pre-processing step takes three consecutive slices as input to remove background. Then in the second step a CNMF-based visual model is applied to obtain discriminant features in the image by finding out W and H matrices. In the third step fuzzy clustering is used on H matrix to get the initial lung segmentation. In the fourth step lung separation is employed to separate the left and right lung lobes. Finally, lung boundary is smooth using Newton-Cottes algorithm.

2.1 Image Pre-processing

The main aim of the pre-processing step is to remove the background of the CT image. Background of the image include the non-anatomical pixels. The air around the body and CT image background contribute towards the background pixels and these pixels are connected to the border of the image. Since the lung tissue pixel values are very similar to that of background, it is very important to remove background pixels to obtain more accurate results. Non-anatomical pixels are removed by applying the OTSU thresholding in combination with 2-D connected component labeling.

2.2 Visual Appearance Model

Visual appearance model of a particular slice is constructed by considering the next and previous slice of the same. Every pixel within the slice is modeled by taking into consideration its closest 26 neighbors as illustrated in Step-2 of Fig. 1. All these 27 elements are stacked together to form a voxel. Data matrix *V* is an array of all such sequential voxels.

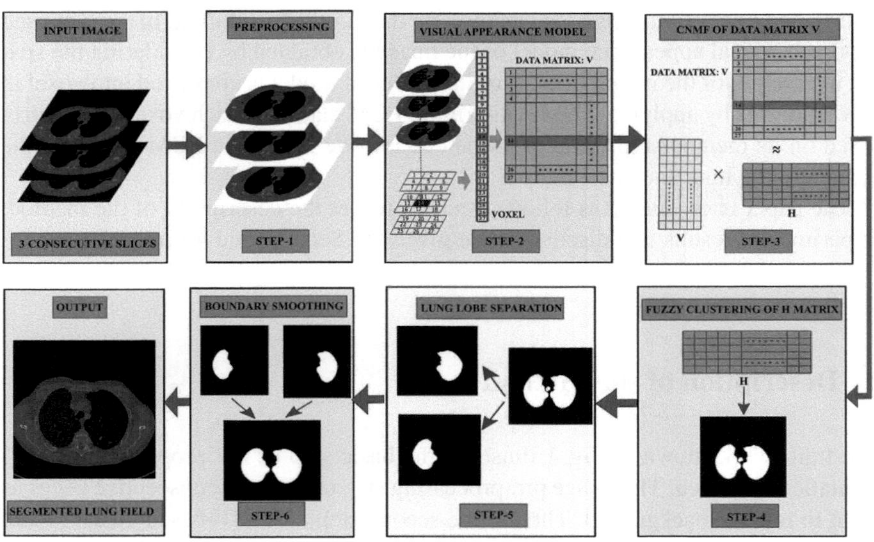

Fig. 1 Outline of automatic lung field segmentation in CT scan images

2.3 CNMF—Based Visual Model

NMF is one of the dimensionality reduction tool used in the machine learning. It has many application in the data mining and pattern reduction. It also used for feature extraction since it is able to discriminate among various object or modalities. NMF can be used for the high dimensional data analysis in which all the elements of the data must be non- negative. This non negative nature of the data with low-rank approximation is not only used for natural interpretation but also to extract the hidden structures of data. To introduce the main concept of the NMF, let us consider a matrix $V \in \mathbb{R}^{M \times N}$. With desired rank $K \ll \min\{M, N\}$. A data matrix V can be factorized into two matrices, a basis matrix $W \in \mathbb{R}^{M \times K}$, and a coefficient matrix $H \in \mathbb{R}^{K \times N}$ such that.

$$V \approx WH \qquad s.t \quad W, H \geqslant 0 \tag{1}$$

To obtain non-negative matrices W and H Eq. 1. can be reformulated as

$$\min_{W,H} \equiv \frac{1}{2} \|V - WH\|_F^2, \qquad s.t \quad W, H \geqslant 0 \tag{2}$$

Since the advent of NMF various types of the optimization techniques have been investigated which comprises multiplicative update rule proposed by Lee and Seung's [14], gradient decent [15] and alternative least square method [16]. In this paper for the lung segmentation we have used the alternative non-negative constrained least means square on active set (ANLS-AS) [17] method. the convergence criteria for ANLS algorithm is given by.

$$\min_{W \geq 0} \left\| H^T W^T - V^T \right\|_F^2, \tag{3}$$

$$\min_{H \geq 0} \|WH - V\|_F^2, \tag{4}$$

The general form of the cost function for the ANLS-AS algorithm given by the

$$\min_{G \geq 0} \|BG - Y\|_F^2 \tag{5}$$

where $B \in \mathbb{R}^{p \times q}$ and $Y \in \mathbb{R}^{p \times l}$ are given and we have to find the optimal value of the $G \in \mathbb{R}^{q \times l}$. By applying ANLS-AS [17] algorithm Eq. 5 can be separated into l independent sub-problems.

$$\min_{G \geq 0} \|BG - Y\|_F^2 \rightarrow \min_{g_1 \geq 0} \|Bg_1 - y_1\|_2^2, \ldots \min_{g_l \geq 0} \|Bg_l - y_l\|_2^2 \tag{6}$$

where $G = [g_1, ..., g_l] \in \mathbb{R}^{q \times l}$ and $Y = [y_1, ..., y_l] \in \mathbb{R}^{p \times l}$ By alternative solving Eqs. 3 and 4 cost function of Eq. 2 can be minimized. At every iteration Eqs. 3 and 4 are alternatively converted into form of Eq. 5 and then solved by using Eq. 6. This process is continued until the convergence criteria is satisfied. Once the Convergence criteria is satisfied then the column of the basis matrix W forms the basis of the visual appearance and each column of the coefficient matrix H expresses each voxel of the image in new feature space. Therefore H matrix gives the set of new visual appearance features.

2.4 Fuzzy Clustering of H Matrix

To obtained the initial lung field estimation, we applied Fuzzy c-means (FCM) clustering on the H matrix generated from the CNMF based visual model. FCM algorithm is used to classify the voxels in the coefficient matrix H into several tissue categories. FCM is a very prominent unsupervised segmentation algorithm for the pixel classification. Each voxel in the H matrix is classified based on its membership values of the tissue classes. Initially, each voxel is randomly classified into six clusters corresponding to six different cluster centroid. Cluster label enjoying the maximum membership value is assigned to each voxel. Then $l^2 - norm$ of the center of the cluster is computed, the maximum value from the center of cluster corresponds to the background. The second largest value corresponds to the chest; smallest value belongs to lung region since the intensity of the lung voxel is low as compared to chest region. Remaining all labels are assigned to the chest region. Finally, an initial lung field estimation is obtained by displaying the lung voxels (Fig. 1).

2.5 Lung Lobe Separation

After the fuzzy clustering step, the main trachea and bronchi are still present in the initial segmentation result (Fig. 2). To remove it from the lung area 3-D connected component labeling is applied. It removes the trachea and bronchi easily when they are not closer to the lungs. However when the bronchi are connected or when it enters into the lung parenchyma, then connected component labeling algorithm treats it as a part of lung itself. So in order to avoid the inclusion of this structure into the segmented result, we used anatomical knowledge of trachea. Since trachea is filled with air and CT value of air is in a fixed range, we used 2-D region growing [18] method to remove this specific region. A seed point is selected based on the centroid of the segmented trachea region. The region growing algorithm uses this seed point to segment the trachea region.

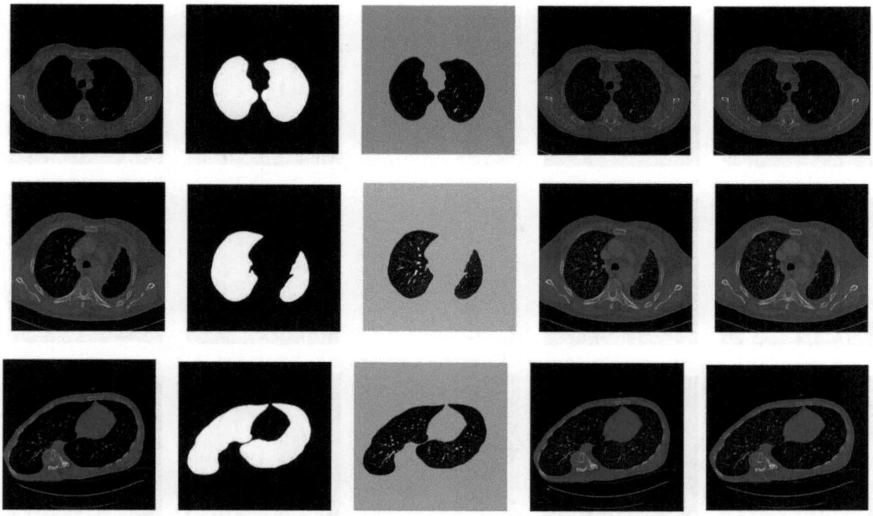

Fig. 2 Segmentation results of proposed method on LOLA11 data set (First Column) original image (Second Column) lung lobe (Third Column) segmented lung (Forth Column) manual segmentaion (Fifth Column) result of proposed method

2.6 Boundary Smoothing

The purpose of boundary smoothing is to obtain smooth and consistent borders for the segmented lung lobe near mediastinum. After the lung lobe separation step, the obtained irregular contour is smooth. To remove noise on the lung boundaries, we have used Newton-Cottes Based Smoothing (NCBS) algorithm from numerical analysis to smooth the lung border. In this step, the noisy pixel is replaced by the new value obtained from the NCBS algorithm.

3 Experimental Results and Discussion

In order to calculate the segmentation accuracy of our proposed CNMF and fuzzy clustering based algorithm. The algorithm was tested on the Lobe and Lung Analysis 2011 (LOLA11) challenge public database and a local database of 20 subjects containing mid range of interstitial lung disease (ILD) provided by the Dr. Jankharia's Imaging Centre, Mumbai. The local database was acquired at Jankaria's imaging Mumbai with SIEMENS SOMATOM Definition Edge scanner with following specifications: slice thickness 1 mm, reconstructed after 0.5 mm, scanning pitch 1.7; 120 KV; 100 MA; and F.O.V 50 cm. The size of each CT image ranges from $512 \times 512 \times 492$ to $512 \times 512 \times 650$.

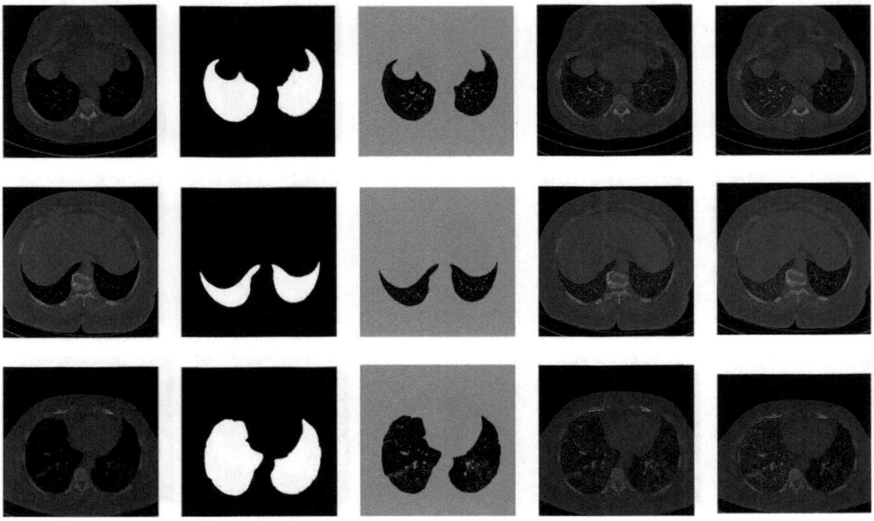

Fig. 3 Segmentation results of proposed method on Local data set **a** original image (Second Column) lung lobe (Third Column) segmented lung (Forth Column) manual segmentaion (Fifth Column) result of proposed method

The projected results of our proposed algorithm on the 2-D axial slices on the selected cases of LOLA11 data set are shown in the Fig. 3 and on the local data set are illustrated in Fig. 3. The quantitative performance of the proposed algorithm is measured by three different performance metrics: Dice Similarity Coefficient (DSC) [19], sensitivity and specificity between automatic segmentation result of proposed algorithm and ground truth. The ground truths were marked by an expert radiologist using Microsoft Surface Pro4 with an active area of 292.10×201.42 mm with a resolution of 2736×1824 and pressure sensitivity of the surface stylus is 1024 levels.

Table 1 summarizes the performance analysis on the two different data sets used in our study. An average DSC of more than 98% for the $(3 \times 3 \times 3)$ neighborhood of size of voxel was obtained by our performance evaluation of over 40 CT scans of patients. To demonstrate the performance of our algorithm, we compared our method against four other segmentation algorithms based on nonnegative matrix factorization (NMF), intensity (I), a combination of intensity and spatial model (IS) and in-

Table 1 Overall performance of the proposed algorithm on the different data sets

Data set	DSC	Sensitivity	Specificity
LOLA11	0.986	0.978	0.998
Local data	0.987	0.975	0.911
Average	**0.987**	**0.977**	**0.954**

Table 2 Qunatitative comparision of the segmentation accuracy using DSC of proposed method with other four algorithm

Segmentation algorithms	DSC
IT [1]	0.816
I [20]	0.632
IS [20]	0.783
NMF [12]	0.966
Proposed method	**0.987**

teractive thresholding (IT). Table 2 demonstrate that our proposed method achieved largest DSC as compared to already existing segmentation methods.

4 Conclusion

In this paper, we proposed a fully automatic method for lung field segmentation based on constrained non-negative matrix factorization (CNMF) and fuzzy clustering. The proposed method consider the spatial interaction of the neighboring voxel to obtain novel image features and these image features were modeled by CNMF algorithm. Then fuzzy clustering is applied on obtained H matrix from the CNMF algorithm to discriminate between lung and chest voxel. The effectiveness of our algorithm was evaluated on forty patients from two different datasets and also compared with other four already existing lung segmentation algorithms. The results shows that proposed algorithm works better than the other segmentation algorithm significantly in terms of DSC performance measure. Currently, the proposed method works well for the minimal to mid range of abnormality. But when the lungs are greatly affected by the pathologies then our method may have reduced performance. Our future work will be focused on the developing segmentation algorithms for severe abnormalities.

Acknowledgements We would like to acknowledge the assistance and support provided by Center of Excellence in Signal and Image Processing, SSGS Institute of Engineering and Technology, Nanded, India and Dr. Jankharia's Imaging Centre, Mumbai for providing the CT image database with the ground truth.

References

1. Hu, S., Hoffman, E.A., Reinhardt, J.M.: Automatic lung segmentation for accurate quantitation of volumetric x-ray CT images. IEEE Trans. Med. Imaging **20**, 490–498 (2001)
2. Kuhnigk, J.-M., Hahn, H., Hindennach, M., Dicken, V., Krass, S., Peitgen, H.-O.: Lung lobe segmentation by anatomy-guided 3D watershed transform. In: Sonka, M., Fitzpatrick, J. M.

(eds.) Medical Imaging 2003: Image Processing of Procspie, vol. 5032, pp. 1482–1490, May 2003
3. Ali, A.M., Farag, A.A.: Automatic Lung Segmentation Of Volumetric Low-dose CT Scans Using Graph Cuts, pp. 258–267. Springer, Berlin Heidelberg (2008)
4. Itai, Y., Kim, H., Ishikawa, S., Yamamoto, A., Nakamura, K.: A segmentation method of lung areas by using snakes and automatic detection of abnormal shadow on the areas
5. Silveira, M., Nascimento, J., Marques, J.: Automatic segmentation of the lungs using robust level sets. In: 2007 29th Annual International Conference of the IEEE Engineering in Medicine and Biology Society, pp. 4414–4417. IEEE (2007)
6. Sluimer, I.C., Niemeijer, M., Van Ginneken, B.: Lung field segmentation from thin-slice CT scans in presence of severe pathology. In: Medical Imaging 2004, pp. 1447–1455. International Society for Optics and Photonics (2004)
7. El-Ba, A., Gimel'farb, G., Falk, R., Holland, T., Shaffer, T.: A New Stochastic Framework for Accurate Lung Segmentation, pp. 322–330. Springer, Berlin Heidelberg (2008)
8. El-Baz, A., Farag, A., Ali, A., Gimel'farb, G., Casanova, M.: A Framework for Unsupervised Segmentation of Multi-modal Medical Images, pp. 120–131. Springer, Berlin Heidelberg (2006)
9. Brown, M.S., Goldin, J.G., McNitt-Gray, M.F., Greaser, L.E., Sapra, A., Li, K.-T., Sayre, J.W., Martin, K., Aberle, D.R.: Knowledge-based segmentation of thoracic computed tomography images for assessment of split lung function. Med. Phy. **27**(3), 592–598 (2000)
10. Prasad, M.N., Brown, M.S., Ahmad, S., Abtin, F., Allen, J., da Costa, I., Kim, H.J., McNitt-Gray, M.F., Goldin, J.G.: Automatic segmentation of lung parenchyma in the presence of diseases based on curvature of ribs. Acad. Radiol. **15**(9), 1173–1180 (2008)
11. Mansoor, A., Bagci, U., Xu, Z., Foster, B., Olivier, K.N., Elinoff, J.M., Suffredini, A.F., Udupa, J.K., Mollura, D.J.: A generic approach to pathological lung segmentation. IEEE Trans. Med. Imaging **33**(12), 2293–2310 (2014)
12. Hosseini-Asl, E., Zurada, J.M., El-Baz, A.: Lung segmentation based on nonnegative matrix factorization. In: 2014 IEEE International Conference on Image Processing (ICIP), pp. 877–881. IEEE (2014)
13. Ahuja, N., Rosenfeld, A., Haralick, R.M.: Neighbor gray levels as features in pixel classification. Pattern Recognit. **12**(4), 251–260 (1980)
14. Lee, D.D., Seung, H.S.: Learning the parts of objects by non-negative matrix factorization. Nature **401**(6755), 788–791 (1999)
15. Lin, C.-J.: Projected gradient methods for nonnegative matrix factorization. Neural Comput. **19**(10), 2756–2779 (2007)
16. Berry, M.W., Browne, M., Langville, A.N., Pauca, V.P., Plemmons, R.J.: Algorithms and applications for approximate nonnegative matrix factorization. Comput. Stat. Data Anal. **52**(1), 155–173 (2007)
17. Kim, H., Park, H.: Nonnegative matrix factorization based on alternating nonnegativity constrained least squares and active set method. SIAM J. Matrix Anal. Appl. **30**(2), 713–730 (2008)
18. Sonka, M., Hlavac, V., Boyle, R.: Image processing, analysis, and machine vision. Cengage Learning (2014)
19. Dice, L.R.: Measures of the amount of ecologic association between species. Ecology **26**(3), 297–302 (1945)
20. Abdollahi, B., Soliman, A., Civelek, A., Li, X.-F., Gimel'farb, G., El-Baz, A.: A novel gaussian scale space-based joint MGRF framework for precise lung segmentation. In: 2012 19th IEEE International Conference on Image Processing, pp. 2029–2032. IEEE (2012)
21. Sluimer, I., Prokop, M., Van Ginneken, B.: Toward automated segmentation of the pathological lung in CT. IEEE Trans. Med. Imaging **24**(8), 1025–1038 (2005)
22. Dubuisson, M.-P., Jain, A.K.: A modified Hausdorff distance for object matching. In: Proceedings of the 12th IAPR International Conference on Pattern Recognition, 1994. Vol. 1-Conference A: Computer Vision & Image Processing, vol. 1, pp. 566–568. IEEE (1994)

Competitive Advantage Through Social Media: A Study of Indian Firms

Pallavi Thacker and H. P. Mathur

Abstract The purpose of this paper is to visit the role of social media in attaining or/and sustaining competitive advantage. Not before some 5–10 years, especially in the Indian context, was social media this active and influential. Internet existed merely as a source of information; and nothing more. Technological and social advances have led internet to give birth to an entirely new phenomenon called "social media". Twitter, Facebook, LinkedIn, and many other web platforms have been established. People share their views, companies promote themselves, all the protests and campaigning is done, very much through social media. In this changed time, companies are opting to stand out among their competitors through the use of social media. In proposing out the new social marketing strategies, this paper contributes to the literature pool as well as opens scope for practicing new ways to gain advantage through social media.

Keywords Competitive advantage · Social media · Social networking sites
Resource-based view · Facebook · Twitter · LinkedIn

1 Introduction

> The reality is that most of your competitors aren't likely doing a very good job with social media (most companies aren't), which gives you the chance to stand out. Also consider the flip side. If you avoid social media, you leave a big opening that allows your competitors to capture your audience. (Chandler [3])

P. Thacker (✉) · H. P. Mathur
Institute of Management Studies, Banaras Hindu University, Varanasi, India
e-mail: pallavithacker@yahoo.com

H. P. Mathur
e-mail: hpmathur@yahoo.com

X.-S. Yang et al. (eds.), *Smart Trends in Systems, Security and Sustainability*, Lecture Notes in Networks and Systems 18,
https://doi.org/10.1007/978-981-10-6916-1_7

To stand out among the competitors is the ultimate motive of any business organization. Now that is the big question of the hour, how to differentiate ourselves from others?

This is an era when copying technology of any other physical/infrastructural asset/advantage is not that difficult. Neither, any of such advantage is actually sustainable. Gone are the times when advantages could be sustained for long. With the fast moving times, advantages too move fast! Very recently, Rita McGrath [8] in her book "The End of Competitive Advantage" names this advantage that firms try to attain as "Transient Competitive Advantage" and not "Sustainable Competitive Advantage". To win in this era of Transient Competitive Advantage is the biggest challenges firms face now days.

There can be many ways through which companies try to attain that edge over competitors. Innovation, leadership, technological advancement, differentiation, cost leadership, cultural advantage and many more. A detailed analysis on factors is presented as we proceed to the section of literature survey. With the changing time, the factors and the influence they hold are changing. Some new factors which did not exist some years back have also joined the queue. This research addresses one such factor, and a very hot area of discussion in marketing these days, *Social Media.*

This research paper addresses the following questions:

i. What are the Sources of Competitive Advantage for firms in today's era?
ii. Does Social Media act as a source to Competitive Advantage of firms?
iii. How companies build their brand images through Social media?

2 How Companies Create Awareness over the Internet

Before directly reaching to the platform of social media, it is important to know how companies create awareness over social media, and also what strategies do they adopt to compete over social media. Three methods lie in front of the companies:

i. Through Web-based strategies
ii. Through Internet-based strategies
iii. Through Social media platforms

In web-based strategies, companies create their own websites and create awareness through their own portals. In internet-based strategies, they can put their advertisements/glimpse over the frequently used websites by people; so that when people open those sites, they may look at the highlighting picture of the company's product/services at one of the strategic corners. Apart from these two, there is another strategy that uses the social networking sites. Here companies create their pages on the social media sites such as Facebook, LinkedIn, Twitter, and the like.

Also, the world is talking of digital transformation. There are three important drivers to this transformation (Fig. 1).

The figure shows that Social connectivity comes first, and then come data and mobility. Social connectivity, or social media, gives a direct link to companies and their leaders to connect directly with all types of current and prospective customers; which is never possible through their products/services or through any other media.

This paper tries to figure out how these social media platforms help firms to create/sustain competitive advantages.

3 Social Media—A Glimpse

Comprised of two terms "social" and "media", social media is a platform that enables people/firms to share their content /updates /information among each other. This is a kind of "digitalized word-of-mouth". People interact on a virtual platform and they have a choice to like /dislike /comment over other people's activities. Social media comes with many benefits. It takes less time to reach a large number of people; in case of products/services, instant feedback and review is possible (Fig. 2).

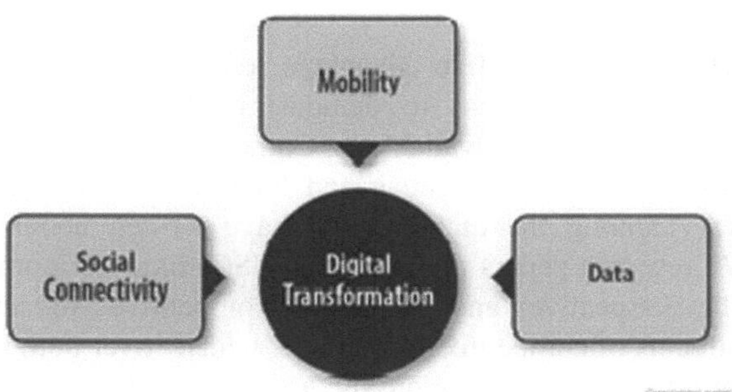

Fig. 1 (Adopted from *The Engaged Leader: A Strategy for your Digital Transformation,* Charlene Li [7])

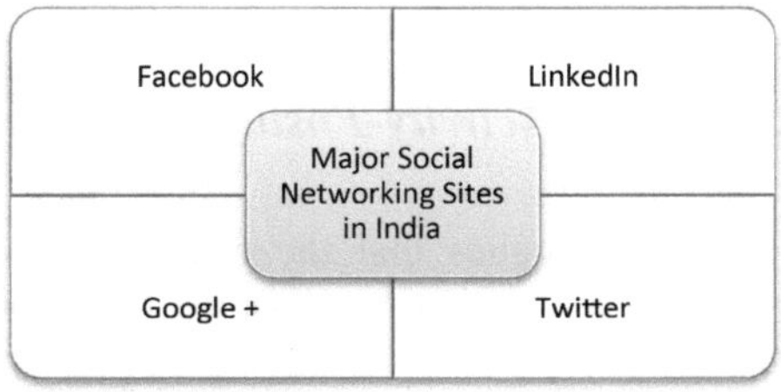

Fig. 2 Source—Compiled by the Author

4 Why is Social Media Important?

The meaning of Marketing has changed of late. The focus of marketing has shifted from products to consumers. Also, the awareness on part of consumers has increased. Social media is no more only about sharing pictures with families and friends or updating videos on Youtube. Social media is now being used widely by business organizations for promotion, improvement and profit-making. Social networking sites also help companies to build and strengthen their images among their customers; and also this helps them to attract prospective customers. Companies now a day hire social media managers who help them maintaining their social advertisements and conversations. While some companies have started this on regular basis, not all companies are yet open to social media. This has been said especially for developing nations.

Also, as per a UN Report, With 356 million 10–24 year-olds, India has the world's largest youth population despite having a smaller population than China, and this generation is *always online*. This fact encourages marketers to design their strategies in such a way as to reach them easily and giving them an easy portal to get answered of all their inquisitiveness about the product. *Social Media* seems to be a solution to the same.

5 Competitive Advantage—A Literature Survey

The term competitive advantage refers to "a set of capabilities that permanently enable the business to demonstrate better performance than its competitors" [2]. The literature available on Competitive Advantage is plenty. Many studies have been conducted on Indian and International organizations; also, the perspectives vary among the authors who have written about the concept. Approximately 75 research papers (belonging to SSCI indexed journals) were referred to find out the perspectives and sources to Competitive Advantage. Majorly, three research streams of competitive advantage, which cover both internal and external attributes of a firm, exist. These are the activity position view, the resource based view, and the relational view.

5.1 The Activity-Position View

This view argues that the firm's superior performance mostly results from its strategic choice that provides the firm a better positioning in the industry structure

Porter [10]. In particular, [11] emphasises that competitive advantage resides in business activities and activity systems, rather than firm's resources.

This view is considered to be the classic view of competitive advantage. And the researchers who brought forward this view are known to be the pioneers of the theory of Competitive Advantage.

5.2 The Resource Based View

After the Activity-position view, another view, i.e., the Resource-based view was named by Birger Wernerfelt in his article A Resource-Based View of the Firm (1984). This view holds that dissimilar resource endowments result in distinctive competitive advantage and different performances between firms. (e.g., [1], [9]) According to this view, the primary resources of the firm and not the activity systems define a firm's competitive advantage. Physical assets, financial capital, human resources, organizational systems, technology and knowledge, and intangible assets (e.g., trademark, patent, copyright and goodwill) are the factors responsible for the edge. This was the first view that held that people can be responsible for gaining the edge and not exactly the facilities of the firm. This is the most popular view as per the literature and business leaders; because directly or indirectly, facilities/activities of the firm are in hands of people, and thus the human-factor is the main driving force behind any success/failure.

5.3 The Relational View

This view of competitive advantage goes beyond the firm's boundaries. It suggests that competitive advantage stems from collaboration or social relations, rather than a firm's distinctive resources or individual activities (Dyer and Singh [4], Lavie [6]).

According to this view, an individual firm alone will never be able to generate competitive advantage; instead, the advantage is determined by dynamic interactions between organizations to create mutual benefits.

This research will take into account, majorly, the Resource-based view and partly, the Relational view of Competitive Advantage as it majorly focuses on the knowledge and people part, which forms the core of this study.

Various scholars have worked upon the possible sources of competitive advantage. Scanning the literature will be incomplete without including some major findings in this area.

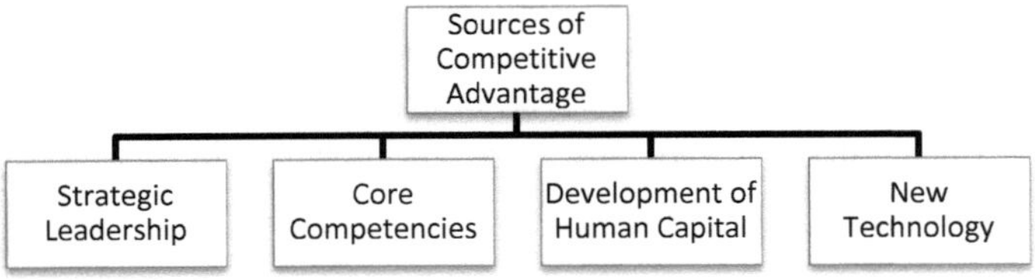

Fig. 3 Source—Compiled by the Author

6 Sources of Competitive Advantage

Focusing on the Relational View, a few sources of Competitive Advantage are discussed below:

Hitt et al. [5] gave six ways to achieve competitive advantage. They said a firm can gain competitive advantage either by exercising strategic leadership (it is a leadership by the top management personnel of the firms), or by building the dynamic core competences of the firm. They can also do so by focusing and developing human capital or by making effective use of new technology. The firms can also gain advantage by engaging in valuable strategies or by developing new organizational structures and culture (Fig. 3).

Another important view point is given by Wen-Cheng et al. [12]. They suggested three major sources of Competitive Advantage in their study, i.e., Technology and Innovation, Human Resources and the Organizational Structure of the firm.

Combining the theories and sources of the competitive advantage, it can be concluded that people and technology both constitute the major factors of determining the advantage. And social media/networking is anytime a mix of two. It is the result of a technology innovation and is managed by people.

7 Methodology and Findings

The profound scanning of literature tells us that there can be various methods to gain and sustain the competitive advantage.

Mainly, three views/theories towards the achievement of competitive advantage exist, Activity-position view, Resource-based view and the Relational view. And among the sources of competitive advantage, human capital and technology are focused within the resource based view. Social media is practically an example of the combination of two, i.e., it is a technology completely depending upon how a person uses it. Studies show that social networking has given a new shape to *leadership*, there are now more ways in which a leader connects to the followers and to the bigger world outside the firm, that too directly.

8 Conclusion

Social media has emerged as one of the strongest ways companies present themselves in front of their present and prospective buyers. Increased awareness level of consumers has made it almost a mandate for firms to come online and interact with them. Easy availability, easy feedback and complaint systems and other advantages make Social Networking sites friendly for consumers when it comes to brand selection.

This study opens a door to further analysis, as to what Indian firms prefer when it comes to making their firms/strategies visible over the social media. If Facebook works better, or Twitter, or some other platform? Globalisation we say is already in our economy and systems; but social networking and marketing is giving an entirely new meaning to globalization by bringing on more ease in connecting and interacting. All technological advancement is a boon, but a curse as well. Social media marketing has limitations too! The challenge is to win over the curse part and use it a tool towards gaining and sustaining competitive advantage.

References

1. Barney, J.B.: Firm resources and sustained competitive advantage. J. Manag. **17**, 99–120 (1991)
2. Bobillo, A., Iturriaga, F., Gaite, T.: Firm performance and international diversification: The internal and external competitive advantages. Int. Bus. Rev. **19**, 607–618 (2010)
3. Chandler: The Hidden Benefits of Social Media Marketing: Why Your Strategy May Be Working Better Than You Think, March 12, 2013. Available at http://www.forbes.com/sites/work-in-progress/2013/03/12/the-hidden-benefits-of-social-media-marketing-why-your-strategy-may-be-working-better-than-you-think/#64c3cf7b1185 (2013)
4. Dyer, J.H., Singh, H.: The relational view: corporate strategy and sources of inter-organizational competitive advantage. Acad. Manag. Rev. **42**(4), 37–43 (2006)
5. Hitt, M.A., Keats, B.W., Demaiie, S.M.: Navigating in the new competitive landscape: Building strategic flexibility and competitive advantage in the 21st century. **12**(4) (1998)
6. Lavie, D.: The competitive advantage of interconneted firms: An extension of the resource-based view. Acad. Manag. Rev. **31**(3), 638–658 (2006)
7. Li, C.: The Engaged Leader: A Strategy for Your Digital Transformation. Wharton Digital Press. Retrieved from https://play.google.com/books/reader?id=wGO9BwAAQBAJ&printsec=frontcover&output=reader&hl=en&pg=GBS.PA1 (2015)
8. McGrath, R.G.: The End of Competitive Advantage: How to keep your strategy moving as fast as your buisness (2013)
9. Peteraf, M.A.: The cornerstones of competitive advantage: A resource-based view. Strateg. Manag. J. **14**, 170–181 (1993)
10. Porter, M.E.: Competitive Strategy. Free Press, New York (1980)
11. Porter, M.E.: Towards a dynamic theory of strategy. Strateg. Manag. J. (Winter Special Issue) **12**, 95–117 (1991)
12. Wen-Cheng, W., Chien-Hung, L., Ying-Chien, C.: Types of competitive advantage and analysis. Int. J. Bus. Manag. **6**(5), 100–104 (2011). https://doi.org/10.5539/ijbm.v6n5p100

A Real-Time Devnagari Sign Language Recognizer (α-DSLR) for Devnagari Script

Jayshree Pansare and Maya Ingle

Abstract Devnagari Sign Language (DSL) is used for communication between dump and deaf users. Our Alphabet Devnagari Sign Language Recognizer (α-DSLR) system is used to translate DSL alphabets into Devnagari alphabets along with speech. Devnagari alphabets comprises fourteen vowels ranging from "A" to "A:" and thirty-three consonants ranging from "k" to "&". Work flow of α-DSLR system emphasizes on sequential phases along with algorithmic approach used in our system. The system works with Single Hand Single Camera approach and applies template based and clustering based algorithms. The detection rate of 97% is accomplished by α-DSLR system against a complex background.

Keywords DSL alphabets • Devnagari alphabets • Single hand single camera approach • *Biggest BLOB* algorithm • *Sim-Temp-Match* algorithm *K-Cluster-Temp-Match* algorithm

1 Introduction

Multiple sign language recognizer have been designed for recognition of various sign languages. Alphabetic HGRS perceives alphabet signs of American Sign Language (ASL) alphabets based on centroid, Euclidian distance, median filter, and morphological operation. The system recognizes static gestures in real-time with 91.19% recognition rate against the complex background [1]. HGRS includes histogram matching to recognize 46 DSL alphabets in static background and achieves 87.82% precision [2]. N-DSLA system recognizes 10 DSL numbers

J. Pansare (✉)
Modern Education Society's College of Engineering, Pune, S.P. Pune University, Pune, India
e-mail: jayshree.pansare23@gmail.com

J. Pansare · M. Ingle
Devi Ahilya Vishwavidyalaya, Indore, India
e-mail: mayaingle22@gmail.com

X.-S. Yang et al. (eds.), *Smart Trends in Systems, Security and Sustainability*, Lecture Notes in Networks and Systems 18,
https://doi.org/10.1007/978-981-10-6916-1_8

Table 1 Alphabets of Devnagari script

अ	आ	इ	ई	उ	ऊ	ए	ऐ	ओ	औ
अं	अः	क	ख	ग	घ	च	छ	ज	झ
ट	ठ	ड	ढ	ण	त	थ	द	ध	न
प	फ	ब	भ	म	य	र	ल	व	श
स	ह	ळ	क्ष	ज्ञ	ऋ	ड			

ranging from 0 to 9 using Exhaustive Template Matching and Single Hand Two Cameras approach. It achieves recognition rate of 97.2% in complex background. We have presented the literature related to research articles from [3] in our previous research article [4]. In this work, we described about ASL, Arabic Sign Language (ArSL), Greek Sign Language (GSL), Chinese Sign Language (CSL) etc. and multiple techniques. We have discussed all techniques comprehensively in [5]. However, there is widespread scope for development of sign language recognizers for DSL alphabet identification and translation in Devnagari alphabets. Devnagari alphabets are represented in Table 1. These invasive DSL alphabets are as represented in Fig. 1a and b are used in α-DSLR system for experimentation. The standard DSL alphabets are depicted in Figs. 1a and 4 new DSL alphabets proposed in α-DSLR are as shown in Fig. 1b. Our prime objective is to design Devnagari hand gesture recognizer that convert 47 DSL alphabets in Devnagari alphabet against complex background. In Sect. 2, we emphasizes on workflow and algorithms for α-DSLR system. Experimental results of α-DSLR system are discussed in Sect. 3. Comparative performance of α-DSLR system is enlightened in Sect. 4. Lastly, we conclude in Sect. 5.

2 Work Flow and Algorithms Used in α-DSLR System

Our α-DSLR system executes sequentially from the image capturing phase to text to speech conversion phase. In this section, work flow associated with aforementioned phases is depicted in Fig. 2. In this view, we emphasize on our major contribution, i.e. development of three algorithms present in the highlighted portion of workflow of α-DSLR system, namely; *biggest BLOB*, *Sim-Temp-Match* and *K-Cluster-Temp-Match* algorithm in detail.

Fig. 1 **a**, **b** Invasive DSL alphabets used in α-DSLR System **a** standard DSL alphabets used in α-DSLR system **b** proposed DSL alphabets used in α-DSLR system

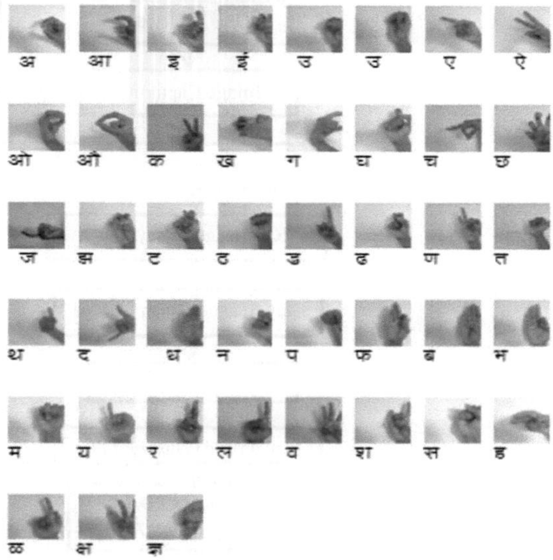

(a) Standard DSL Alphabets used in α-DSLR System

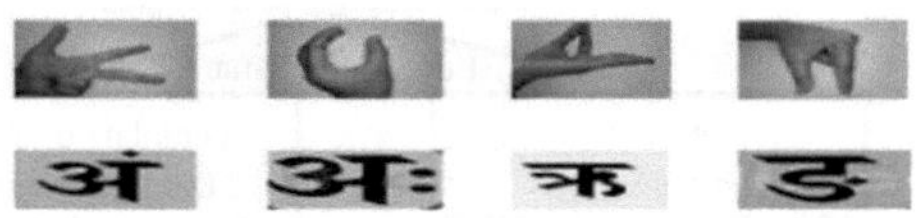

(b) Proposed DSL Alphabets used in α-DSLR System

In image capturing phase, we capture the input image of DSL alphabet in RGB color format from the distance of 25 cm. Fixed position camera of resolution 8 Mega Pixel is mounted on the monitor of computer is used to capture snapshots against a complex background. Image pre-processing phase consists of conversion of RGB image to gray image and gray image to a binary image. Further, noise removal is applied on binary image of DSL alphabet followed by morphological operations such as erosion and dialation. Furthermore, 8-connected Binary Linked Object (BLOB) of size 80 × 60 is extracted using *biggest BLOB* algorithm (Algorithm I). Image post-processing phase includes feature extraction using template-based tracking technique and *Sim-Temp-Match* algorithm (Algorithm II) for feature matching. Feature vector of running input template image and training dataset template images are provided by feature extraction phase. Our proposed Algorithm II that is based on template matching technique is applied for matching these feature vectors effectively. Further, *K-Cluster-Temp-Match* algorithm (Algorithm III) and SVM are used for clustering and classification purpose respectively followed by text to speech conversion phase. The aforementioned algorithms are presented in subsequent section.

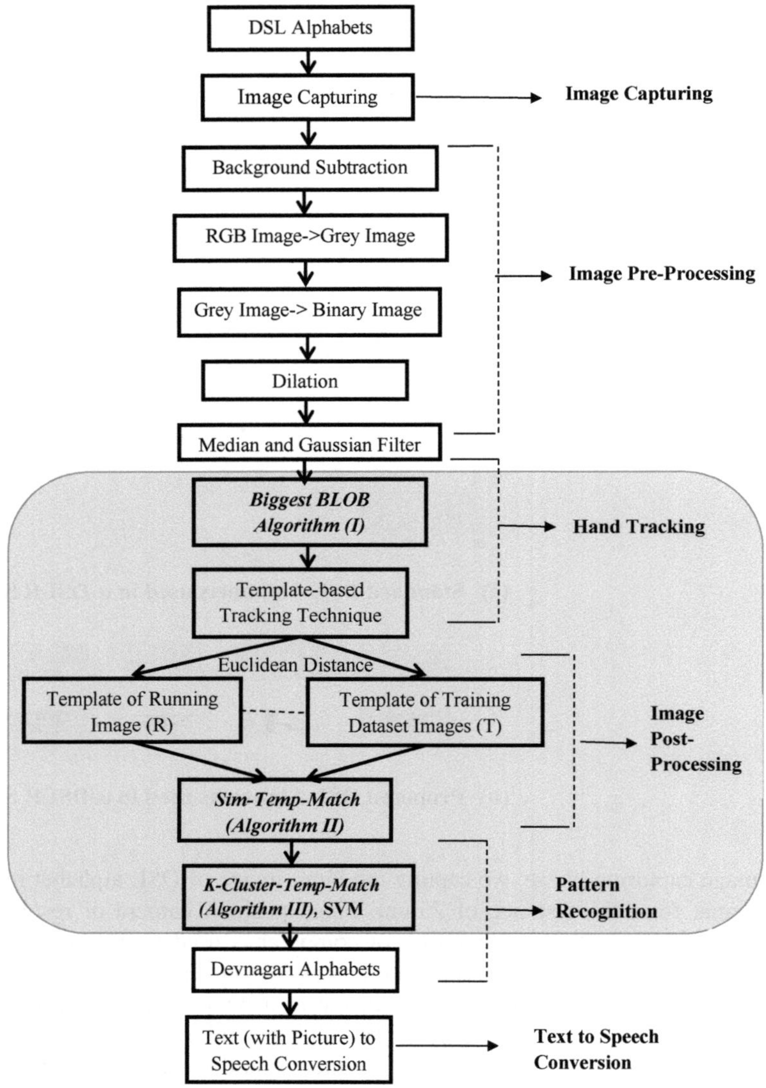

Fig. 2 Work flow for α-DSLR system

2.1 Biggest BLOB Algorithm

In Image processing, BLOB detection process is responsible to find a region of interest in the image that differs in properties such as brightness, color and area than other regions. We present Biggest-*BLOB* algorithm.

Algorithm- I: *Biggest-BLOB* Algorithm for α-DSLR System

```
Input:   Noise recovered two dimension image matrix.
Output:  Extracted hand two dimension image matrix (a largest region).
Begin
  1. temp = zeroes(size(input image));   /* Black image with the same size as input image */
  2. [index total_lablels]= labeling(input image,8); /*8 component connectivity Elements are labeled */
     if(total_labels>0) // At least one label must present for further processing//
  3.   for i=1 to total_labels //For each label//
  4.      area[i] =length (i);  // calculate length of each label//
  5.   end
  6.   large=Max(area) ; // select largest length label from area set//
       temp(large)=1;     /* 1 represents white pixel value, the largest label assign into temp image */
  7.
  8. end                  // return output image contains only a largest label//
  9.    return temp;
End
```

2.2 *Sim-Temp-Match Algorithm*

Enriched template matching techniques are applied in this algorithm and presented as follows:.

Algorithm- II: *Sim-Temp-Match* Algorithm for α-DSLR System

Input: $accu$, X_i $\{TD \mid X_i \in TD, i = 1, 2 \ldots N\}$
/* TD: Training Dataset for DSL Alphabets; N: Total No. of Samples of DSL Alphabets; *accu*: accuracy expected*/
Output: W_i // Recognized Patterns as Devnagari script

Begin

1. *for i = 1 : N*
2. *Begin*
3. *get R* // *R* : Running Image of DSL Alphabet
4. $T_i = TM(R, X_i)$ /* *TM*: Template Matching function that matches
   $\mu_i = T_i . \omega$ /* μ_i measure of match
5.
6. $PR_{i,p} = find\ (min\ (\mu_i))$ // $PR_{i,p}$ is useful for storing mean
7. *Begin*
8. *find* $PRi_{,k}$ // Pattern Recognition
9. *If* $PR_k \geq accu$
10. *Begin*
11. *get R of* $PRi_{,k}$
12. $g_j(R) = min\ (d_{cham,\tau}(R, X_i))$ // eq. 4.17 used to find similarity
13. $W_i = j$ // j^{th} value recognized as Devnagari script
14. *end*
15. *end*
16. *end*

End.

2.3 K-Cluster-Temp-Match Algorithm

In α-DSLR system, we employ K-means clustering data grouping scheme. Using K-means clustering partitions, a set of data is partitioned into *k* subsets. On the basis of above clustering scheme, we introduce Algorithm II-*K-Cluster-Temp-Match* algorithm in α-DSLR system.

Algorithm-III: *K-Cluster-Temp-Match* Algorithm for α-DSLR System

Input: X_i $\{T \mid X_i \in T, i = 1, 2 \ldots N\}$ // T: Training Dataset for DSL Alphabets
// N: Total No. of Samples for DSL Alphabets
Output: $PR_{i,p}$ // $PR_{i,p}$: Recognized Pattern in Cluster
Begin
for i = 1 : m // m=Total No. of Samples in each Cluster
Begin
for j= 1 : n // n= Total No. of Clusters
Begin
get c_j, x_i // c_j : No. of clusters, x_i : Samples in each cluster
calculate $m(c_j|x_i)$ // Calculate Membership Function for each data x_i
calculate $w(x_i)$ // Calculate Weight for each data x_i
$C_j = \frac{\sum_{i=1}^{n} m(c_j|x_i)w(x_i)x_i}{\sum_{i=1}^{n} m(c_j|x_i)w(x_i)}$ // Compute the Center of the Cluster
$\mu_i = C_j$ // Initialize the Center of the Clusters
if l≠i
Begin
for k= 1:p
Begin
ci= $d(xj, \mu i) \leq d(xj, \mu l)$ // Identification of Cluster
$\mu i = |ci| * sum\ (j)$ /* Position of cluster */
$KM(T, \mu i) = \sum_{i=1}^{n} min_{j \in \{1..k\}} \|x_i - c_j\|^2$ /* KM: K-Means function for Minimization of Squared Distance
$KM\ (T, \mu i) = PR_k$ // Assign Resultant to Pattern
$PR_{i,p} = find\ (min\ (PR_{i,k}))$ // Recognized Pattern in Cluster
end
end
end
end
End.

3 Experimental Results of α-DSLR System

Our α-DSLR has designed for translation of DSL alphabets into Devnagari alphabets. It consists of we camera of resolution 8 Mega Pixel, input image of size 160 × 120, distance of posed 25 cm from fixed position camera. The proposed α-DSLR system is implemented using C# and *.NET*. In α-DSLR system, 47 orientation groups for canny edges are formed and hence eight separate distance maps are calculated per image. Similarly, the template points are also divided into 47 groups (i.e. Cluster). In this context, 100 templates for 47 DSL alphabets are collected and designed 47 classes as 47 clusters in α-DSLR system. The aforementioned input running image and training dataset is constructed with 4700 template images of DSL alphabets. Our system includes diversified methods related to consequent phases such as image pre-processing, hand tracking, image post-processing and pattern recognition. *Sim-Temp-Match* algorithm and *K-Cluster-Temp-Match* algorithm are applied in α-DSLR system. *Sim-Temp-Match* algorithm is applied for finding the highest measure of match between input running template

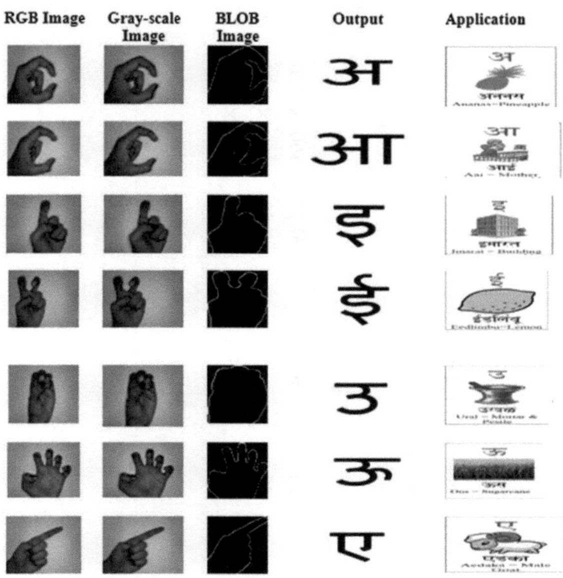

Fig. 3 Outcome of α-DSLR system

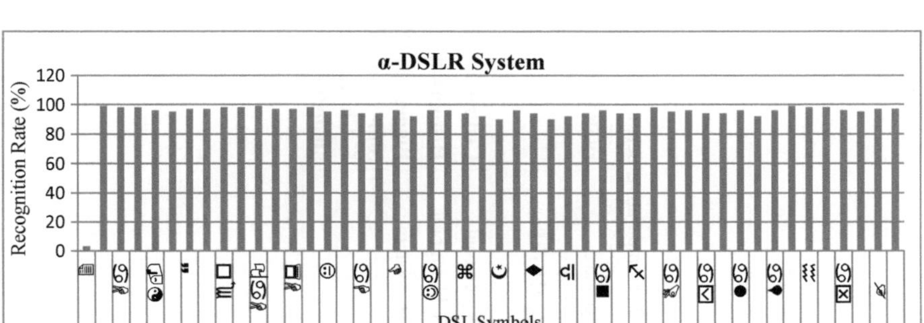

Fig. 4 Experimental results of α-DSLR system

Table 2 Comparative performance of DSLR system with other sign language recognizers

Hand gesture recognition systems (HGRSs)	HGRS techniques	No. of gestures	Rec. time (sec/frame)	Rec. rate (%)	Frame resolution	No. of training dataset images	Background	Lighting condition
HGRS_centroid	**Centroid of BLOB**	26	**0.7**	90.19	160 × 120	2600	Complex	Natural
HGRS_histogram	**Histogram**	46	0.5	87.82	160 × 120	4600	Static	Mixed
α-DSLR system	**Biggest BLOB, Sim-Temp-Match, K-Cluster-Temp Match algorithm**	**47**	**0.5**	**97**	**160 × 120**	**4700**	**Complex**	**Mixed**

image and training dataset template images. *K-Cluster-Temp-Match* algorithm forms 47 clusters in the training dataset for respective DSL alphabets. It represents best suitable pattern and displays Devnagari alphabet as text with picture and further converted to speech. Outcome of α-DSLR system and experimental result in graphical form are as depicted in Figs. 3 and 4 respectively.

4 Comparative Performance of α-DSLR System with Existing Sign Language Recognizers

The comparative performance of α-DSLR system with other recognizers is presented in this Section. It is observed from Table 1 that α-DSLR system detects 47 DSL alphabets in complex background with a recognition rate of 97%. The system outperforms using *Sim-Temp-Match* algorithm as compared to existing HGRS recognizes 46 DSL alphabets using histogram matching in static background achieves a recognition rate of 87.82%. Moreover, on the basis of some vital factors, as a comparative study is presented as shown in Table 2. However, Table 2 proved that α-DSLR system is superior as compared to existing sign language recognizers based on vital factors.

5 Conclusion

Our α-DSLR system is applied for detection of 47 DSL alphabets. DSL alphabets are converted into Devnagari alphabets using Single Hand Single Camera approach. It works robustly in complex background with mixed lighting condition. It accomplishes detection rate of 97% and identification time of 0.5 s and uses 4700 samples in training dataset. The experimental results of α-DSLR are compared with existing HGRS and measured performance has shown superior results in α-DSLR system. The outcome of DSLR system is text (in picture form), word and sentence formation along with speech as well.

References

1. Pansare, J., Ingle, M., Gawande, S.: Real-time static hand gesture recognition for American sign language (ASL) in complex background. J. Sig. Inf. Process. **3,** 364-367 (2012). https://doi.org/10.4236/jsip.2012.33047
2. Pansare, J.R., et al.: Real-time static Devnagri Sign Language translation using histogram. Int. J. Adv. Res. Comput. Eng. Technol. **2**(4), 1–5 (2013)
3. Pansare, J.R., et al.: Real-time static hand gesture recognition system in complex background that uses number system of Indian sign language. Int. J. Adv. Res. Comput. Eng. Technol. **2** (3), 1086–1090 (2013)

4. Pansare, J.R., Ingle, M.: 2D hand gesture numeric Devnagari Sign Language analyzer based on two cameras. In: International Conference on Intelligent Human Computer Interaction 2016 (IHCI 2016), pp. 148–160. Springer LNCS 10127 (2017). https://doi.org/10.1007/978-3-319-52503-7_12
5. Pansare, J.R., Ingle, M.: Comprehensive performance study of existing techniques in hand gesture recognition system for sign languages. Int. J. Comput. Sci. Inf. Technol. **7**(3), 1343–1347 (2016)
6. Hu, J., et al.: Bare-fingers touch detection by the button's distortion in a projector—camera system. IEEE Trans. Circuits Syst. Video Technol. **24**(4), 566–575 (2014)
7. Kanwar, A.: An Appearance Based Approach For Gait Identification Using Infrared Imaging, pp. 719–724. IEEE (2014)
8. Ghafouri, S., Seyedarabi, H.: Hybrid Method for Hand Gesture Recognition Based on Combination of Haar-Like and HOG Features, pp. 0–3. IEEE (2013)
9. Abid, M.R., et. al.: Dynamic sign language recognition for smart home interactive application using stochastic linear formal grammar. IEEE Trans. Instrum. Meas. 1–10 (2014)
10. Tofighi, G., et. al.: Hand pointing detection using live histogram template of forehead skin. In: Proceedings 19th International Conference on Digital Signal Processing, pp. 383–388 (2014)
11. Yao, Y., et al.: Contour model-based hand-gesture recognition using the kinect sensor. IEEE **24**(11), 1935–1944 (2014)
12. Pansare, J.R., et. al.: Real-time static hand gesture recognition system in complex background that uses number system of Indian sign language. Int. J. Adv. Res. Comput. Eng. Technol. **2** (3), 1086–1090 (2013)
13. Pansare, J.R., et al.: Real-time static hand gesture recognition system using HCI for recognition of numbers. Int. J. Adv. Res. Comput. Sci. **4**(4), 258–262 (2013)
14. System, R., et al.: Novel FPGA implementation of hand sign. IEEE Trans. Circuits Syst. Video Technol. **25**(1), 153–166 (2015)
15. Zabulis, X., et. al.: Vision-Based Hand Gesture Recognition For Human-Computer Interaction, pp. 1–56 (2005)
16. Zeng, B., et al.: A hand gesture based interactive presentation system utilizing heterogeneous cameras. Tsinghua Sci. Technol. **17**(3), 329–336 (2012)
17. Alvi, A.K., et al.: Pakistan sign language recognition using statistical template matching. Int. J. Inf. Technol. **1**(1), 1–4 (2004)
18. Yao, Y., et al.: Contour model-based hand-gesture recognition using the kinect sensor. IEEE Trans. Circuits Syst. Video Technol. **24**(11), 1935–1944 (2014)

Novel Strategy for Fairness-Aware Congestion Control and Power Consumption Speed with Mobile Node in Wireless Sensor Networks

Sagar B. Tambe and Suhas S. Gajre

Abstract The power issue in wireless sensor network (WSN) stays one of the real barriers keeping the complete abuse of this technology. The WSN is a dense network of sensors which sense the environmental conditions, process and propagate that data towards sink node. Limited battery life of sensor node and unbalanced utilization of that energy can affect the lifetime of the entire sensor network. In proposed work, mobile nodes are used to transmit the data nearby the area where the power consumption of the nodes is more. The mobile node reduces the workload and congestion of the nodes which is controlled by adjusting the reporting rate (RR) according to the buffer occupancy level. In this paper, we have correlated the existing Ad hoc on demand vector (AODV) routing protocol with our new approach to the delivery of power consumption. The proposed work evaluate the performance of the lifetime and energy consumption of the WSN with and without mobile nodes and results achieved by adjusting the number of mobile nodes, location and the speed of mobile node. The RR is also adjusted to control the buffer occupancy of each node and mitigate the congestion that occurs in the sensor network. Based on this simulation results, this proposed work increases the lifetime of the nodes whose power consumption speed is high and enhances the life of the entire network. The use of a dual threshold for buffer and mobile node can reduce the congestion and the waiting time for data reducing the delay.

Keywords Wireless sensor network · Energy · Network lifetime · Sensors
Energy-aware systems · Mobile node

S. B. Tambe (✉)
Department of Information Technology, SGGSIE&T, Nanded, Maharashtra, India
e-mail: tambesagar@sggs.ac.in

S. S. Gajre
Department of Electronics and Telecommunication Engineering, SGGSIE&T, Nanded, Maharashtra, India
e-mail: ssgajre@sggs.ac.in

X.-S. Yang et al. (eds.), *Smart Trends in Systems, Security and Sustainability*, Lecture Notes in Networks and Systems 18,
https://doi.org/10.1007/978-981-10-6916-1_9

1 Introduction

Nowadays, improvements in integrated circuit innovation have made possible the development of a large number of modest and low-power sensor hubs with on wireless communication, signal preparing, and remote sensing abilities [1, 2]. A WSN is a highly distributed system of small, lightweight wireless nodes, deployed in large number to monitor the environment or system by the estimation of physical parameter [3]. The sensor network can be described as a collection of tiny sensor nodes which can provide access to information anytime, anywhere by gathering, preparing and investigating of data [4]. These devices or nodes called sensors which consist of sensing, local data processing, and communicating components like controller, transceiver also onboard storage and power source i.e. battery. The typical sensor network arrangement as shown in Fig. 1.

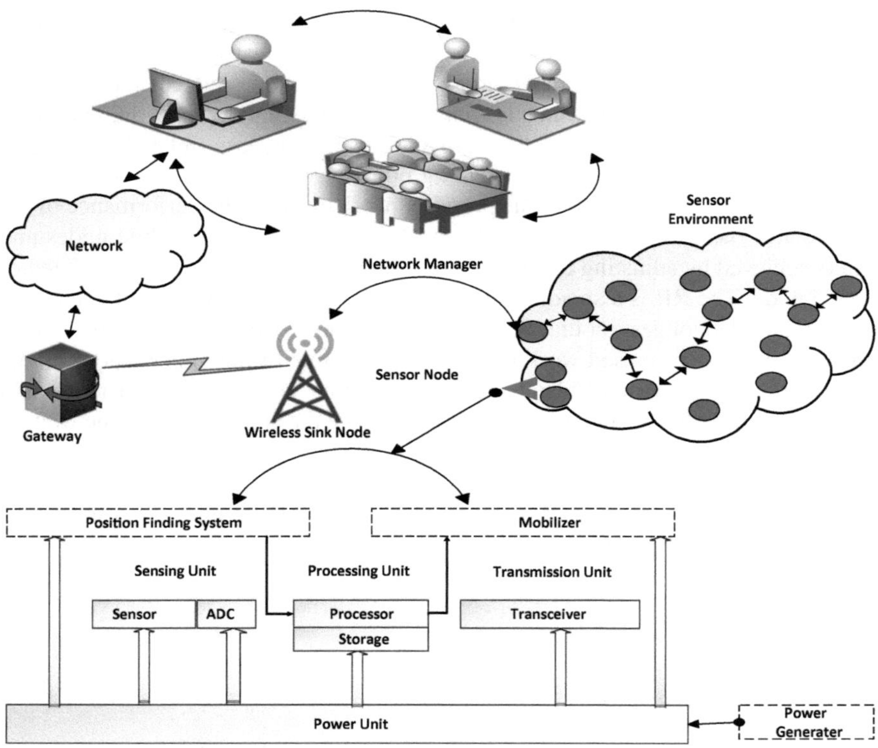

Fig. 1 Typical sensor network arrangement

Generally a WSN consist of seismic sensor used for measuring motion of the ground, infrared sensor for detecting characteristic and movements of surrounding [5], The WSN of these special sensor can be utilized in application such as forest fire control [6], thermal sensor for measuring temperature of surrounding [7, 8], humidity in air [9, 10], vehicle surveillance [11, 12], noise levels [13], military [14], healthcare environment [15, 16], chemical processing scenarios [17], and so on. Each sensor node is made up of transceivers, micro-controller, external memory, power sources [18] and defined as:

Controller: The controller controls the functionality of other components in the sensor node performs tasks and processes data. While the most common controller is a microcontroller, a general purpose desktop microprocessor, digital signal processors, FPGAs [19] and ASICs [20] can be used as controller.

Transceiver: The transceivers function as both transmitter and receiver. The operational states are transmitted, receive, idle, and sleep. The transceivers in its receiving mode consume almost same energy as that of its idle mode. Whenever the node is not transmitting or receiving the data, it's better to shut it down than to keep it idle.

External memory: In sensor nodes, the important factor is the memory. The on-chip memory of a microcontroller is mostly used. Flash memories are used due to their cost and storage capacity. Two categories of memory based on the purpose of storage are: user memory used for storing application related or personal data and program memory used for programming the device.

Power source: The important factor in any electronic or mechanical device is the power or the battery. The sensor node consumes power for sensing, communicating and data processing. Two power saving policies are being used. These are the Dynamic Voltage Scaling (DVS) policy and the Dynamic Power Management (DPM). The DVS scheme is based on varying the power level based on the workload. DPM helps conserve power by shutting down those part of the sensor node that is idle.

In WSN, the limited battery life of sensor node and unbalanced consumption of that energy can affect the coverage of the network, connectivity, reliability as well as lifetime of the whole sensor network. In the WSN, there are many different issues like flow and congestion control, loss recovery, quality of service, fairness, reliability, energy efficiency, etc. To address these different issues, a lot of work has been done. Among them, energy is a most critical issue because of the battery power with the sensors and the applications where sensors are used. It is almost impossible to change the battery of the sensors. To control or avoid the congestion as well as to save the bandwidth and energy of the nodes, the data sending rate of the nodes close to the source should be controlled. So to adjust the sending rate of each flow as early as possible and save the scarce resource at the nodes close to the sink node, all intermediate sensor nodes are categorized into near-source nodes and near sink node. The probability of the congestion is high to the nodes those are near to sink due to convergent nature of the WSN. Based on this simulation results, this proposed work that will increase the life of the entire network by balancing energy consumption speed and reducing the congestion.

The rest of the paper is organized as follows: In Sect. 2, we introduce related work. In Sect. 3, we present several related methodologies of power consumption and congestion control in WSN. In Sect. 4, presents the proposed system design and methodology. In Sect. 5, we present a proposed system and algorithm. Experimental setup, result, discussion and comparison analysis are described in Sect. 6, and finally we conclude the paper in Sect. 7.

2 Related Work

The major issue in the WSN related to flow and congestion control is studied in detail [21–27]. The loss recovery are discussed in [28, 29], whereas the quality of service issues are focused in [30, 31]. The fairness factor in WSN is addressed in [32] and reliability matters are investigated in [33, 34]. The energy efficiency [35] being the most critical issue in WSN due to limited battery power is discussed in [1, 36, 37].

The congestion control saves the bandwidth as well as the energy of the nodes. In order to speed-up the transmission of packets, two types of node namely near-source node and adjacent sink node are introduced in the network. The probability of the congestion is high to the nodes those are near to sink due to convergent nature of the WSN. In [32], the authors proposed the fairness aware congestion control (FACC) scheme in which the queue at each node is maintained with two thresholds Q_l and Q_h. If the queue occupancy is less than Q_l, the packet is accepted. On the other hand, if the queue occupancy exceeds Q_h, the arriving packets will be dropped, which indicates that the traffic is overwhelming, and rate of all passing flows should be reduced. A Warning Message (WM) consisting of a flow ID along with node ID is generated by the sink node once the congestion occurs.

Like FACC, ECODA [21] also maintains a queue dual buffer thresholds and weighted buffer difference for congestion detection. When congestion occurs, packets are dropped to alleviate congestion. The total number of packets are N, Two thresholds Q_{min} and Q_{max} are used. If $0 \leq N \leq Q_{min}$, all incoming packets are buffered because queue utilization is low. If $Q_{min} \leq N \leq Q_{max}$, some packets with low dynamic priority are dropped or overwritten by subsequent packets with high dynamic priority and the expected average buffer length increases. If $Q_{max} \leq N \leq Q$, some packets with high dynamic priority are dropped or overwritten, then the expected average buffer length increases. Once the source node receivers a backpressure message it immediately adjusts its transmission rates in line with receiver or multihop receivers if exit.

In [38] AODV protocol, the route from source to destination is discovered only when the source has data to send. If no information is available, it broadcasts route request (RREQ) packet to find the path to the destination. When any node receives the RREQ packet, it also checks in its own routing table. If the route is available, then the node replies back by sending route reply packet (RREP) with the path to reach the destination. The main drawback of AODV protocol is multiple packets are needed to be transferred. In response to a single request packet, multiple reply packets are generated. So the modes in the path loss their energy which imbalance the network power consumption. The high energy consumption rate is controlled by comparing energy consumption speed with different nodes.

The use of mobile node i.e. either sink or sensor node helps to reduce the number of transmissions. Authors [39] proposed a new technique for data distribution using mobile sink groups C-SGM. The total sensing field F is divided into the coarse-grained grid. If two different nodes send a notice at the same time, the node with the lowest identifier wins and becomes the header. When the energy of the header node goes below the certain threshold, it reports the header node reselection inside its cluster.

In [37], the authors proposed a new strategy in which all intermediate nodes in between sink and source node cooperate in the packet forwarding. When any node in the intended path fails to receive the data packet and send the acknowledgment to the sender. i.e. failing node and the sender node will work as a cooperative node. This cooperative node forwards the data packet to the next hop node and sends the acknowledgment to the sender of the packet. Each node multicast the packet to all other nodes that are within the communication range of that node. The other nodes sense the channel which is silent, so the cooperative node sends the acknowledgment to the sender and forward the packet to next hop node. It is possible that there can be many nodes which act as a cooperative node or not even a single one. If there are many nodes then to decide that which node will reply back and act as a cooperative node, a timer is set with each node. This timer is called as backoff timer. Then the node having shortest backoff timer will reply back. Other nodes which are currently sensing the channel turn off their timer by overhearing the acknowledgment (Table 1).

3 Sensor Networks

Some of the features of sensor networks include the following: (1) Sensor nodes are prone to failures; (2) Mobility of nodes; (3) The topology of a sensor network changes very frequently; (4) Sensor nodes are densely deployed; (5) Ability to withstand harsh environmental conditions; (6) Dynamic network topology; (7) Sensor nodes are limited in power, computational capacities, and memory; (8) Sensor nodes may not have global identification because of the large amount of overhead and the large number of sensors.

Table 1 Comparison with existing solutions

No	Proposed idea	Approach	Strength	Weakness
1	FACC [27]	Queue maintained with two thresholds Q_l and Q_h	(1) Compared to no congestion scheme and backpressure, provides higher throughput, less packet loss, less power consumption; (2) Improve channel utilization, reduces the interference; (3) The starving problem for the long flows is resolved, it achieves better fairness	(1) Backpressure message increases traffic in the network
2	ECODA [21]	Maintains a queue dual buffer thresholds and weighted buffer difference for congestion detection	(1) Provides higher throughput with reducing packet loss than the CODA; (2) CODA, there in no too many ACKs, so the energy consumption is less; (3) Compared to CODA, the end to end delay is less; (4) It uses the priority of the data packets and provides weighted fairness	(1) Buffer is in reject state, drop the higher priority packets or may be delayed by newly arrived packets
3	AODV [32]	Broadcasts route request (RREQ) packet to find the path to the destination	(1) Increase the average lifetime of the nodes by putting then off; (2) Balance the energy consumption of the network	(1) Node is turned off due to high energy consumption speed, then again to find the new route (a) Increase number of packets to transfer; (b) Can increase delay; (c) Can stop the data transfer of other nodes; (d) Can cause congestion (2) If no route available, data is buffered. If buffer goes full, can cause packet loss
4	C-SGM [33]	New technique for data distribution using mobilesink groups C-SGM	(1) All the sensors consume equal battery power as the role of the header is changed continuously i.e. balanced energy consumption is achieved; (2) Data is distributed via almost the shortest path	(1)As the data is forwarded by towards the header node,it is possible to increase the number of hops; (2) Number of messages are required to send for header advertisement each time when header change and location update loss

(continued)

Table 1 (continued)

No	Proposed idea	Approach	Strength	Weakness
5	EECC [31]	All intermediate nodes in between sink and source node cooperate in the packet forwarding	(1) This EECC protocol reduces the number of retransmissions as the cooperative nodes help in the packet retransmission; (2) The energy consumption is also reduced by reducing the retransmission; (3) The network performance is improved; (4) The overall delay is minimized	(1) If cooperative nodes are available, else it is pure retransmission; (2) All cooperative nodes have to receive and store the data from all sensors till acknowledgment does not reach the destination
6	AODV with Mobile node	This proposed work increases the lifetime of the nodes whose power consumption speed is high and enhances the life of the entire network	(1) To detect and reduce the congestion (2) To control the node wise traffic flow (3) To reduce packet loss ratio (PLR) (4) To increase the network lifetime (5) To increase packet delivery ratio(PDR) (6) To increase throughput (7) To balance the energy consumption of all nodes in the network (8) To reduce energy consumption	(1) This protocol is suitable only for event-driven WSN (2) The protocol needs extra nodes i.e. mobile nodes (3) Extra overhead of maintain the mobile nodes (4) Need to check that the intermediate node or source node is really a part of our network (that is security challenge)

3.1 Protocol Stack of Sensor Networks

The sensor networks follow the general rules of networks concerning their protocol with the following layers: Application, Transport, Network, Data-Link and Physical [40]. Though, due to specific characteristics of these networks especially power constraints or specific application tasks, some of these layers could merge in one. Moreover, some of the functionalities of each layer, as it is explained below, can be very different than classical Transmission Control Protocol/Internet Protocol (TCP/IP) protocol stack. The major functions of the physical layer include a selection of transmission frequency, generation of a carrier frequency, detection of a signal in the environment, modulation of data as well as it encryption. All these functions are carried out in such a way that the total energy consumption is minimized. Data link layer deals with the multiplexing of data streams, data frame detection, medium access and error control. Due to power constraints, WSNs implement specific MAC protocols that take in mind the power conservation and data—centric routing. The design of MAC protocols embodies the achievement of at least two targets. The first objective is the creation of a specific network infrastructure that can self-organize and, also, to be able to establish connections among thousands of nodes. The second target is the fair resource allocation between all nodes. Unfortunately, existing MAC protocols fail to meet these objectives because power conservation is a secondary concern in their development. These MAC protocols are not designed to provide self organization of the network during the initial deployment or in a case of nodes failure, due to power limitations. Two efforts for MAC protocols that meet these requirements are the Sensor-MAC (SMAC) and CSMA MAC [41]. Network layer design is based on the power considerations of the system. WSNs embody attributed based addressing and location awareness. The simplest and mostly spread network protocol is flooding. Each node broadcast its data to all the other nodes in the network until these data reach the destination. The message in terms of packets from network layered is segmented into manageable blocks in the transport layer at source subnet while the same block is reassembled into packets at destination subnet. Transport layer makes use of TCP as a flexible whereas User Datagram Protocol (UDP) as flexible protocols. TCP here is flexible in terms of adjusting the data transmission rate a sender side and the same is not possible in UDP being nonflexible. Application layer protocol is of paramount importance as it addresses the implementation issues of some major functions such as data management, data fusion, positioning, and clock synchronization.

3.2 Congestion and Energy Consumption

The congestion control is one of the major issues in WSN. Some of the nodes transmit data at a higher rate than receivers capacity. Thus the buffers at the receiver may overflow which results in congestion. All of the sensors get active when an event is detected; and all sensors start sensing, processing and forwarding the data to the sink node. The many to one nature that is the convergent character of the WSN results in the congestion to the nodes those are near to sink node. Limited bandwidth, high data sending rate are also significant factors in the congestion. Congestion results in buffer overflow, a loss of packets, comprehensive queuing management [42] and reduction in overall system throughput. The power with the sensors is very limited and the lifetime of the network depends on the battery having with the sensors. If the energy of the sensors within a particular area of the network drains, the network connectivity may be lost. Congestion causes packet loss which results in retransmission ultimately lead to energy consumption. The transport layer plays a vital role in controlling the congestion by adjusting an appropriate transmission rate within limit from sender to receiver. For this congestion control; it is essential to detect first where congestion has occurred. Accordingly, the sender and receiver are notified the transmission rate is then adjusted according to receivers capacity and sender is notified to transmit within this limit. The event driven nature of the WSN is the greatest challenge in the congestion control mechanism.

3.3 Effect of Energy Consumption on Sensor Network

The energy consumption of a single node affects the entire network. It directly affects the network lifetime, performance of the network, reliability, connectivity as well as coverage of the network. The total energy of the sensor is utilized to sense the signals in the environment capture and process the data for transmitting to the sink node. Some of the energy is also used to forward some of the packets to other nodes in the network. The sensor may run out of energy after a lot of workloads, and thus it gets disconnected from the network. Due to this disconnected node, the workload on the other nodes increases with more data traffic from the nearest sink node. The node with high data traffic utilized more energy consumption rate and are declared as critical nodes. These critical nodes may limit the overall the lifetime of the network. [1]. To improve the network lifetime, energy of all nodes should be consumed equally. In a case of the fixed nodes, connectivity between two nodes is stable, which is changing in moving sensors. Energy depletion of a single node can also cause to break the connectivity of the network. By reducing the amount of data to be transmitted i.e. RR, reducing the number of reporting sensors, shortening communication range, proper clustering [36] or mobility [39, 43], we can avoid unnecessary energy consumption. Proper deployment of sensors [44] or good transmission [1] and routing policies [45] also helps to increase the network lifetime. To increase the network lifetime, we have to pay attention towards both reductions of energy consumption of single node as well as entire network.

3.4 Clustering

All the node are organized in a clustered structure each having a cluster head. Each node in the cluster transmits its data to sink node. The redundancy in the data collected from the nodes is decreased and then exchanged with the base station by the cluster head. The reduced redundancy results in less overhead and faster communication useful to achieve balanced energy consumption. In [46], authors proposed the energy efficient, fair clustering scheme. In [36], authors proposed an energy efficient clustering scheme. Some sink nodes in the network can result in the faster communication.

4 System Design and Methodology

It is assumed that the sensor in the network is topologically arranged in a random manner. Each sensor has some spatial limit to measure a rate of passing the data to the sink nodes. This spatial limit is predefined in such a way that the transmis-

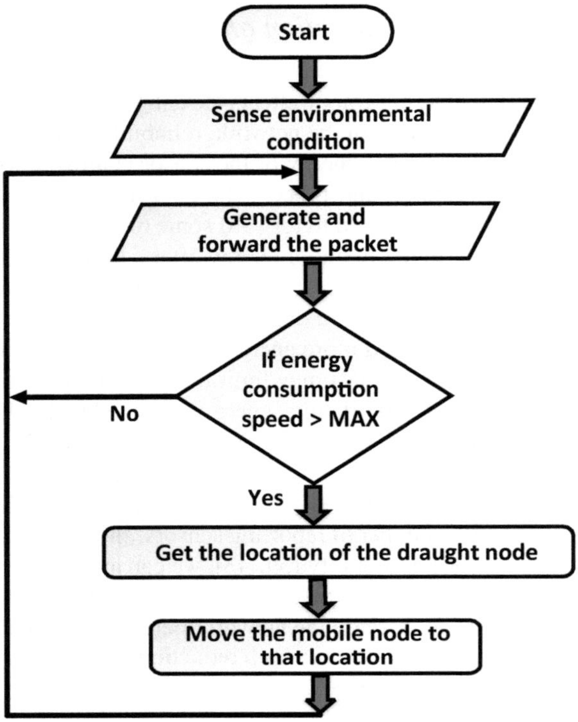

Fig. 2 Flowchart for the mobile node scheme

sion energy is minimized. Which in turn extends the overall lifetime of the sensor (Figs. 2 and 3).

4.1 Single Hop Mode

At the point when the sensor nodes use single hop communication, there is no transferring of packets. Every node straightforwardly transmits its packet to the cluster head. Since the communication is direct between the sensor nodes and the cluster head, one and only node should forward at once, and a dispute less MAC is favored and accepted. The sensor node at the farthest distance from its cluster head (at a distance a) in a single hop network consumes the highest energy than other nodes. The battery parameter and energy requirement are confirmed in line with the worst case energy consumption of the sensor nodes. Hence to ensure a lifetime of at least T cycles, we require that the battery energy of the sensor nodes in the single hop communication system E_s be

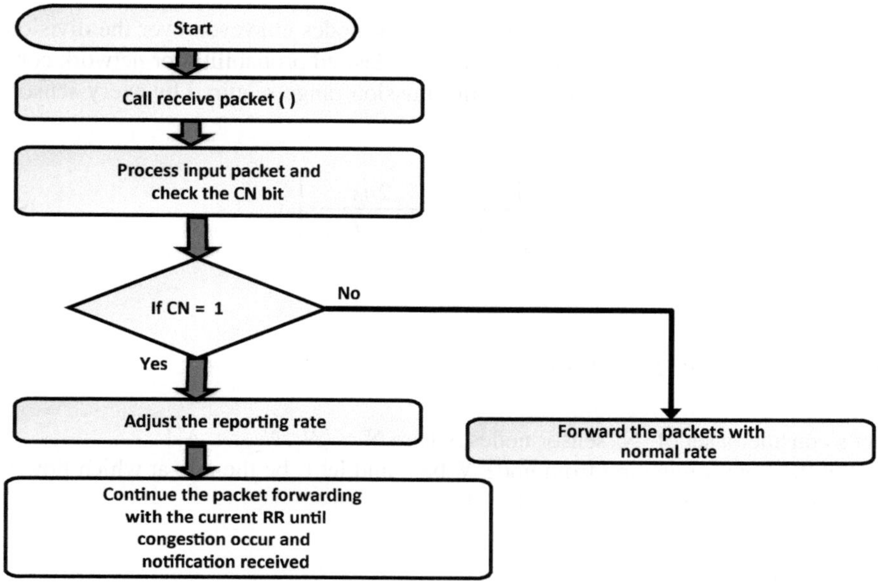

Fig. 3 Flowchart for AFRC algorithm

$$E_s = T(l + \mu a^k) \tag{1}$$

Nodes may utilize power control to save energy and to reduce interference with the neighboring clusters. However, this has no effect on the problem of battery dimensioning which needs to represent the most pessimistic scenario in energy expenditure. Since the area of the region is πA^2, we can approximate each cluster to be a circular region of area $\frac{\pi A^2}{n_1}$, i.e., of radius $\frac{A}{\sqrt{n_1}}$ When single hopping is utilized within the cluster, using (1), the required battery energy of a type 0 node E_0^s 0 is [47]

$$E_0^s = T(l + \mu(\frac{A}{\sqrt{n_1}})^k) = T(l + \frac{\mu A^k}{n_1^{k/2}}) \tag{2}$$

4.2 *Network Connectivity*

We consider a WSN where different sensors are arbitrarily and reliably passed on over the system zone. Two critical measurements for WSNs are connectivity and coverage. For the nodes to effectively utilize multihop communication, it is important to guarantee that any event the condition for node connectivity is met. For *n* sensor nodes randomly conveyed over a unit area, each having transmission range *r*(*n*), the probability of node connectivity is given by [1]

$$pr(Connectivity) \geq 1 - ne^{-\pi n r^2(n)} \tag{3}$$

Normalizing the above relation for 'n' sensor nodes conveyed over the division having area $\frac{1}{2}\theta R^2$ rather than unit area and the desired probability for network connectivity being at least p_{con}, the base transmission range required by every sensor, signified as r_{con} (=r(n)), is given by

$$r_{con} \geq R\left[\frac{\theta}{2n\pi} log \frac{2n\pi}{\theta(1 - p_{con})}\right]^{\frac{1}{2}} \tag{4}$$

4.3 Congestion Detection

Let's consider a set of 'N' sensor nodes, where N = $\{N_1, N_2, \ldots, N_i\}$

Let the flow originating from node N_i be f_i and let r_i be the rate at which flow f_i is admitted into the network [22] (Fig. 4).

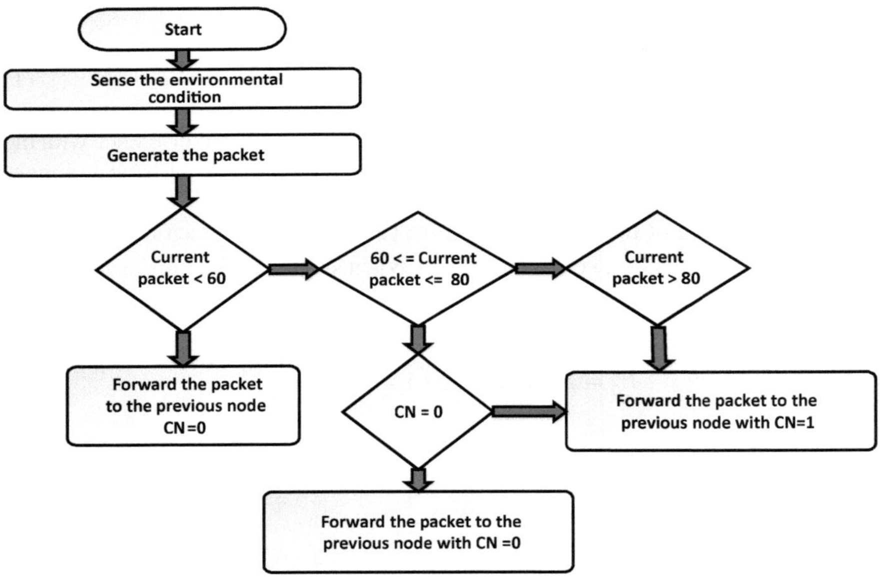

Fig. 4 Flowchart for Congestion detection algorithm

Adaptively assign flow f_i, rate r_i based on queuing theory (M/M/1 Model).

$$BS^{i}_{unoccupancy} = BS^{i}_{max} - BS^{i}_{occupancy} \tag{5}$$

where, $BS^{i}_{occupancy}$ = Current queue length of node N_i

BS^{i}_{max} = Maximal buffer size of node N_i

Node N_i detects the queue length and accordingly sets the values of congestion notification(CN) bit.

$$CN = \begin{cases} 0, & ifBO(t-1) < BO(t) \\ 1, & ifBO(t-1) > BO(t) \end{cases} \tag{6}$$

where, BO (t – 1) is buffer occupancy of previous node and BO (t) buffer occupancy of current node.

$$CN = \begin{cases} 0, & ifBO < \alpha \\ 1 & BO > \alpha < \beta \\ 1 & ifBO \geq \beta \end{cases} \tag{7}$$

Let Buffer Occupancy (BO) at node N_i is currentpr.

If currentpr ≥ 80 then set CN = 1 and decrease RRs.

Else increase RRs.

Now, relation between buffer occupancy and RRs by using linear regression analysis is as follows:

$$\begin{cases} y = 51.56.ln(x) - 85.56 \\ i.e.x = exp(y + 85.56/51.56) \end{cases} \tag{8}$$

where, x = RR, y = Buffer occupancy

Equation (8) is obtained from theoretical results of queuing theory for finding the different values of buffer occupancy by varying RRs. Then theoretical results are compared with results of a simulation. From results, the confidence value is calculated which is good in case of a logarithmic equation. Hence, this equation is considered in Adaptive Flow Rate Control (AFRC). For Differed Reporting Rate (DRR) algorithm,

$$r_i(\mathrm{t}+\mathrm{F}) = r_i(\mathrm{t}) + \mathrm{A}, \text{If CN} = 0$$
$$r_i(\mathrm{t}+\mathrm{F}) = r_i(\mathrm{t}) - \mathrm{A}, \text{If CN} = 1$$

Fig. 5 Flowchart for proposed protocol

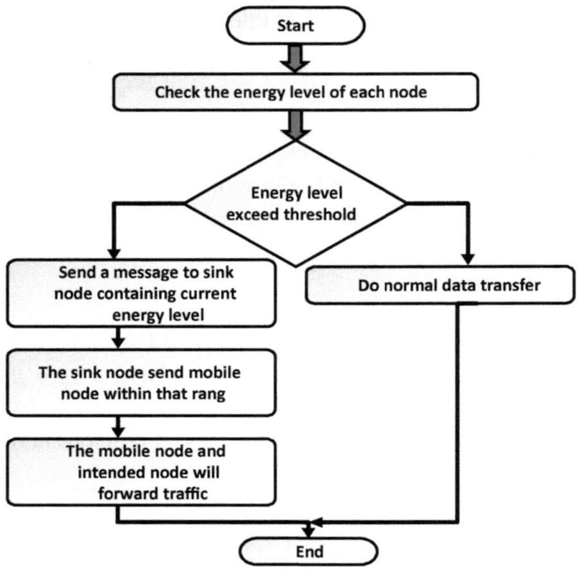

where, F is the frequency of rate update and depends on the network status. ‘A’ is the amount of additive increase or decrease. Total rate, r_i(t), is based on the CN of nodes (Fig. 5).

4.4 Power Dispersion Technique

Assume that $'InitEng'$ represent the initial energy of the node, $'RemEng'$ indicates remaining energy, and time indicates the period that node takes to consume energy. The energy consumption speed $'ConsSpeed'$ of each node is computed using the equation (9) as below.

$$ConsSpeed = \frac{InitEng - RemEng}{Time} \tag{9}$$

The total period during which the node keeps on receiving and transmitting packets without fail is known as it lifetime. Which is calculated using the formula given in equation (10) as below.

$$Lifetime = \frac{InitEng}{ConsSpeed} \tag{10}$$

To find remaining energy $'RemEng'$ of the node can be calculated by following formula.

$$RemEng = InitEng - ((PktT.TEng) + (PktR.REng)) \tag{11}$$

where, $PktT$ = Number of packets transmitted by the node, $PktR$ = Number of packets received by the node, $TEng$ = The amount of energy required to transmit a packet, $REng$ = The amount of energy needed to receive the packet.

$$RemainTime = \frac{RemEng}{ConsSpeed} \tag{12}$$

The above formula is to calculate the remaining lifetime left to the node. Total consumed energy can be calculated as

$$TotalConsumedEnergy = InitEng - RemEng \tag{13}$$

and average energy per hop is

$$AverageEnergyPerHop = \frac{TotalConsumedEnergy}{N_h} \tag{14}$$

where, N_h = Number of nodes per hop

5 Description of Proposed Algorithm

Our previous work on congestion control and minimization of power congestion speed [48] is further extended in this paper (Table 2).

Algorithm 1 Congestion Detection Algorithm:

Input: I_p, C_{pr}, C_p
Output: D_c
1. Require: Watch buffer occupancy consistently;
2. W_b
3. **if** $I_p = C_{pr}$, **then** $C_{pr} = C_b$;
4. **if** $C_{pr} >= 80$, **then** set CN bit;
5. **Send** C_p to P_n with α;
6. $P_n < F_r$;
7. When $C_{pr} <= 60$;
8. Reset CN bit;
9. $P_n > F_r = D_c$;
10 **Repeat** step 3 - 9;
11. Apply **DRR**;

In this algorithm, to detect congestion in the sensor network, buffer occupancy method is used. If buffer occupancy at sensor node is greater than the specified higher

Table 2 Mathematical notations

No	Symbol	Explanation
1	I_p	Incoming flow of packet
2	C_{pr}	Current packet ratio
3	C_p	Choke packet
4	D_c	Detected congestion
5	W_b	Watch buffer occupancy consistently
6	C_b	Current buffer occupancy
7	P_n	Previous node
8	F_r	Flow rate
9	N_h	Number of hops
10	N	Sensor node
11	R_{ca}	Rate control adjustment
12	N_{fr}	New flow rate
13	C_{ql}	Current queue length
14	P	Packet
15	U_{fr}	Updated flow rate
16	C_{ql}	Current queue length
17	O_r	Output rate
18	$ConsSpeed$	Energy consumption speed
19	T	Threshold
20	m_n	Mobile node
21	d_n	Drought node

threshold value, then CN bit is set, and explicit notification is sent to previous nodes. After receiving this packet, all sensor nodes have to decrease the flow rate. If buffer occupancy crosses the lower threshold value, then CN bit is reset and sensor nodes can increase the flow rate. In DRR algorithm, the sensor node which is near to sink has minimum RR, and it will go on increasing as we go away from the sink. For this, first, the distance of every sensor node from the sink is calculated. Then, the RRs for every sensor node are estimated. To estimation of RR for each sensor node, AFRC algorithm is used.

Congestion control by using AFRC algorithm mainly run on each node including source and sink of the network with minimal functionality at the source nodes. More precisely, sensor nodes only need the following functionalities. The major events such as detection of a signal, transmission, reception of data packets, processing the data, etc. are generated by the source node in line with the frequency band provided to it. Thus the data packets are received, processed and forward to the next node via hops if the link to the destination is free. Congestion may occur due to speed mismatch of the sender with a destination node. As a result of congestion, data packets

Algorithm 2 DRR Algorithm:

1. Requir: Hop-count in the path from source to sink;
2. Compute N_h;
3. Estimate RR;
4. **foreach** N;
5. Set N near to sink RR;
6. **Each** N = R_{ca};
7. **Repeat** step 1 - 4;

Algorithm 3 AFRC Algorithm:

1. Requir: Receiving packet P;
2. **if** CN bit is set, **then** N_{fr} (Packet/Sec);
3. $C_q l$ = P;
4. Calculate N_{fr}, by using y = 51.56 ln(x)–85.56,
5. where, x = U_{fr}, y = C_{ql};
6. O_r set with N_{fr};
7. N_{fr} maintained up to next CN will be sensed;

are lost, and more energy is consumed with no productive work. So to avoid it this, proposed algorithm runs continuously on the node, and whenever the buffer occupancy crosses a certain threshold value, algorithm sends the message to its previous node i.e. alert message of congestion is going to occur in the network. And as per the flow rate control, the bit is set in the packet; it will decrease the flow rate till the buffer occupancy reaches to the desired level. The mobile node is nothing but an extra node in the network which is put near the sink node. This mobile node moves to the node where the power consumption speed is more. The sensor nodes sense and generate packets. These packets are forwarded towards the sink node. After a particular time, the power consumption speed of all the nodes in the network is compared with max. If any node having higher energy consumption speed, then a message with the location of the mobile node is forwarded to the sink node. The sink moves toward the intended node location. Some of the traffic at the intended node is forwarded via the mobile node. This reduces the traffic as well as the energy consumption speed of the intended node. The proposed algorithm work on (1) To detect and reduce the congestion; (2) To control the node wise traffic flow; (3) To reduce packet loss ratio (PLR); (4) To increase the network lifetime; (5) To increase packet delivery ratio (PDR); (6) To increase throughput; (7) To balance the energy consumption of all nodes in the network; (8) To reduce energy consumption.

Algorithm 4 Algorithm for mobile node scheme:

1. Require: Check *ConsSpeed* for all N;
2. Compare *ConsSpeed* with T;
3. **if** *ConsSpeed* > T = d_n;
4. Set $m_n = d_n$;
5. Move m_n toward the location;
6. **Repeat** step 2 - 4;

6 Result and Discussion

The proposed algorithm intends to minimize the power consumption by controlling the congestion in the network. This is achieved by tuning of energy consumption parameter which has substantial impact on total energy consumption. The various parameters considered as node density i.e. number of nodes in the system, RR—the number of packets transmitted every second, packets size, etc. The Table 3, shows the simulation parameters used for energy consumption. Simulation has been carried out using a powerful network simulator-2 (NS-2) to measure the performance of the proposed method. By varying these parameters, energy consumption in the network is noted. The Fig. 6a shows the graph of total energy consumption on the

Table 3 Experimental setup

Simulation parameters	Network area 500 × 500	Network area 1000 × 1000
Topoloogy	Random	Random
MAC protocol	802.11	802.11
Routing protocol	AODV	AODV
Number of sensor nodes	50	26
Queue length	100	100
Initial energy	2 J	2 J
Simulation time	50 s	10 s
Packet size	50	50
cbr start time	1.0 s	1.0 s
cbr stop time	49.0 s	10.0 s
Packet interval	0.1	0.025
Number of sink node	01	01
Position of sink node [(m, m)]	(500, 500)	(800, 800)
Radio signal range	15 m	25 m
Mobile sink velocity	5 m/s	10 m/s
Payload size of data packet	1000 Bytes	1000 Bytes
Channel bandwidth	1 mbps	1 mbps
Transmission range	150 m	250 m
Sensing range	90 m	550 m
Antenna type	OmniAntenna	OmniAntenna
Radio-propagation model	TwoRayGround	TwoRayGround
Interface queue type	Queue/droptail/priqueue	Queue/droptail/priqueue
Idle power	0.10	0.10
rxpower	0.04	0.05
txpower	0.08	0.10
Sleep power	0.000015	0.000015

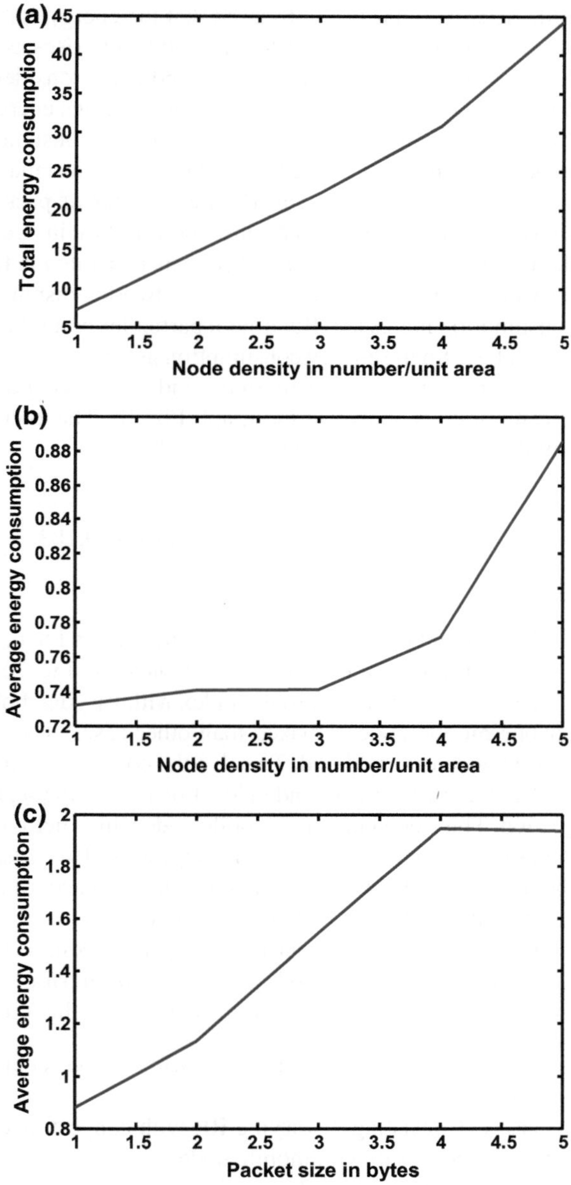

Fig. 6 **a** Total energy consumption versus node density; **b** Average energy consumption versus node density; **c** Average energy consumption versus packet size

node density. It has been observed that the energy consumption gradually increases as node density varied for 1 to 4 and suddenly rises as node density changes from 4 onwards. The same scenario is observed if we considered average energy consumption as shown in Fig. 6b. Figure 6c shows average energy consumption versus packet size. It is to be noted that the average energy consumption is linearly varies for the packet size changes from 2 to 4 and thereafter it remains constant at 1.92 for packet size more than 4. Change in RR i.e. number of packets transmitted per second is also one of the factors which cause the variation in a rate of energy consumption. As the RR increases, the number of packets generated in the network also increases. To forward that number of packets, the network consumes more energy, so the energy consumption increased. But after a particular point, the increase in RR increases the packet loss. So the energy consumption gets reduced as shown in Fig. 7a. By keeping all the simulation parameters same, and just varying the simulation time, the energy consumption is shown in the graph Fig. 7b. The energy consumed is linear as the simulation time increases and is given by

$$y = 0.170x + 0.183$$

As the number of mobile nodes changes from 1 to 6, the fluctuations in the average energy consumption is observed as depicted in Fig. 7c.

The Fig. 8 shows the fairness index with varying mobile nodes. The fairness index for one mobile node is maxed than other cases. Therefore, for this work, only one mobile node is considered. To solve the congestion problem, an extra node is generated and moved to the intended location nearby the sender node concerned with congestion. This node known as mobile node which helps the sender node to forward the data to the intended receiver with congestion. This intended location is detected with the use of congestion detection algorithm and accordingly to the receivers capacity, an appropriate mobile node is generated.

The mobile nodes are placed nearby the sink node to avoid the loss of energy due to distance between two nodes. The simulation parameter mentioned above are used to compute the PDR, PLR, and throughput for the four different cases as:

- Case 1: Describes the basic scenario without congestion control and no use of mobile nodes.
- Case 1: Consisting of varying RR with congestion control mechanism.
- Case 1: Scenario with mobile nodes.
- Case 1: Consists of mobile nodes and congestion control mechanism.

All these four cases are elaborated in Fig. 9. The PDR is found to be maximum in case of scenarios with congestion control mechanism. The scenarios with mobile nodes result in improved PDR than otherwise. The global scenario depicts that when the PLR increases, then PDR decreases. In the scenarios, the congestion control mech-

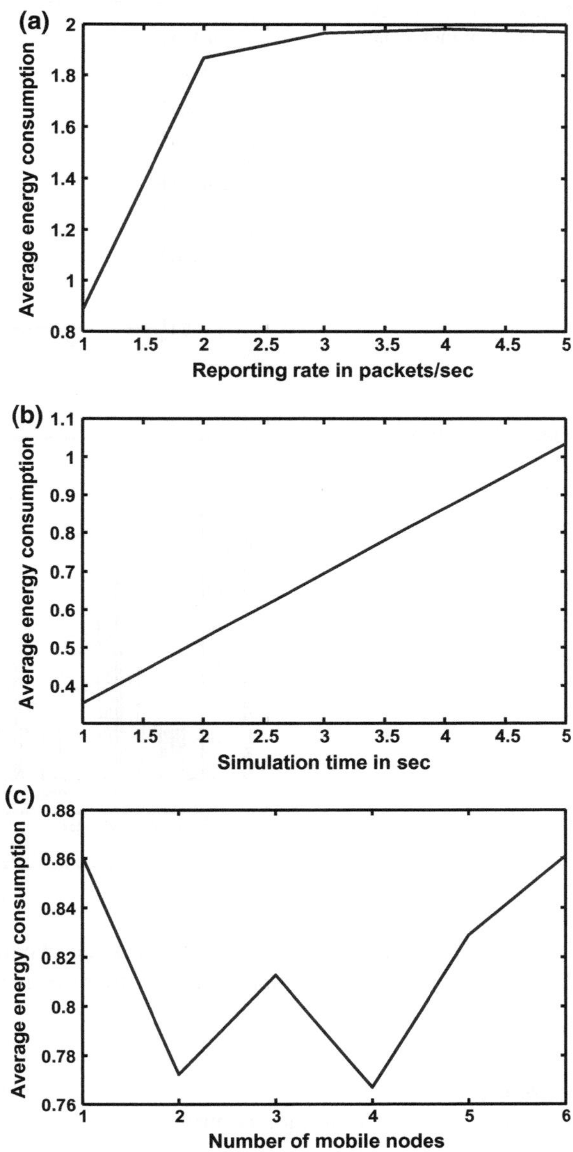

Fig. 7 **a** Average energy consumption versus RR; **b** Average energy consumption versus simulation time; **c** Average energy consumed versus no of mobile node

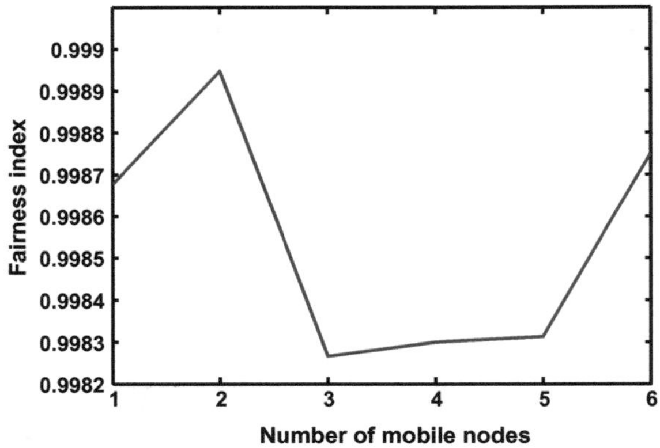

Fig. 8 Fairness index versus no of mobile nodes

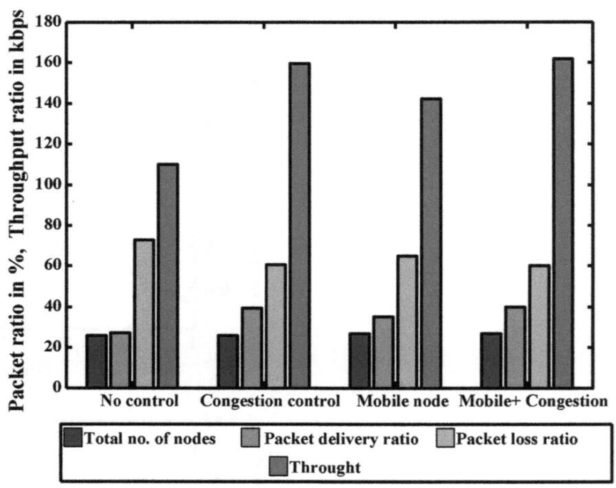

Fig. 9 PDR, PLR, throughput for 4 distinct test cases

anism, and mobile node together improves the overall throughput of the network. Furthermore,energy consumption rate for four distinct cases are studied as:

- No control.
- Congestion.
- Mobile node.
- Mobile node + Congestion control.

and shown in Fig. 10. The energy consumption rate is maximum in the no control scenario i.e. more energy is consumed in that case. This energy consumption rate is decreased with the use of congestion control algorithm. The rate of energy consumption is mostly reduced with the use of mobile nodes. The node which consumes the

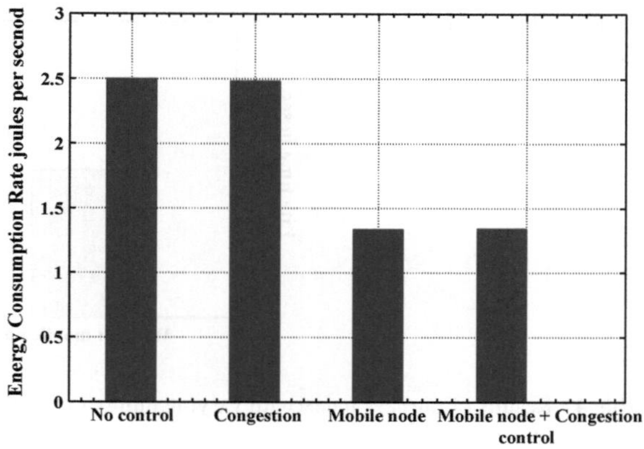

Fig. 10 Rate of energy consumption in 4 distinct test cases

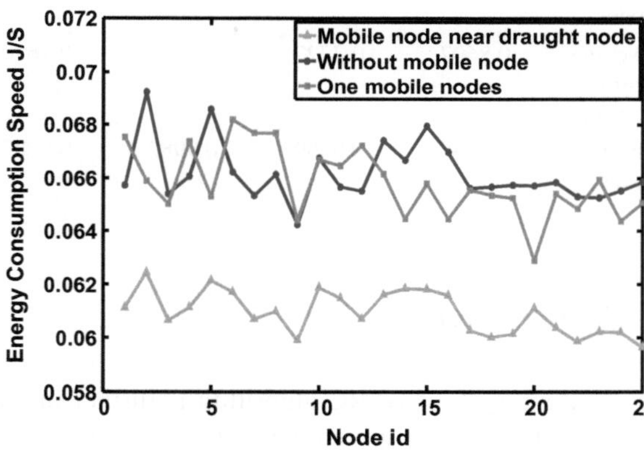

Fig. 11 Energy consumption speed of each node

energy at the higher rate is likely to die earlier than other. Figure 11 depicts the speed of power consumption by each node. Without mobile nodes, the consumption speed variation is more than the use of mobile nodes. The use of mobile nodes somewhere balances the energy consumption speed. The node number 2 is having energy consumption speed more than other nodes. Therefore, when mobile nodes reach to the node 2, the energy consumption speed of node 2 gets decreased. Also, the energy consumption speed of all nodes is reduced. A lifetime of node 2 is increased with the help of mobile nodes. The Fig. 12 shows the graph of a lifetime of the intended node i.e. node 2, which confirms the usefulness of the proposed method (Table 4).

Fig. 12 Lifetime of the intended node

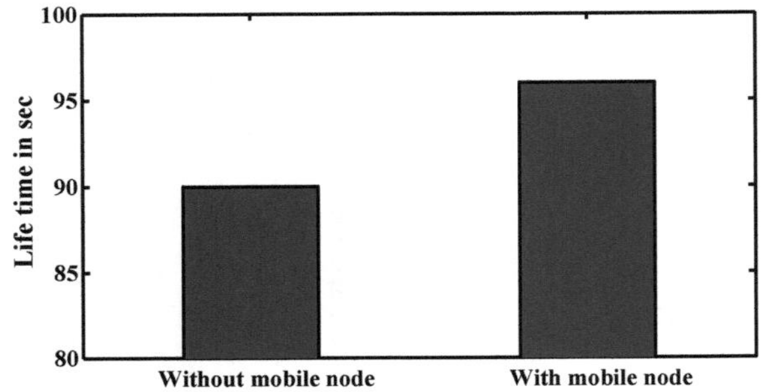

Table 4 Results of fairness index with varying number of mobile nodes

Number of mobile nodes	0	1	2	3	4	5
Average energy consumed	0.8607	0.7722	0.8128	0.7670	0.8288	0.8609
Fairness index	0.9986	0.9989	0.9982	0.9983	0.9983	0.9987

7 Conclusion

The real strength of AFRC algorithm is further investigated to leverage its potential benefits in this proposed work. The novel strategy presented here demonstrates the use of mobile nodes for effective congestion control as well as reducing energy consumption in the node. For evaluation purpose, we have performed simulation analysis using NS-2 and showed the results are as follows. Using the Congestion control scheme, the PDR increases by 44.80%, PLR decrease by 83.19% and the throughput increases by 44.80%. The mobile nodes in the network, the energy consumption rate decreases by 53.4%, so the lifetime of a node increases by 5.52%. Also the PDR increases by 29.27%, PLR decrease by 89.02% and the throughput increases by 29.15%.

References

1. Azad, A.K.M., Kamruzzaman, J.: Energy-balanced transmission policies for wireless sensor networks. IEEE Trans. Mob. Comput. **10**(7), 927–940 (2011)
2. Alippi, C., Anastasi, G., Di Francesco, M., Roveri, M.: Energy management in wireless sensor networks with energy-hungry sensors. IEEE Instrum. Measur. Mag. **12**(2), 16–23 (2009)

3. GowriDurga, A., Prakash, B.: A zigbee sms alert system with trust mechanism in wireless sensor networks. In: 2013 International Conference on Information Communication and Embedded Systems (ICICES), pp. 1010–1014. IEEE (2013)
4. Tubaishat, M., Madria, S.: Sensor networks: an overview. IEEE Potentials **22**(2), 20–23 (2003)
5. Suraiya, Ta, et al.: Energy conservation challenges in wireless sensor networks: a comprehensive study. Wireless Sens. Netw. **2**(06), 483 (2010)
6. Hefeeda, M., Bagheri, M.: Wireless sensor networks for early detection of forest fires. In: IEEE International Conference on Mobile Adhoc and Sensor Systems, 2007. MASS 2007, pp. 1–6. IEEE (2007)
7. Opasjumruskit, K., Thanthipwan, T., Sathusen, O., Sirinamarattana, P., Gadmanee, P., Pootarapan, E., Wongkomet, N., Thanachayanont, A., Thamsirianunt, M.: Self-powered wireless temperature sensors exploit RFID technology. IEEE Pervasive comput. **5**(1), 54–61 (2006)
8. Reindl, L., Shrena, I., Kenshil, S., Peter, R.: Wireless measurement of temperature using surface acoustic waves sensors. In: Proceedings of the 2003 IEEE International Frequency control symposium and PDA exhibition jointly with the 17th European Frequency And Time Forum, 2003, pp. 935–941. IEEE (2003)
9. Garzón, C.A.L., Riveros, O.J.R.: Temperature, humidity and luminescence monitoring system using wireless sensor networks (wsn) in flowers growing. In 2010 IEEE ANDESCON, pp. 1–4. IEEE (2010)
10. Luomala, J., Hakala, I.: Effects of temperature and humidity on radio signal strength in outdoor wireless sensor networks. In: 2015 Federated Conference on Computer Science and Information Systems (FedCSIS), pp. 1247–1255. IEEE (2015)
11. Umit Bas, C., Coleri Ergen, S.: Ultra-wideband channel model for intra-vehicular wireless sensor networks beneath the chassis: from statistical model to simulations. IEEE Trans. Veh. Technol. **62**(1), 14–25 (2013)
12. Hung, C.-C., Peng, W.-C.: Model-driven traffic data acquisition in vehicular sensor networks. In: 2010 39th International Conference on Parallel Processing (ICPP), pp. 424–432. IEEE (2010)
13. Segura-Garcia, J., Felici-Castell, S., Perez-Solano, J.J., Cobos, M., Navarro, J.M.: Low-cost alternatives for urban noise nuisance monitoring using wireless sensor networks. IEEE Sens. J. **15**(2), 836–844 (2015)
14. Bekmezci, I., Alagöz, F.: Energy efficient, delay sensitive, fault tolerant wireless sensor network for military monitoring. Int. J. Distrib. Sens. Netw. **5**(6), 729–747 (2009)
15. Mascareñas, D., Flynn, E., Farrar, C., Park, G., Todd, M.: A mobile host approach for wireless powering and interrogation of structural health monitoring sensor networks. IEEE Sens. J. **9**(12), 1719–1726 (2009)
16. Bhuiyan, M.Z.A., Wang, G., Cao, J., Wu, J.: Deploying wireless sensor networks with fault-tolerance for structural health monitoring. IEEE Trans. Comput. **64**(2), 382–395 (2015)
17. Mendis, C., Skvortsov, A., Gunatilaka, A., Karunasekera, S.: Performance of wireless chemical sensor network with dynamic collaboration. IEEE Sens. J. **12**(8), 2630–2637 (2012)
18. Pattnaik, P.K., Mall, R.: Fundamentals of Mobile Computing. PHI Learning Pvt. Ltd., (2015)
19. Kilts, S.: Advanced FPGA design: Architecture, Implementation, and Optimization. Wiley (2007)
20. Madurawe, R.U.: Alterable application specific integrated circuit (asic), June 20 2006. US Patent 7,064,579
21. Tao, L.Q., Yu, F.Q.: Ecoda: enhanced congestion detection and avoidance for multiple class of traffic in sensor networks. IEEE Trans. Consum. Electron. **56**(3), 1387–1394 (2010)
22. Wang, G., Liu, K.: Upstream hop-by-hop congestion control in wireless sensor networks. In: 2009 IEEE 20th International Symposium on Personal, Indoor and Mobile Radio Communications, pp. 1406–1410. IEEE (2009)
23. Kafi, M.A., Ben-Othman, J., Ouadjaout, A., Bagaa, M., Badache, N.: REFIACC: reliable, efficient, fair and interference-aware congestion control protocol for wireless sensor networks. Comput. Commun. (2016)

24. Kafi, M.A., Djenouri, D., Othman, J.B., Ouadjaout, A., Bagaa, M., Lasla, N., Badache, N.: Interference-aware congestion control protocol for wireless sensor networks. Procedia Comput. Sci. **37**, 181–188 (2014)
25. Esmaeelzadeh, V., Hosseini, E.S., Berangi, R., Akan, O.B.: Modeling of rate-based congestion control schemes in cognitive radio sensor networks. Ad Hoc Netw. **36**, 177–188 (2016)
26. Ghaffari, A.: Congestion control mechanisms in wireless sensor networks: a survey. J. Netw. Comput. Appl. **52**, 101–115 (2015)
27. Heikalabad, S.R., Ghaffari, A., Hadian, M.A., Rasouli, H.: Dpcc: dynamic predictive congestion control in wireless sensor networks. IJCSI Int. J. Comput. Sci. Issues **8**(1) (2011)
28. Shih, H.-C., Ho, J.-H., Liao, B.-Y., Pan, J.-S.: Fault node recovery algorithm for a wireless sensor network. IEEE Sens. J. **13**(7), 2683–2689 (2013)
29. Zou, Z., Bao, Y., Li, H., Spencer, B.F., Ou, J.: Embedding compressive sensing-based data loss recovery algorithm into wireless smart sensors for structural health monitoring. IEEE Sens. J. **15**(2), 797–808 (2015)
30. Sobrinho, J.L., Krishnakumar, A.S.: Quality-of-service in ad hoc carrier sense multiple access wireless networks. IEEE J. Sel. Areas Commun. **17**(8), 1353–1368 (1999)
31. Tati, R., Ahmadi, F., Rashidy, R., Ashkoti, F.: Designing and simulation of a distributed algorithm for quality of service in wireless sensor networks. In: International Conference on Application of Information and Communication Technologies, pp. 1–5. IEEE (2009)
32. Xiaoyan, Y., Xingshe, Z., Rongsheng, H., Yuguang, F., Shining, L.: A fairness-aware congestion control scheme in wireless sensor networks. IEEE Trans. Veh. Technol. **58**(9), 5225–5234 (2009)
33. Emad, F., Chang-Gun, L., Eylem, E.: Mmspeed: multipath multi-speed protocol for qos guarantee of reliability and timeliness in wireless sensor networks. IEEE Trans. Mob. Comput. **6**, 738–754 (2006)
34. Mukhopadhyay, S., Schurgers, C., Panigrahi, D., Dey, S.: Model-based techniques for data reliability in wireless sensor networks. IEEE Trans. Mob. Comput. **8**(4), 528–543 (2009)
35. Ghaffari, A.: An energy efficient routing protocol for wireless sensor networks using a-star algorithm. J. Appl. Res. Technol. **12**(4), 815–822 (2014)
36. Wei, D., Jin, Y., Vural, S., Moessner, K., Tafazolli, R.: An energy-efficient clustering solution for wireless sensor networks. IEEE Trans. Wirel. Commun. **10**(11), 3973–3983 (2011)
37. Fang, W., Liu, F., Yang, F., Shu, L., Nishio, S.: Energy-efficient cooperative communication for data transmission in wireless sensor networks. IEEE Trans. Consum. Electron. **56**(4), 2185–2192 (2010)
38. Abusaimeh, H., Yang, S.-H.: Balancing the power consumption speed in flat and hierarchical wsn. Int. J. Autom. Comput. **5**(4), 366–375 (2008)
39. Park, S., Lee, E., Jin, M.-S., Kim, S.-H.: Novel strategy for data dissemination to mobile sink groups in wireless sensor networks. IEEE Commun. Lett. **14**(3), 202–204 (2010)
40. Fahmy, H.M.A.: Protocol stack of WSNs. In: Wireless Sensor Networks, pp. 55–68. Springer (2016)
41. Ye, W., Heidemann, J.: Medium access control in wireless sensor networks. In: Wireless Sensor Networks, pp. 73–91. Springer (2004)
42. Ahammed, G.F., Banu, R.: Analyzing the performance of active queue management algorithms. arXiv:1003.3909 (2010)
43. Karenos, K., Kalogeraki, V.: Traffic management in sensor networks with a mobile sink. IEEE Trans. Parallel Distrib. Syst. **21**(10), 1515–1530 (2010)
44. Cheng, Z., Perillo, M., Heinzelman, W.B.: General network lifetime and cost models for evaluating sensor network deployment strategies. IEEE Trans. Mob. Comput. **7**(4), 484–497 (2008)
45. Ren, F., Zhang, J., He, T., Lin, C., Ren, S.K.D.: Ebrp: energy-balanced routing protocol for data gathering in wireless sensor networks. IEEE Trans. Parallel Distrib. Syst. **22**(12), 2108–2125 (2011)
46. Park, S., Lee, W., Cho, D.: Fair clustering for energy efficiency in a cooperative wireless sensor network. In: 2012 IEEE 75th Vehicular Technology Conference (VTC Spring), pp. 1–5. IEEE (2012)

47. Mhatre, V., Rosenberg, C.: Design guidelines for wireless sensor networks: communication, clustering and aggregation. Ad Hoc Netw. **2**(1), 45–63 (2004)
48. Tambe, S.B., Thool, R.C., Thool,V.R.: Power consumption and congestion control of rendezvous node for wireless biosensor network. In: Proceedings of International Conference on ICT for Sustainable Development, pp. 647–655. Springer (2016)

Two Stage Wireless Sensor Node Localization Using Firefly Algorithm

Eva Tuba, Milan Tuba and Marko Beko

Abstract Locations of the sensors in wireless sensor networks usually have to be known. Wireless sensor networks contain large number of sensors so installing global position system device in each of them is not acceptable. Few anchor nodes with known positions are usually used. Estimating sensor nodes positions from radio strength signal index is a hard optimization problem. In this paper we proposed a two stage algorithm with semi-mobile anchors that uses swarm intelligence firefly algorithm for optimization. The proposed algorithm was tested by simulation of standard benchmark wireless networks where it obtained better results compared to other approaches from literature.

Keywords Wireless sensor networks · Localization problem · Firefly algorithm Swarm intelligence · Metaheuristics

1 Introduction

Wireless sensor networks (WSN) are widely used in many different areas. WSN consist of large number of sensor nodes that have ability to collect various data such as

E. Tuba · M. Tuba (✉)
Graduate School of Computer Science, John Naisbitt University, Bulevar umetnosti 29, 11070 Belgrade, Serbia
e-mail: tuba@ieee.org

E. Tuba
e-mail: rtuba@acm.org

M. Beko
Computer Engineering Department, Universidade Lusòfona de Humanidades e Tecnologias, Lisbon, Portugal
e-mail: marko@isr.ist.utl.pt

X.-S. Yang et al. (eds.), *Smart Trends in Systems, Security and Sustainability*, Lecture Notes in Networks and Systems 18,
https://doi.org/10.1007/978-981-10-6916-1_10

temperature, vibration, sound, etc. from the environment. Besides collecting data, these nodes are able to communicate with other nodes and to do some computation.

One of the problems that affect WSN is nodes localization. In most cases collected data are meaningful only with the information of position from where the data were taken. Global positioning system (GPS) module added to each node would facilitate position determination. However, this solution is not acceptable because of the price, size and energy requirements. Better solution is to make known positions only for a few nodes named anchor nodes. They can have GPS or their positions can be known because the sensors are placed manually at exact position. For determination of positions of other sensor nodes some ranging techniques, along with knowledge of positions of anchor nodes are used. In this paper we used radio signal strength indicator (RSSI) for sensor nodes localization. Localization of sensor nodes in this way represents a hard optimization problem for which different solutions can be found in literature. In this paper we propose using novel swarm intelligence algorithm, firefly algorithm, for solving this hard optimization problem.

The rest of the paper is organized as follow. In Sect. 2 literature review is given. In Sect. 3 mathematical model for localization problem is given. Section 4 describes our proposed method, while in Sect. 5 simulation results are presented. Conclusion is in Sect. 6.

2 Related Work

Ranging techniques are necessary for range based algorithms. Some of the most common ranging techniques are time of arrival (TOA), time difference of arrival (TDOA), radio signal strength indicator (RSSI) and angle of arrival (AOA). They are used to estimate the distances or angles between sensor and anchor nodes [16].

A distributed method where nodes themselves estimate their positions using radio signal angle of arrival and least square triangulation was described in [1].

In [19] a sequential greedy algorithm for localization problem was presented. In the proposed framework both range-based and range-free methods were implemented.

A weighted least squares based algorithm that converts distance measurements obtained from time of arrival information into linear equations for WSN node localization was proposed in [6]. The basic algorithm was based on approximation of the maximum likelihood estimator when time differences of arrival estimation errors are small and additionally method for position estimation with less than three neighbors and incorporated boundary and communication range constraints was proposed.

Different guiding functions and different metaheuristics for localization problem were investigated in [17] using two-stage search where error norm was minimized while maximum likelihood was maximized.

Some approaches from literature used bio-inspired algorithms for solving localization problem in wireless sensor networks, however most of them used particle swarm optimization (PSO). Reason for introducing those algorithms was that they are easier to implement and also they are energy and time saving because of the fast convergence and less computing resources [15].

Weighted centroid algorithm for initial estimate and genetic algorithm for further refinement were proposed in [10]. RSSI was used for ranging.

Particle swarm optimization algorithm was used to minimize mean square error as objective function in [8]. Experimental results proved the proposed method better than simulated annealing from literature.

PSO as well as bacterial foraging algorithm were used for ranging-based localization in [13]. Iterative procedure was proposed that used localized nodes as references for localization of other nodes.

In [14] H-best particle swarm optimization (HPSO) as well as biogeography based optimization (BBO) were proposed for localization problem optimization. Proposed distributed algorithm successfuly minimized error function.

Recent paper [2] used fireworks algorithm while [9] implemented cuckoo search algorithm in order to estimate the location of sensors. The proposed method was compared with particle swarm optimization and biogeography based optimization (BBO) and showed that proposed algorithm outperformed both variants.

3 Mathematical Model for Localization Problem

In this paper localization of unknown nodes is done using ranging by RSSI. Radio signal strength index describes relationship between received and transmitted power of radio signals and the distance between nodes.

RSSI propagation models in the WSN can include free-space model, ground bidirectional reflectance model and log-normal shadow model [24]. Free-space model is used in the cases when the transmission distance is much larger compared to antenna size and the radio wavelength λ and when no obstacles stand between sensors. In these cases the following expression can be used for distance determination:

$$P_r(d) = \frac{P_t G_t G_r \lambda^2}{(4\pi)^2 d^2 L} \tag{1}$$

where P_r is the received power, P_t is the transmitted power, G_t and G_r represent anchor and sensor antenna gains, while L refers to system loss factor. Usual values for G_t, G_r and L are 1 [5].

The distances estimation between unknown node and anchor nodes in this paper is calculated as:

$$d = \sqrt{\frac{P_t G_t G_r \lambda^2}{4\,\pi^2 P_r L}} \tag{2}$$

Real locations of sensor nodes in simulations are known, so the real distances between nodes i and j could be calculated. Fitness function that should be minimized is:

$$f(x, y, z) = \sum_{i=1}^{N} \sum_{j=1}^{M} \left| r_{ij} - d_{ij} \right|. \tag{3}$$

where N is number of unknown nodes and M is the number of anchor nodes, r_{ij} is the real distance between sensor i and anchor node j, while d_{ij} is estimated distances by RSSI between sensor node i and anchor node j. Each sensor node will be localized by minimizing the following functions (square is used instead of absolute value):

$$f_i(x, y, z) = \sum_{j=1}^{M} \left(r_{ij} - d_{ij}\right)^2, \quad i = 1, 2, \ldots, N. \tag{4}$$

4 The Proposed Algorithm

We consider WSN 3-D localization model as described in [18], with anchors positioned at the corners of the area of interest. In [21] we proposed an algorithm for WSN sensor localization with semi-mobile anchors that can move once from their initial positions for another set of measurements from new positions. For optimization we used bat algorithm. The algorithm contains three main steps.

The first step is to localize sensor nodes using anchor nodes positioned at the corners of the network by minimizing functions in Eq. 4.

The second step is to move anchor nodes to better locations for determination of positions of sensor nodes. Trilateration is an algorithm used for localization. In order for localization to be more precise, angles between two anchor nodes and one sensor node should be as large as possible. On the other hand, localization will be more precise if the distance between anchor and sensor node is as small as possible since RSSI then introduces smaller error. Based on these facts, positions of anchor nodes can be adjusted with aim to localize sensors more precise. Maximization of angles between sensors and all possible pairs of anchor will place anchors in optimal position. This leads to the objective function that need to be maximized:

$$\max f(x, y, z) = \sum_{i=1}^{N} \sum_{k=1}^{M-1} \sum_{l=k+1}^{M} \angle(A_k T_i A_l) \tag{5}$$

where T_i represents location of sensor node, A_k and A_l are locations of anchor nodes. Instead of angles, we used their cosines.

Optimal positions for anchor nodes are determined using firefly algorithm for solving the optimization problem given by Eq. 5.

At the end, the same procedure as in the first step is done, but with updated positions of anchor nodes. One of the problems that can occur during this process is the so-called flip ambiguity problem. If some neighbor nodes are nearly collinear, one node can be reflected across the line and still connecting the neighbors and also satisfying the distance constraint [11].

To deal with the flip ambiguity problem in [21] we developed two different approaches. In this paper we used the more successful one. In order to avoid flip problem for the final sensors localization, previous distance estimations are included in the fitness function. Thus, in the second localization of sensor nodes instead of 4 anchors, 8 anchor nodes are used. Flipping problem is avoided because of the first position of anchors. When anchor nodes are in the corners of the area of wireless sensor network, flipping problem does not occur.

In this paper for solving the optimization problems in all three steps we used firefly algorithm. Firefly algorithm was proposed by Yang [25] and later different modifications and hybridizations were proposed [7, 12, 22]. Firefly algorithm was successfully used in numerous applications such as multi-objective RFID network planning [23], support vector machine parameter tuning [20], multilevel image thresholding [4], portfolio optimization problem [3], etc.

5 Simulation Results

The proposed algorithm was implemented in Matlab version R2016a and its quality was tested by comparing the results with other algorithms from literature. In [18] Nuo et al. proposed using particle swarm optimization for location estimation. In order to make our results comparable with [18] experimental conditions were set to be the same. We also included comparison with our method proposed in [21] where bat algorithm was used for node localization.

The task is to localize 20 unknown sensor nodes in 3-D area. Size of that area was $50 \times 50 \times 50$ m. Anchor nodes were set at corners of the area in the following positions: A_1 (50, 0, 0), A_2 (0, 50, 50,), A_3 (50, 50, 50), A_4 (0, 0, 50). Parameters for the firefly algorithm were determined empirically. Number of fireflies was 40 and number of iterations was set to 1000. Initial parameters were $\beta_0 = 0.97$, $\gamma = 0.25$ and $\alpha_0 = 0.15$. For parameter adjustments recommendations from [26] was used.

As mentioned before, distances between anchor and sensor nodes are estimated by RSSI. In reality, this estimations contain some error. Nuo et al. in [18] simulated ranging errors by Gaussian distribution $\mathcal{N}(0, \sigma^2)$. Simulations were done for seven different error levels, where σ^2 was set to be 0%, 5%, 10%, 20%, 30%, 40%, 50% of the distance. For each experiment average as well as minimum and maximum error was recorded.

In Table 1 comparison of our proposed algorithm with dynamic anchors and static anchors is presented (better results are in bold). Static anchors were positioned at the corners of the network. It can be seen that with dynamic anchors localization error is significantly smaller that in the case of static anchors. Dynamic anchors are more

Table 1 Localization error comparisons of our proposed algorithm with dynamic anchors and static anchors

Rang	4 anchors		6 anchors		8 anchors		4+4 anchors	
err.%	Avg	Std	Avg	Std	Avg	Std	Avg	Std
0	0.000	0.000	0.000	0.000	0.000	0.000	0.000	0.000
5	2.041	1.072	1.941	0.781	1.736	0.792	**1.363**	**0.614**
10	3.196	1.856	2.584	1.417	2.017	0.838	**1.882**	**0.702**
20	4.912	2.867	3.560	1.659	3.255	1.603	**3.042**	**1.335**
30	5.859	2.593	4.450	1.698	3.697	1.732	**3.102**	**1.521**
40	8.635	5.891	5.472	2.592	4.335	1.845	**4.097**	**1.777**
50	8.976	6.197	6.296	2.849	5.604	1.781	**4.729**	**4.340**

Table 2 Localization error comparisons of our proposed algorithm with [18] and [21] with ranging errors from 0 to 50%

Rang	TLP			BA			FA		
err.%	Avg	Min	Max	Avg	Min	Max	Avg	Min	Max
0	0.212	0.011	0.463	0.022	0.002	0.050	**0.000**	**0.000**	**0.000**
5	4.980	1.290	9.670	1.444	0.620	3.228	**1.363**	**0.514**	**2.978**
10	9.660	3.980	17.259	2.259	0.688	5.367	**1.882**	**0.602**	**3.166**
20	19.670	6.120	26.592	3.098	1.705	**4.707**	**3.042**	**0.543**	6.144
30	27.921	14.780	39.257	3.517	0.887	5.927	**3.102**	**0.862**	**3.102**
40	30.343	15.458	41.364	5.152	**1.260**	9.538	**4.097**	1.406	**8.159**
50	32.691	17.665	41.292	5.239	**1.407**	11.807	**4.729**	2.234	**8.366**

expensive, but it can be concluded that even if they were twice as expensive as the static ones they are the better choice because with four of them precision is better then with 8 static anchors.

Comparison results are shown in Table 2. Our proposed method with relocation of anchor nodes successfully increased accuracy of localization of sensor nodes compared to the algorithm proposed in [18] named TLP. In [18] PSO was used, while we proposed using one of the recent swarm intelligence algorithms, firefly algorithm. The first three columns in the table show the results from [18]. The second triplet of columns represents results reported in [21] while the last triplet represents results of our proposed algorithm.

The first column in Table 1 shows results for our proposed algorithm with 4 static anchors which is equivalent to the method proposed in [18]. Based on these results we can conclude that just using the firefly algorithm gives much better results compared to proposed PSO in [18]. It can also be noticed that with firefly algorithm results were slightly better then the results obtained by using the bat algorithm (better results are in bold).

6 Conclusion

In this paper algorithm for localization of sensor nodes in wireless sensor networks was proposed. The proposed algorithm for localization calculates the position of unknown nodes two times: the first time with 4 anchor nodes placed at the corners of the network and the second time with anchors in optimal positions using also distances calculated in the previous step. For solving the optimization problem one of the latest swarm intelligence algorithm, firefly algorithm, was used. By comparing the results with algorithms from literature it has been shown that our proposed algorithm significantly improves accuracy of localization. Moreover, it has been shown that just using firefly algorithm instead of PSO improves results. Future research may include more benchmark networks and comparison with other algorithms from literature as well as additional firefly algorithm adjustments and modifications.

Acknowledgements M. Tuba acknowledges support by the Ministry of Education, Science and Technological Development of Republic of Serbia, Grant No. III-44006.

References

1. Amundson, I., Sallai, J., Koutsoukos, X., Ledeczi, A., Maroti, M.: RF angle of arrival-based node localisation. Int. J. Sens. Netw. **9**(3/4), 209–224 (2011)
2. Arsic, A., Tuba, M., Jordanski, M.: Fireworks algorithm applied to wireless sensor networks localization problem. In: Congress on Evolutionary Computation (CEC). pp. 4038–4044. IEEE (2016)
3. Bacanin, N., Tuba, M.: Firefly algorithm for cardinality constrained mean-variance portfolio optimization problem with entropy diversity constraint. The Scientific World Journal, Article ID721521 pp. 1–16 (2014)
4. Brajevic, I., Tuba, M.: Cuckoo search and firefly algorithm applied to multilevel image thresholding. In: Cuckoo Search and Firefly Algorithm: Theory and Applications, LNCS, vol. 516, pp. 115–139. Springer (2014)
5. Bulusu, N., Heidemann, J., Bychkovskiy, V., Estrin, D.: Adaptive beacon placement. In: 21st International Conference on Distributed Computing Systems. pp. 489–498 (2001)
6. Chan, F., So, H.: Accurate distributed range-based positioning algorithm for wireless sensor networks. IEEE Trans. Sig. Process. **57**(10), 4100–4105 (2009)
7. Fister, I., Yang, X.S., Brest, J.: Modified firefly algorithm using quaternion representation. Expert Syst. Appl. **40**(18), 7220–7230 (2013)
8. Gopakumar, A., Jacob, L.: Localization in wireless sensor networks using particle swarm optimization. IET Int. Confer. Wireless Mob. Multimed. Netw. **2**, 227–230 (2008)
9. Goyal, S., Patterh, M.S.: Wireless sensor network localization based on Cuckoo search algorithm. Wireless Person. Commun. **79**(1), 223–234 (2014)
10. Huanxiang, J., Yong, W., Xiaoling, T.: Localization algorithm for mobile anchor node based on genetic algorithm in wireless sensor network. In: International Conference on Intelligent Computing and Integrated Systems, pp. 40–44 Oct (2010)
11. Kannan, A., Mao, G., Vucetic, B.: Simulated annealing based wireless sensor network localization with flip ambiguity mitigation. In: IEEE 63rd Vehicular Technology Conference. vol. 2, pp. 1022–1026 (2006)

12. Karthikeyan, S., Asokan, P., Nickolas, S., Page, T.: A hybrid discrete firefly algorithm for solving multi-objective flexible job shop scheduling problems. Int. J. Bio-Inspired Computat. **7**(6), 386–401 (2015)
13. Kulkarni, R.V., Venayagamoorthy, G.K., Cheng, M.X.: Bio-inspired node localization in wireless sensor networks. In: IEEE International Conference on Systems, Man and Cybernetics. pp. 205–210 (2009)
14. Kumar, A., Khosla, A., Saini, J., Singh, S.: Meta-heuristic range based node localization algorithm for wireless sensor networks. In: International Conference on Localization and GNSS (ICL-GNSS). pp. 1–7 (2012)
15. Li, D., Wen, X.B.: An improved PSO algorithm for distributed localization in wireless sensor networks. Int. J. Distrib. Sens. Netw. **11**(7), 1–8 (2015)
16. Maoa, G., Fidanb, B., Anderson, B.D.: Wireless sensor network localization techniques. Comput. Netw. **51**, 2529–2553 (2007)
17. Molina, G., Alba, E.: Location discovery in wireless sensor networks using metaheuristics. Appl. Soft Comput. **11**(1), 1223–1240 (2011)
18. Nuo, W., Qiang, G., Ming-Lei, S., Jia-liang, L., Ming, Y.: Three-dimensional localization algorithm of wireless sensor networks base on particle swarm optimization. J. China Universit. Posts Telecommun. **91**(1), 7–12 (2012)
19. Shi, Q., He, C., Chen, H., Jiang, L.: Distributed wireless sensor network localization via sequential greedy optimization algorithm. IEEE Trans. Sig. Process. **58**(6), 3328–3340 (2010)
20. Tuba, E., Mrkela, L., Tuba, M.: Support vector machine parameter tuning using firefly algorithm. In: 26th International Conference Radioelektronika. pp. 413–418. IEEE (2016)
21. Tuba, E., Tuba, M., Simian, D.: Range based wireless sensor node localization using bat algorithm. In: 13th ACM Symposium on Performance Evaluation of Wireless Ad Hoc, Sensor, and Ubiquitous Networks. pp. 41–44 (2016)
22. Tuba, M., Bacanin, N.: Improved seeker optimization algorithm hybridized with firefly algorithm for constrained optimization problems. Neurocomputing **143**, 197–207 (2014)
23. Tuba, M., Bacanin, N., Alihodzic, A.: Firefly algorithm for multi-objective RFID network planning problem. In: 22nd Telecommunications Forum Telfor. pp. 95–98 (2014)
24. Xu, J., Liu, W., Lang, F., Zhang, Y., Wang, C.: Distance measurement model based on RSSI in WSN. Wireless Sens. Netw. **2**(08), 606–611 (2010)
25. Yang, X.S.: Firefly algorithms for multimodal optimization. In: Stochastic Algorithms: Foundations and Applications, LNCS, vol. 5792, pp. 169–178. Springer (2009)
26. Yang, X.S., He, X.: Firefly algorithm: recent advances and applications. Int. J. Swarm Intell. **1**(1), 36–50 (2013)

Detection of Brain Tumor from MR Image Sequence Using Image Segmentation and Blob's Centroid

Thanawit Chaisuparpsirikun and Varin Chouvatut

Abstract This paper proposes a method to search for a probable tumor in magnetic resonance (MR) images of a human brain. Typically, a tumor can be found in some contiguous images of the MR sequence and positions of its appearance in such contiguous images usually have similar centroid thus their corresponding projections should be able to be detected automatically in order to support a user or a doctor for further diagnosis. Once region of a probable tumor is detected, matched checking between a pair of contiguous MR images can be done and relabeled to indicate the same area of the tumor amongst sequential images. Any regions without match between contiguous images are initially considered as irrelevant components and will not be analyzed further unless the doctor indicates otherwise. Then, ratio of tumor to brain is calculated to support as an initial diagnosis of tumors appeared in an MR image sequence.

Keywords Magnetic resonance · MRI · Tumor · Brain · Region Blob

1 Introduction

Magnetic Resonance Imaging (MRI) is done for the purpose of risk reduction for a patient from irradiation damage whereas images resulting from MRI provide as similar details as those resulting from x-ray imaging. Furthermore, applying image processing techniques computing by a modern personal computer helps improve

---

T. Chaisuparpsirikun
Faculty of Science and Graduate School, Computer Science Department,
Chiang Mai University, Chiang Mai, Thailand

V. Chouvatut (✉)
Faculty of Science, Computer Science Department,
Chiang Mai University, Chiang Mai, Thailand
e-mail: varinchouv@gmail.com; varin.ch@cmu.ac.th

X.-S. Yang et al. (eds.), *Smart Trends in Systems, Security and Sustainability*, Lecture Notes in Networks and Systems 18,
https://doi.org/10.1007/978-981-10-6916-1_11

medical analysis known as medical image processing. The main objective of medical image processing is for increasing performance of an expert doctor or specialist in disease diagnosis. There have been many researches using various techniques in medical image processing recently.

Researchers in [1] separated brain portion in an MRI-brain image using image segmentation. In order to segment the brain portion, researchers demonstrated global, local, and adaptive thresholding. Thresholding values applied to the segmentation were selected from some methods including histogram, interactive, Otsu, and clustering. Some different methods were applied for comparison purpose.

Researchers in [2] detected tumor in MR images of a human brain by checking whether color of the brain's surface is normal. To extract segment of the detected tumor, researchers used the seeded region growing method. The method was also used in [3] to get boundary of cervical cancer. In this method, the seed pixel or threshold value of color intensity identifying the searched region must be initialized or fixed.

MR images of a human brain were also researched in [4]. Researchers extracted the brain region out of the images automatically using a two-stage method. The brain region is considered as the region of interest (ROI) in the method of feature extraction. Since the background of MR images is uniform (the whole dark region), a threshold value was thus used for the ROI extraction. Then, the obtained binary image was used to separate the scalp part. After that, the segmentation is used further.

Some researchers such as in [5] used statistical data in order to detect and segment the brain tumor from MR images. To segment the tumor region, Gabor wavelet feature extraction was used. Unfortunately, working with statistical data usually requires high computation due to complex equations, e.g. exponential equations used in their method.

Image segmentation could be done using some classification techniques such as those used in [6] and [7]. Using a classification technique requires training stage where a large data set should be available on hand in order to gain high accuracy in the test stage of the classification.

Most methods worked on segmentation in each MR image frame by frame separately and relationship of the detected or segmented tumor among frames or images in the MRI sequence have not yet been concerned. Thus, this paper aims to proposed a methodology where several image processing techniques are applied to MR images in order to segment a human brain's tumor region(s) appeared in the give image sequence. Contributions of the proposed method are that even tumor whose region is smaller than those found in typical cases can still be detected, segmented, and furthermore mapped between sequential images. Also, the techniques used in this paper require no complex equation, where a long processing time or a large data set for initialization is unavoidable.

2 Methodology and Experimental Results

Methodology for the purpose of finding tumor region(s) appeared in an MR image sequence includes Six Sections; 2.1 Input Images, 2.2 Image Enhancement, 2.3 Multi-thresholding, 2.4 Blob Analysis, 2.5 Removal of Irrelevant Components, and 2.6 Computation of Tumor's Area.

2.1 Input Images

File extension of input magnetic resonance (MR) images is.dcm or DICOM (Digital Imaging and Communications in Medicine) file type which is a medical standard type for store and distribution purposes. An example of sequence of five MR images of a human brain with a tumor is shown in Fig. 1.

2.2 Image Enhancement

There can be several noisy pixels in an original MR image, an unsharp masking filter is applied to eliminate such noises and also emphasizes all edges in the image

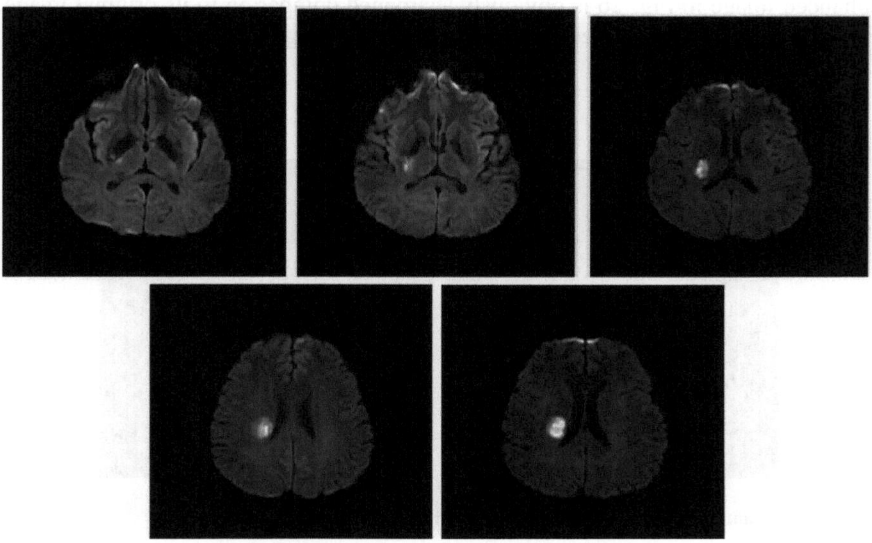

Fig. 1 An example MR image sequence with appearance of a tumor

to increase detection performance of an object in it. To enhance the image using an unsharp masking filter, Eqs. (1)–(3) are used.

First of all, the image should be blurred using a box filter with size of 3 × 3. Let $m(a,b)$ be coefficients of the box filter where a and b are in ranges of [−1, 1], the blurred or filtered image results from $m(a,b)$ filter can be obtained using

$$\bar{f}(x,y) = \frac{\sum_{a=-1}^{1}\sum_{b=-1}^{1} m(a,b)f(x+a,y+b)}{\sum_{a=-1}^{1}\sum_{b=-1}^{1} m(a,b)}. \quad (1)$$

where $f(x,y)$ is intensity value of pixel's coordinates (x, y) in the input image of r rows and c columns, x and y are in ranges of [0, $r-1$] and [0, $c-1$], respectively, and $\bar{f}(x,y)$ is intensity value of the corresponding pixel's coordinates (x, y) in the filtered image.

After the image was blurred, a mask for unsharp masking $s_{mask}(x,y)$ can be calculated using

$$s_{mask}(x,y) = f(x,y) - \bar{f}(x,y). \quad (2)$$

Then result image obtained from unsharp masking can be calculated from

$$r(x,y) = f(x,y) + s_{mask}(x,y). \quad (3)$$

Figure 2 shows an example of result image obtained from unsharp masking. The enhanced image in Fig. 2b is obviously sharpened comparing to its original image, Fig. 2a. That is, details of edges in the sharpened image can be more easily seen.

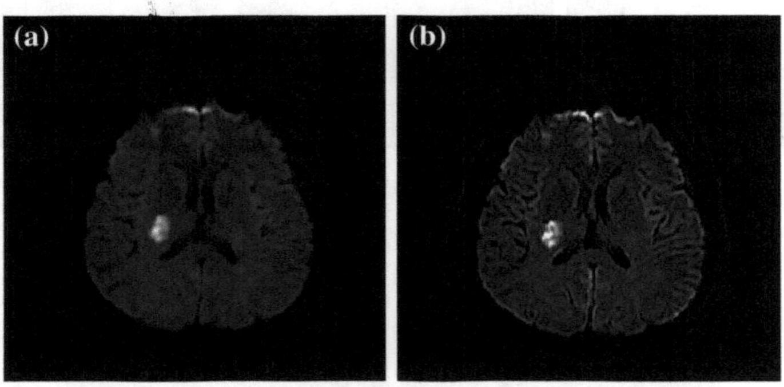

Fig. 2 An example of enhanced image (b) results from the input image (a)

2.3 Multi-thresholding

After enhancing the input image, the tumor region should be separated from other brain's regions in the image. Since the tumor region is typically brighter than other regions of the brain in an MR image and the background region which is not part of the brain is typically darker than the brain itself, thus image segmentation based on discontinuity (or on edges) can be done using multi-thresholding. The multi-thresholding used in this paper has two threshold values, called *T*1 and *T*2, for the significantly different ranges of intensities in an MR image.

$$b(x,y) = \begin{cases} 255 & \text{, if } f(x,y) \geq T1 \\ 0 & \text{, otherwise} \end{cases} \tag{4}$$

From (4), $b(x,y)$ which is the intensity value at coordinates (*x*, *y*) in the segment of the whole brain will be converted to a white color (intensity value of 255) and other parts out of the brain's region will be converted to black (intensity value of 0).

After the whole brain segment was separated from the background, we will get all coordinates specifying the boundary of the whole brain. The boundary coordinates of the whole brain will then be used to scope region for the second threshold value, *T*2, to extract the tumor out of segment of the whole brain from the original MR input image using

$$t(x,y) = \begin{cases} 255 & \text{, if } f(x,y) \geq T2 \\ 0 & \text{, otherwise} \end{cases} \tag{5}$$

where $t(x,y)$ is the intensity value at coordinates *(x, y)* in the tumor segment appeared in the separated segment of the whole brain. Thus, the tumor segment typically brighter than the other parts of brain will be converted to white with intensity value of 255 while the not-tumor segments will be converted to black with intensity value of 0.

In conclusion, the first threshold value, *T*1, is used to separate segment of the whole brain out of the background and the second threshold value, *T*2, is then used to separate the tumor segment out of the segment of the whole brain obtained from the first thresholding.

The following Figs. 3 and 4 show an example form for adjusting the two threshold values in an MR image's histogram and an example result of brain and tumor segments obtained from the multi-thresholding process, respectively.

From Fig. 3, the first threshold value, *T*1, is set to 25 (the left green vertical-line) and the second threshold value, *T*2, is set to 108 (the right red vertical-line).

From Fig. 4b, there are two segments detected as probable tumors at the first stage since they are the brighter segments found in the segment of whole brain. However, these segments of probable tumors will later be mapped to the contiguous MR image in the sequence.

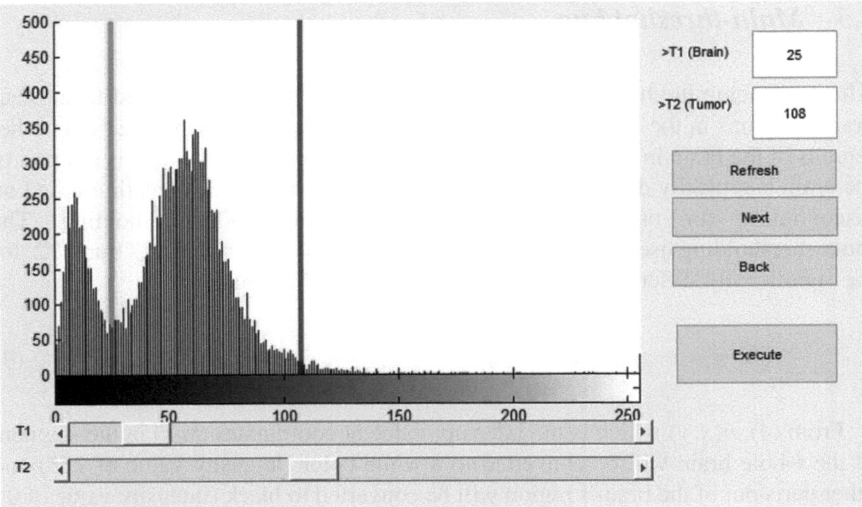

Fig. 3 An example of threshold value adjustment using two scroll bars for the histogram of an input MR image

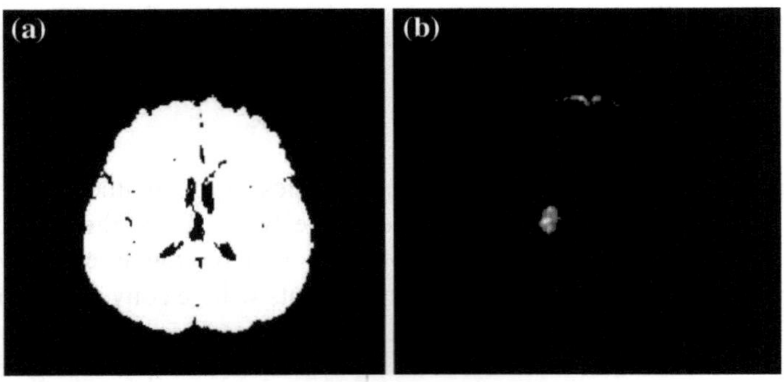

Fig. 4 An example result of separating segment of the whole brain (a) and its corresponding segment of the tumor (b) using multi-thresholding values, *T*1 and *T*2, respectively

2.4 Blob Analysis

In order to map the same segment in contiguous MR images, the segment's centroid should be computed first. Blob coloring analysis can be used for the purposes of finding centroid of an object (or an image segment) and the object's size. There are two types of neighboring pixel consideration, four-connected pixels and eight-connected pixels. An example result of applying the blob analysis to the detected segment(s) of probable tumor(s) in five contiguous MR images of a human brain is as Fig. 5. With this method, even a small region still can be detected.

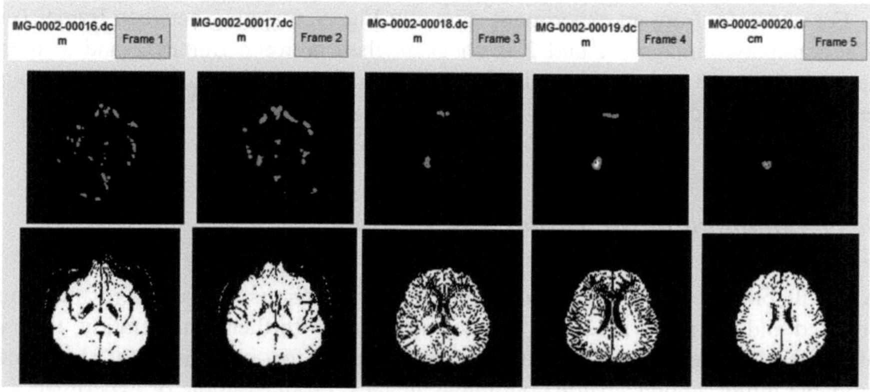

Fig. 5 An example of five contiguous MR images each of which has different numbers of blobs detected, the top row shows separate blob(s) (or segment(s)) of the corresponding MR image in the bottom row

2.5 Removal of Irrelevant Components

Once all probable tumor(s) in an MR image sequence are detected and segmented in blob(s) from Sect. 2.4, the same segment appeared all along the given MR image sequence should be mapped using the segment's centroid obtained from the blob analysis. A detected segment with no match in any other contiguous MR image is initially considered as not a tumor except the doctor investigates otherwise.

To support a doctor in mapping the same segment appeared in a certain contiguous MR images, each segment's centroid in an image is measured. Since the region boundary of a tumor appeared in even contiguous MR images may be different due to its naturally free-formed shape, distances among centroids of segments (or blobs) in any two contiguous images must be measured to find out if they are centroids of the same blob. Euclidean distance, $d(p,q)$, between any pair of centroids' coordinates, p and q, each of which is from one of two contiguous images can be calculated by

$$d(p,q) = \sqrt{(x_p - x_q)^2 + (y_p - y_q)^2} \qquad (6)$$

where (x_p, y_p) is coordinates of the centroid p and (x_q, y_q) is coordinates of the centroid q. Once all distances between any pair of centroids each of which gained from different images (of two contiguous MR images), a threshold distance, $T3$, will be used to eliminate some irrelevant or not matched blobs between the two contiguous images. That is, any $d(p,q) \leq T3$ will be accepted as matched centroids. Any centroid with no match will be rejected and thus removed from further analysis.

For example, from an experiment shown in Fig. 5, mapping between frames (or MR contiguous images) 1 and 2 is checked, then between frames 2 and 3, then between frames 3 and 4, and finally between frames 4 and 5. In total, there are four sets of mapping checks.

2.6 Computation of Tumor's Area

After all blobs (or segments) in each pair of contiguous images were mapping checked and the irrelevant blobs were removed from further consideration, theory of intersection set from (7) is applied to searched for appearance of the same probable tumor among several contiguous images. From example of experiment in Fig. 5, let four sets of mapping checks explained in Sect. 2.5 be denoted as f_1 to f_4. This process of finding intersection set of blobs is required because the blob number labeled by blob coloring analysis in each MR image can be different for the same region of segment. The different labels in different images result from the different numbers of blobs in each image. Note that, relabeling of the same blob in all MR images where its appearance is detected is essential in order to identify the number of probable tumors in the given image sequence. After relabeling, the same blob appeared in different images should have the same label finally.

$$f_1 \cap f_2 = \{l: l \in f_1 \wedge l \in f_2\} \tag{7}$$

From (7), let l be a blob label appeared in both mapping set 1, f_1, and set 2, f_2. Thus, if a blob is a matched blob between sets f_1 and f_2, the blob's label l should be identical in both sets.

Once all blobs of probable tumors are mapped and relabeled, ratio of probable tumors' portion and the whole brain's portion extracted from all MR sequential images can be estimated from

$$Tumor\colon Brain = \frac{\sum_{i=1}^{N} Tumor\_Portion_i}{\sum_{j=1}^{N} Brain\_Portion_j} \tag{8}$$

where $Tumer\_Portion_i$ is the area of all probable tumors detected in all intersected frame sets and appeared in image i, $Brain\_Portion_j$ is the area of the whole brain extracted from image j using method in Sect. 2.3 as an example shown in Fig. 4a, and N is the number of the given sequential images. The $Tumer\_Portion_i$ and $Brain\_Portion_j$ can be calculated using (9) and (10), respectively. Equation (8) can be used to estimate ratio of tumor portion to brain portion since the image sequence as input to this research can be considered as layers of cross-sections of a human brain imaging from top to bottom (or from left to right) where one or more volumes of probable tumors are appeared in a certain number of layers.

$$Tumor\_Portion = \sum_{i=1}^{M} \sum_{j=1}^{N} Tumor\_Coordinates_{i,j} \tag{9}$$

where M is the number of matched tumor labels in an MR image, N is the number of pixels of the currently considered tumor label i, and $Tumor\_Coordinates_{i,j}$ is equal to 1 if the pixel belongs to the currently considered tumor label i in the image and equal to 0 otherwise.

$$Brain\_Portion = \sum_{i=1}^{N} Brain\_Coordinates_{i} \tag{10}$$

where N is the number of pixels of the whole brain area in the currently considered image and $Brain\_Coordinates_i$ is equal to 1 if the pixel is of the brain area and 0 otherwise.

Table 1 below shows some example results of original brain images shown in Fig. 6, calculated from (8)–(10).

Table 1 Three example results obtained from computation of tumor's area of three cases shown in Fig. 6

Case	Tumor portion	Brain portion	Tumor: brain (%)
1	21,521	444,474	4.84
2	18,552	581,699	3.19
3	29,202	470,675	6.20

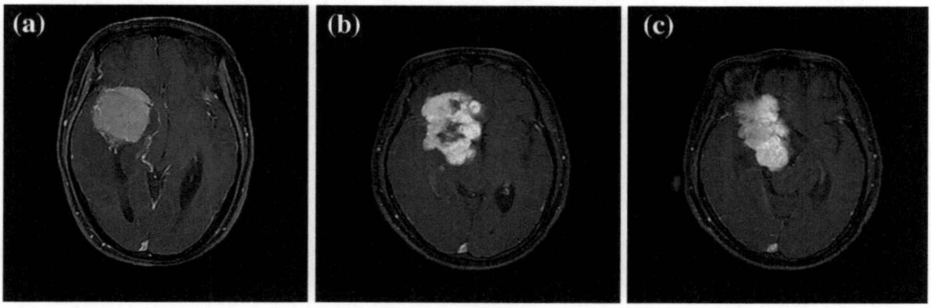

Fig. 6 Three example original images of three human brains

3 Discussion and Conclusion

From an MR image sequence where one or more tumors may be seen in only some images in the sequence, it would be helpful to a doctor in diagnosis of tumor disease. Since a tumor typically appears in a certain number of contiguous images of an MR sequence. Before searching for a tumor's region in images, the MR image must be enhanced first and for this purpose, an unsharp masking filter is used. Next, the segments of the whole of human brain and the probable tumor(s) must be separated from the background and this can be done using multi-thresholding technique in which the image's histogram is needed. Then, blob analysis is applied to the segments of the probable tumors to gain centroid of each blob. Thus, centroid of each blob is compared to each other in contiguous images to match the same tumor which may be appeared in any two contiguous images and to remove any irrelevant segments detected from further consideration. Finally, the ratio of tumor portion to the whole brain portion of all images in the given MR sequence can be computed for diagnostic support for a doctor or a medical specialist.

Acknowledgements We would like to thank you Maharaj Nakorn Chiang Mai Hospital, Faculty of Medicine, Chiang Mai University, Thailand for example MR image sequences for this research.

References

1. Evelin, G., Sujji, Y.L.: MRI brain image segmentation based on thresholding. Int. J. Adv. Comput. Res. 2249–7277 (2013)
2. Mukesh Kumar, K.K.: A texture based tumor detection and automatic segmentation using seed region growing method. IJCTA. 855–859 (2011)
3. Mashor, Othman, Mat Isa: Seeded region growing features extraction algorithm: its potential use in improving screening for cervical cancer. Int. J. Comput. Int. Manag. 61–70 (2005)
4. Somasundaram, K., Kalaiselvi, T.: Automatic brain extraction methods for T1 magnetic resonance images using region labeling and morphological operations. Comput. Biol. Med. 716–725 (2011)
5. Nooshin, N., Miroslav, K.: Brain tumors detection and segmentation in MR images: Gabor wavelet vs. statistical features. Comput. Biol. Med. 286–301 (2015)
6. Anitha, V., Murugavalli, S.: Brain tumor classification using two-tier classifier with adaptive segmentation technique. IET J. 9–18 (2015)
7. Wei, H., Kab, L.C., Jiayin, Z.: Region-based nasopharyngeal carcinoma lesion segmentation from mri using clustering and classification-based methods with learning. Imag. Informat. Med. 472–482 (2012)

A Project Metric Model for Assessing ICT Project Success/Failure

Ezekiel U. Okike and Ofaletse Mphale

Abstract Project success metrics play a vital role in project control as they focus on project deliverables as justification of success. However, this traditional view is not sufficient for measuring the degree of success/failure with consideration of the resulting Return on Investment (ROI). This study presents a model project management metric for assessing project success or failure which sufficiently accounts for necessary expectations from the user community. The model was evolved from existing project management metrics by adding features to justify precise measurement of degrees of project success/failures.

Keywords ICT project · Project management · Metrics
Assessment model · Success/failure

1 Introduction

Information and Communication Technology/Information Technology (ICT/IT) project deployment in public and private institutions around the world is on the increase. However, in some cases, a number of the Information System (IS) projects implemented have not been reportedly successful due mainly to the neglect of software project management best practices by organizations, and other factors identified in [6, 9, 16]. Project failures as defined in [6] occur when there is cost overrun, late delivery, poor quality and the product is not useful. Although ICT deployment aids human capital empowerment, and improves the quality of life of human-beings in any domain (Agriculture, Aviation, Banking, Education, Health etc.), neglecting to apply appropriate project management principles and techniques could spell doom for many projects. Therefore, in general, the application of

E. U. Okike (✉) · O. Mphale
Department of Computer Science, University of Botswana, Gaborone, Botswana
e-mail: euokike@gmail.com

O. Mphale
e-mail: ofaletse_offie@hotmail.com

X.-S. Yang et al. (eds.), *Smart Trends in Systems, Security and Sustainability*, Lecture Notes in Networks and Systems 18,
https://doi.org/10.1007/978-981-10-6916-1_12

software project management best practice should lead to successful projects outcomes.

This paper proposes a hybrid project metric based on existing models for assessing ICT project success or failure.

1.1 Statement of the Problem

IT senior managers are often faced with the challenge of ascertaining the exact measures of project success/failure and which specific metric to apply. Therefore, there is need to define a metric based guideline to measure the degree of a project's success or failure.

1.2 Aims and Objectives

This paper aims to examine the main factors contributing to ICT project success or failure, and to evolve a project metric framework to guide the process of ascertaining the degree of success/failure contributed by identified factors.

The objectives are as follows:

- To identify the factors contributing to success of ICT projects
- To identify the factors contributing to failure of ICT projects
- To develop a metric for ascertaining the degree of success or failure contributed by the identified factors

The rest of the paper counts of 5 sections. Section 2 presents a review of the literature. Section 3 presents the methodology of the study and metrics development approach. Section 4 presents the result of our study, and Sect. 5 the conclusion.

2 Literature Review

Organizations and enterprises invest in ICT projects to assist them manage their business processes. It is suggested in [7] that ICT investment also give organizations competitive advantage over rivals. Furthermore, the history of ICT projects indicate that there could be significant challenges in successfully attaining a return on many of the IT investment projects [18, 19] if appropriate project management is neglected.

Many research reports such as [4, 5, 9, 18, 19] identified software project success factors from many perspectives such as "on time, on budget with satisfactory result" e.t.c. A study in [18] reported the percentages of successful projects at 29% in 2011, 27% in 2012, 31% in 2013, 28% in 2014, and 29% in 2015; while percentages of failed projects from the same study were 22% in 2011, 17% in 2012, 19% in 2013, 17% in 2014, and 19% in 2015. According to [19] successful projects could be due to the following factors: Improved project environment processes, effective project methods, skilled personnel, effective project costing, tools, decisions, optimization, addressing of the project internal and external influences and effective team chemistry. In addition, according to [10], IT projects are successful when they are executed within scope, schedule, budget, goal and there is value added. Other factors critical to project success are discussed and summarized in [5]. These include management issues, communication issues, training and education, team composition, available resources, stakeholder involvement, software development issues, implementation strategies etc.

In terms of failures, IT project fail when the IT system does not deliver required expectations within the expected time and expenditure [3, 9, 19].

The key reasons for ICT projects failure have been identified in [1, 11–14, 20]. A study in [1] suggested that the main causes of IT project failure are deeply rooted in factors involving project management issues, top management issues, technology issues, organizational factors, project complexity factors, and process factors.

3 Methodology and Metric Development

3.1 Methodology

In this study, we employed document analysis of relevant literarture in order to appraise the main factors leading to software project successes and failures, and hence propose a metric for assessing ICT project success/failures. The model was empirically tested with data from Botswana to demonstrate its workability. A flowchart of the approach of this study is shown in Fig. 1.

3.2 Metric Development

A study in [17] classified IS benefits reported in IT project success stories into four dimension namely: operational, managerial, infrastructural, and organizational dimensions. In terms of metrics, IS project success metrics may be aligned to three success levels namely: Project management success, Project success and Consistent project success [8, 17]. Using approaches from many sources such as [2, 8, 10, 15, 19],

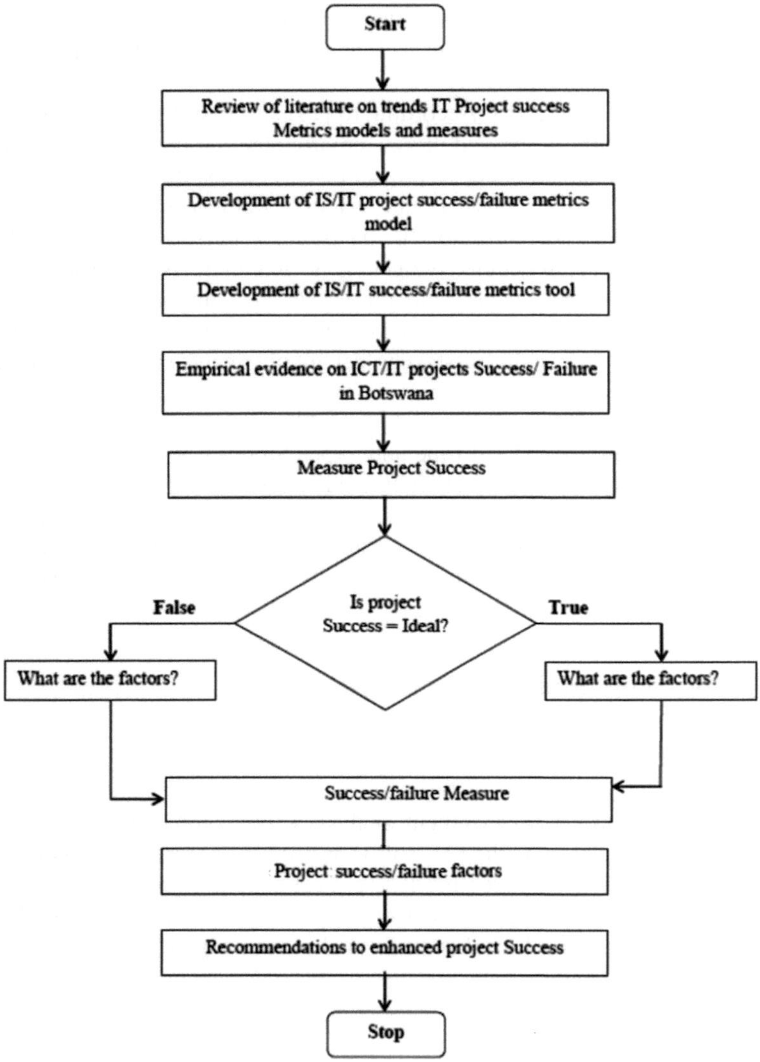

Fig. 1 Study approach flowchart

a hybrid project success metric and measurement has been developed. However, using the approach of [7, 19], a study in [6] suggested that the metric did not account for usability and functionality factors in order to be a useful measure of project success. Hence, by incorporating these factors into the concept of an Information System (IS), we propose a model to measure project success or failure.

4 Proposal of a Project Success/Failure Metric

Following the trend of project success or failure factors earlier identified, we define and demonstrate a measure of project success or failure based on the ideal components of an IS, and whose scope precision quantifies the Return-On-Investment (ROI) and project justification. This establishes new quality management process for the organization and delivers the processes that address the specific business needs as the real measure of success.

The initial phase of the proposed success metric model, analyzed and integrated all useful IS/IT success factors from various scholars in literature. The resulting IT project success metric at the initial stage is shown in Fig. 2.

This model may be represented mathematically as

$$y_{ij} = \beta_0 + \beta_1 x_{1j} + \beta_2 x_{2j} + \beta_3 x_{3j} + \beta_4 x_{4j} + \ldots \beta_n x_{nj} \tag{1}$$

where y_{ij} represents a dependent variable, and $\beta_0 + \beta_1 x_{1j} + \beta_2 x_{2j} + \beta_3 x_{3j} + \beta_4 x_{4j} + \ldots \beta_n x_{nj}$ the independent variables, in this case, the chosen metrics from various sources.

The second phase of our proposal relates the project metrics model with the major components of Information System (IS) as identified in [19] namely: hardware, software, data, procedures, and people.

Using Eq. (1) a mathematical model for an IS which satisfies the form:

$$y_{ij} = \beta_0 + \beta_1 x_{1j} + \beta_2 x_{2j} + \beta_3 x_{3j} + \beta_4 x_{4j} + \cdots \beta_n x_{nj}$$

where y_{ij} represents a dependent variable, $\beta_0 + \beta_1 x_{1j} + \beta_2 x_{2j} + \beta_3 x_{3j} + \beta_4 x_{4j} + \ldots \beta_n x_{nj}$ the independent variables may be defined as follows:

$$\text{IS} = \text{Procedures} + \text{Data} + \text{Network} + \text{Hardware} + \text{Software} + \text{People} + \text{Organization} \tag{2}$$

Equation (2) may be refined to

$$\text{IS} = \text{Information}\{\text{Procedures} + \text{Data}\} + \text{Network} + \text{Hardware} + \text{Software} + \text{People} + \text{Organization} \tag{3}$$

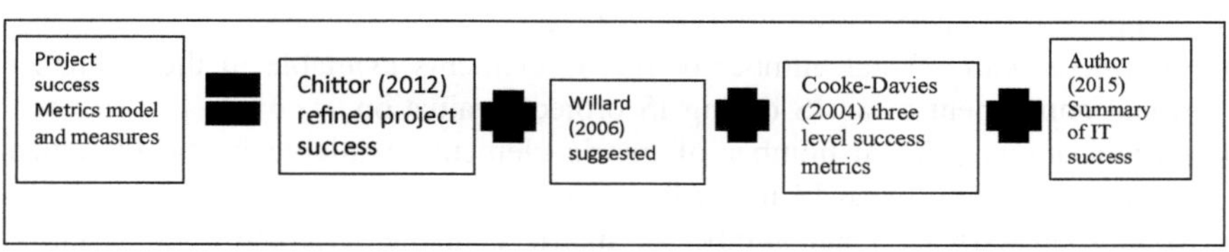

Fig. 2 Proposed project metric model

Incorporating other IS components in Eq. (3):

$$\begin{aligned} &\text{IS} = \\ &\text{technology}\{\text{Information} + \text{Network} + \text{Hardware} + \text{Software} + \text{People} + \\ &\text{Organizational issues}\{\text{Organization}\} + \text{Human resource}\{\text{People}\} \end{aligned} \tag{4}$$

The final model defines IS as shown in Eq. 5:

$$\text{IS} = \text{technology} + \text{organizational issues} + \text{Human resource} \tag{5}$$

4.1 Integrating IS Components in the Metric

The third phase of the IS metric model development was to categorise the metrics measures based on the IS components as shown in Table 1.

4.2 Definition of IS Project Success/Failure in the Refined Metric

In order to make IS project metrics model more inclusive, we define project success or failure measurement and its evaluation by modifying Eq. (4) *above:*

- Success/failure of an IS project may be estimated by the rate of IS components to the Total IS components elements.

$$|\text{SUCCESS/FAILURE}| = \frac{(Techonology_1 + Organisational_1 + Human\_Resource_1) * 100\%}{\text{Total\_origional\_}metric_{\text{-}0}} \tag{6}$$

Individual components of Eq. (6) are explained as follows:

- *$Technology_1$*: Total number of metric elements available in the Technology component of an IS during IS project success evaluation
- Technology_0: Total number of metric elements available in the Technology component of an IS in the original metric.
- *$Organisational_1$*: Total number of metric elements available in the Organizational component of an IS during IS project evaluation
- Organisational_0: Total number of metric elements available in the Organizational component of an IS in the original metrics
- *$Human\_Resource_{1:}$* Total number of metric elements available in the Human resource component of an IS during an IS project evaluation

Table 1 Refined IS/IT project success metrics and measures

Category	Metrics (measures)
Technology	IT functionality/Capabilities
	Ease of use/quantity of use
	Happiness/willingness of end users
	Technology and technological issues
	Software development
	Software prototyping and testing
	IT vendor capabilities
	IT outsourcing strategy
	IT implementation strategy
	IT solved problem(s) that was intended to solve
	Software quality improvements
	Safety (if applicable)
Organisational	Top management support
	Project Schedule
	Project Time
	Project cost
	Project accuracy (Specificationsmet)
	Management of requirements
	Change management
	Cultural management
	Quality management
	Business process re-engineering
	Financial resource
	Management of expectations
	Business plan and vision
	Leadership style
	Stakeholder involvement
	Security strategy
	Benefit(s) to the organization
	Un-intentional improvements
	Reduction of manual intervention/process
	Improved operating efficiencies
	Issues recorded since implementation
	Resource management improvement
	Support business growth
Human resource	Use of consultants
	IT project management
	User training, education and support

(continued)

Table 1 (continued)

Category	Metrics (measures)
	IT project champion
	Commitment
	Cooperation
	Enhanced productivity
	Empowerment
	Expanding/Improving core competency
	Increased flexibility
	Empowerment

- $Human\_Resource_0$: Total number of metric elements available in the Human resource component of an IS in the original metric
- $Total\_Origional\_Metric\_{_0}$: Total number of the metric elements in the original metric
- $Total\_Origional\_Metric\_{_0} = Technology_0 + Organisational_0 + Human\_Resource_0$

Assumptions:

- For a successful IS/IT Project all the IS components and their metric measurement must be available.
- If some metric elements or some components are missing then the concept of acceptable failure is used.
- Acceptable failure is when some components are missing and their absence considered insignificant by the project manager.

4.3 Metric Components Weighting

In order to evaluate and measure the IS/IT project success, some weights are assigned to the developed IS project success metric in line with [8]. For instance, in the technology category of IS metrics, there are 12 metric measures (Table 1, Sect. 4.1), hence its weight value is 12 (Table 2, Sect. 5). For the organisational component, there are 23 metric measures (Table 1), and hence a weighting of 23 (Table 2). Similarly, the human resource component has a weighting of 11 (Table 2). The total weight of metric is 12 + 23 + 11, giving the total weight of 46 (Table 2, Sect. 5).

Table 2 Technology versus organizational and human resource

Categories	Total number of metrics elements varied											
Technology	1	2	3	4	5	6	7	8	9	10	11	12
Organisational	23	23	23	23	23	23	23	23	23	23	23	23
Human resource	11	11	11	11	11	11	11	11	11	11	11	11
Score in (%)	76.09	78.26	80.43	82.61	84.78	86.96	89.13	91.30	93.48	95.65	97.83	100.00
Diff. in scores	2.17		2.17		2.17		2.17		2.17		2.17	

4.4 Critical Scores Evaluation

Each metric measure is a critical score. If during project evaluation some metric measures are available, a value 1 is assigned, otherwise 0 is assigned to symbolise unavailability of a metric measure. Assuming the Technology component of IS has 11 metric measures, then the critical score is 11.

The following assumptions define the success measure of an IS into two major categories:

- Success = 100% critical score (all metric measures available); this is the ideal case category of success
- Failure = less than 100%, but greater than 0% critical score (Partial metric measures available).

4.5 Acceptable Failure Definition and Categories

An acceptable failure is when the user is aware and understands that the IS/IT project success is in a failure category but is still satisfied with the level of success to carry on with the project.

$$\text{Acceptable failure} = \text{Success} - \mathbf{n} \tag{7}$$

where **n** equals partial metrics measures available (not 100% metric elements).

4.6 Acceptable Failure Categories

Acceptable failure may be categorized into two broad categories which are:

- Acceptable tolerance = less than 100% metric measures, but greater than 50% of the metrics measures.
- Unacceptable tolerance = greater than 0% metric measures, but less than 50% of the metric measures.

Assumption:

- Acceptable failure cannot be equal to 0% otherwise you have not implemented an IS system in your organization.

Acceptable failure categories are shown in Fig. 3.

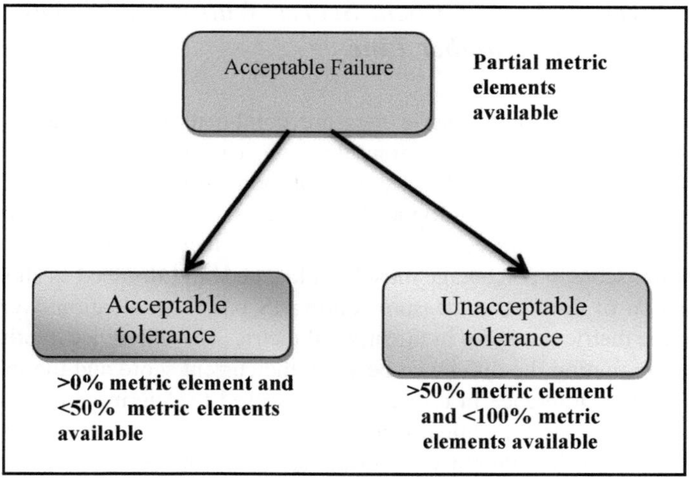

Fig. 3 Acceptable failure main categories

4.7 Success Measurement Number Line

An IS project success measurement is further illustrated in success measurement number line in our study. This is to exhibit and clarify different categories of success and the measures that satisfy them. Figure 4 shows the detailed success measurement number line and its categories.

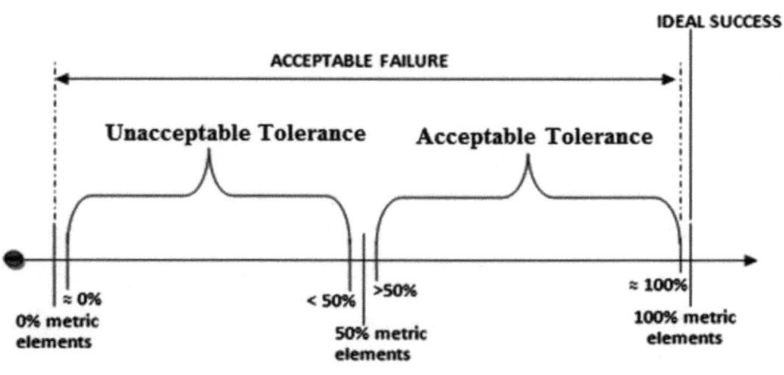

Fig. 4 Success measurement number line

4.8 Discovering More Critical Scores Within the Success Measurement Number Line

In order to explore various metric measure combinations with the intention to discover possible unique combinations that can lead to discovering new critical values, we investigate the possible presence of more critical values which represent critical scores that further sub-divide the IS success measurement number line to enhance its accuracy and precision. To determine if there are more critical values, first the metric score in percentage must be calculated, and summed as shown in (8): the summation of the metric measures during IS project evaluation over the total number of the metric measures in the original metric. The difference in critical score is obtained by finding the difference between the current score and the initial score as shown in (9). Difference in critical scores is used to determine the consistency between the scores and the critical values. A critical value is a metric score that has no relative pair and its difference is not the same as other scores.

$$\textbf{Score} = \sum(\text{Technology metric elements, Organisational metric elements, Human resource metric elements})/\text{Total metric elements from the original metric}*100\text{ per cent}) \tag{8}$$

$$\textbf{Diff. in scores} = \text{current score} - \text{previous score} \tag{9}$$

Assumption:

Scores which have relatively insignificant difference may be combined and represented by only one score. A score that does not belong to a pair is the critical point.

5 Result and Discussion

Applying our metric approach gives the results shown in Table 2.

From Table 2 above, the scores increased at a constant rate. The constant rate is shown by the difference in scores which is 2.17 throughout the entire experiment. This implies that success is balanced. Hence no critical value was discovered.

i. Varying the IS Organizational component while keeping the Technology and Human resource IS components constant, we obtain the result shown in Table 3.

From Table 3 above no critical value is discovered. The difference in the critical score is constant, thus; 2.17. This implies that IS success is still balanced at this level of experiment. Even though 100% did not belong to a pair, it is ignored because it is mentioned in the IS success measures assumptions.

ii. Varying the IS Human resource component while keeping the Organizational and Technology IS components constant yields the result shown in Table 4.

Table 3 Organizational versus technology and human resource

Categories	Total number of metrics elements varied																					
Technology	12	12	12	12	12	12	12	12	12	12	12	12	12	12	12	12	12	12	12	12	12	12	12
Organisational	1	2	3	4	5	6	7	8	9	10	11	12	13	14	15	16	17	18	19	20	21	22	23
Human Resource	11	11	11	11	11	11	11	11	11	11	11	11	11	11	11	11	11	11	11	11	11	11	11
Score in (%)	52.17	54.35	56.52	58.70	60.78	63.04	65.22	67.39	69.57	71.74	73.91	76.09	78.26	80.43	82.61	84.78	86.96	89.13	91.30	93.48	95.65	97.83	100
Diff. in Scores	2.17		2.17		2.17		2.17		2.17		2.17		2.17		2.17		2.17		2.17		2.17		

Critical value

Table 4 Human resource versus organisational and technology

Categories	Total number of metrics elements varied										
Technology	12	12	12	12	12	12	12	12	12	12	12
Organisational	23	23	23	23	23	23	23	23	23	23	23
Human Resource	1	2	3	4	5	6	7	8	9	10	11
Score in (%)	78.26	80.43	82.61	84.78	86.96	89.13	91.30	93.48	95.65	97.83	100
Diff. in Scores	2.17		2.17		2.17		2.17		2.17		

Critical value

From Table 4 no critical value is discovered. The 100% which is observed is ignored, as it is already stated in the success measurement assumptions. The success is still balanced at this level of the experiment variation.

5.1 Variations of One IS Components with Relative Percentage Proportions

i. Varying IS components in ratios of 2.5%

Table 5 represent the results of varying metric components in the ratios of 2.5% as shown.

Table 5 Varying IS components by the ratio of 2.5%

Variation in (%) / Categories	2.5	5	7.5	10	12.5	15	17.5	20	22.5	25	27.5	30	32.5	35	37.5	40	42.5	45	47.5	50
Technology	1	1	1	1	2	2	3	3	3	3	4	4	4	5	5	5	6	6	6	6
Organisational	1	2	2	3	3	4	4	5	6	6	7	7	8	8	9	10	10	11	11	12
Human Resource	1	1	1	2	2	2	2	3	3	3	3	4	4	4	5	5	5	5	6	6
Scores in (%)	6.52	8.70	8.70	15.22	15.22	17.39	19.57	23.91	26.09	26.09	30.43	32.61	34.78	36.96	41.30	43.48	45.65	47.83	50.00	52.17
											2.17		2.17		2.17		2.17		2.17	

Combining same values to be represented by a single value →

Refine Score in (%)	6.52	8.70	15.22	17.39	19.57	23.91	26.09
Diff. in Scores (%)	2.17		2.17			2.17	

↓ Critical value

Variation in (%) / Categories	52.5	55	57.5	60	62.5	65	67.5	70	72.5	75	77.5	80	82.5	85	87.5	90	92.5	95	97.5	100
Technology	7	7	7	8	8	8	9	9	9	9	10	10	11	11	11	11	12	12	12	12
Organisational	12	13	14	14	15	15	16	17	17	18	18	19	19	20	21	21	22	22	22	23
Human Resource	6	6	7	7	7	8	8	8	8	9	9	9	10	10	10	10	11	11	11	11
Scores in (%)	53.35	56.52	60.87	63.04	65.22	67.39	71.74	73.91	73.91	78.26	80.43	82.61	82.61	89.13	91.30	91.30	97.83	97.83	97.83	100.00
Diff. in Scores (%)	2.17		2.17		2.17															

Combining same values to be represented by a single value →

Refined Scores in (%)	71.74	73.91	78.26	80.43	82.61	89.13	91.30	97.83	100.00
Diff. in Scores (%)	2.17			2.17			2.17		

↓ Critical value (78.26) ↓ Critical value (89.13) ↓ Critical value (100.00)

From Table 5, the following critical values are discovered; 19.57, 78.26 and 89.13%. The value of 100% is ignored as it is already given in the success measure assumptions. This implies that IS success is not balanced at these stated critical value measures. Therefore those critical values are noted.

ii. Varying IS components in relative ratio of 1.25%.

The metric elements were varied with relative ratios of 1.25%. The results are presented in Table 6.

From Table 6, no new critical values are discovered. Although 19.57 and 100% are appear as critical values, there are ignored because they are already known. This implies that success at this level of experiment is balanced.

5.2 Discovered Critical Values and Categorisation

The discovered critical scores from our empirical study using data from Botswana is summarized in Table 7.

Table 6 Varying IS components by ratio of 1.25%

Variation in (%) / Categories	1.25	2.5	3.75	5	6.25	7.5	8.75	10	11.25	12.5	13.75	15	16.25	17.5	18.75	20	21.25	22.5	23.75	25	26.25	27.5	28.75	30
Technology	1	1	1	1	1	1	1	2	2	2	2	2	2	3	3	3	3	3	3	3	4	4	4	4
Organisational	1	1	1	2	2	2	2	3	3	3	4	4	4	4	5	5	5	6	6	6	6	7	7	8
Human Resource	1	1	1	1	1	1	1	2	2	2	2	2	2	2	2	3	3	3	3	3	3	3	4	4
Score in (%)	6.52	6.52	6.52	8.70	8.70	8.70	8.70	15.22	15.22	15.22	17.39	17.39	17.39	19.57	21.24	23.91	23.91	26.09	26.09	26.09	28.26	30.43	32.61	32.61
Combining same values to be represented by a single value →																								
Refined Scores (%)	6.52			8.70				15.22			17.39			19.57	21.24	23.91		26.09			28.26	30.43	32.61	
Diff. in scores	2.17							2.17							2.17			2.17				2.17		

Critical value

Variation in (%) / Categories	31.25	32.5	33.75	35	36.25	37.5	38.75	40	41.25	42.5	43.75	45	46.25	47.5	48.75	50	51.25	52.5	53.75	55	56.25	57.5	58.75
Technology	4	4	4	5	5	5	5	5	5	6	6	6	6	6	6	6	7	7	7	7	7	7	7
Organisational	8	8	8	8	9	9	9	10	10	10	10	11	11	11	12	12	12	12	13	13	13	14	14
Human Resource	4	4	4	4	5	5	5	5	5	5	5	5	5	6	6	6	6	6	6	6	7	7	7
Score in (%)	34.78	34.78	34.78	36.96	41.30	41.30	41.30	43.48	43.48	45.65	45.65	47.83	47.83	50.00	52.17	52.17	54.35	54.35	56.52	56.52	58.70	60.87	60.87
Combining same values to be represented by a single value →																							
Refined Scores (%)	34.78			36.96	41.30			43,48		45.65		47.83		50.00	52.17		54.35		56.52		58.70	60.87	
Diff. in scores (%)	2.17				2.17					2.17				2.17			2.17				2.17		

Variation in (%) / Categories	60	61.25	62.5	63.75	65	66.25	67.50	68.75	70	71.25	72.5	73.75	75	76.25	77.5	78.75	80	81.25	82.5	83.75	85	86.25
Technology	8	8	8	8	8	8	9	9	9	9	9	9	9	10	10	10	10	10	10	10	11	11
Organisational	14	14	15	15	15	16	16	16	17	17	17	17	18	18	18	19	19	19	19	20	20	20
Human Resource	7	7	7	7	8	8	8	8	8	8	8	8	9	9	9	9	9	9	9	10	10	10
Score in (%)	63.04	63.04	65.22	65.22	67.39	69.57	71.74	71.74	73.91	73.91	73.91	76.09	78.26	80.43	80.43	82.61	82.61	82.61	82.61	86.96	89.13	89.13
Combining same values to be represented by a single value →					2.17							2.17										
Refined Scores (%)	63.04		65.22				71.74		73.91					50.00		82.61				86.96	89.13	
Diff. in scores (%)	2.17						2.17							2.17						2.17		

Variation in (%) / Categories	87.50	88.75	90	91.25	92.50	93.75	95	96.25	97.50	98.75	100
Technology	11	11	11	11	12	12	12	12	12	12	12
Organisational	21	21	21	21	22	22	22	22	22	22	23
Human Resource	10	10	10	10	11	11	11	11	11	11	11
Score in (%)	91.30	91.30	91.30	91.30	97.83	97.83	97.83	97.83	97.83	97.83	100
Combining same values to be represented by a single value →											
Refined Scores (%)	91.30				97.83						100
Diff. in scores (%)	2.17										

Critical value

Table 7 Discovered critical values

Experiment no:	Experiment name	Critical values (%)
1	Technology versus organisational and human resource	No new critical value
2	Organisational versus technology and human resource	No new critical value
3	Human resource versus technology and organisational	No new critical value
4	Variation of IS components by 2.5% relative ratios	19.57, 78.26, 89.13
5	Variation of IS components by 1.25% relative ratios	No new critical value

Table 8 Success categorization and acceptability levels

Category	Acceptability	Critical values (%)
Ideal success	Ideal	100
Acceptable tolerance	Best acceptable	>= 89.13, and < 100
	Least acceptable	>= 78.26, and < 89.13
Unacceptable tolerance	Worse acceptable	<78.26, and >=19.57
	Unacceptable	>0, and < 19.57

From Table 7, the following success categories are defined; the ideal success, best acceptable, least acceptable, worse acceptable and unacceptable failures. The success categorisation strategy follows the Success vs. Acceptable failure diagram illustrated earlier in Fig. 2. Table 8 shows success categories and its acceptability levels.

5.3 *Refined Success Measurement Number Line*

Using results from Tables 2–8, we define the success measurement number line with its boundaries (Fig. 5/3.5) which highlights the precise success categories and acceptability levels as shown in Table 9.

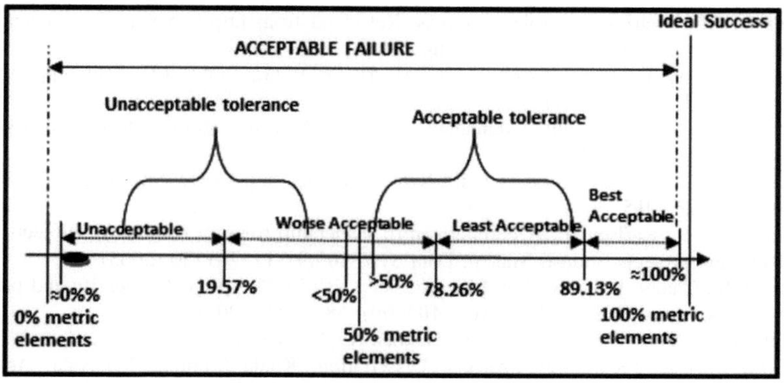

Fig. 5 Refined success measurement number line

Table 9 Success categorization and acceptable levels

Category	Acceptability	Critical values (%)
Ideal success	Ideal	100
Acceptable tolerance	Best acceptable	>= 89.13, and < = 100
	Least acceptable	>= 78.26: and < 89.13
Unacceptable tolerance	Worse acceptable	<78.26, and = 19.57
	Unacceptable	>0, and < 19.57

6 Conclusion

The model as empirically tested in the study can be used to measure project management outcomes and the degree to which success or failure can be expected in ICT projects.

References

1. Al-Ahmad, W.: Taxonomy of an IT project failure: root causes. Int. Manag. Rev. **5**, 93–106 (2009)
2. Basili, V.R., Caldiera, G., Rombach, H.D.: The goal question metric approach. In: Encyclopedia of Software Engineering, 2nd ed. pp. 528–532 (2001)
3. Beales, R.: Where IT starts to go wrong. Retrieved from http://www.pinkelephant.co.za/it-starts-go-wrong/, 10 May 2015
4. Boettcher, J. A.: Ten Steps To IT Project Success, Retrieved from https://www.projectkickstart.com/downloads/IT, 3 July 2014
5. Brocke, H.F., Uebernickel, F., Brenner, W.: Success factors in IT projects to provide customer value propositions. In: Proceedings of the 20th Australian Conference on Information systems (ACIS) 24. http://aisel.aisnet.org/acis2009/24
6. Charette, R.N.: Why software Fails, IEEE Spectrum, September, (2005)

7. Chittoor, R.: Metrics for project success. Retrieved from http://www.project-management.com/metrics-for-project-success/, 27 Aug 2015
8. Cooke-Davies, T.: Project success. In: Morris, P W.G., Pinto, J.K. (eds.) pp. 99–122. Hoboken, NJ, Wiley (2004)
9. Flowers, S.: Software failure: Management failure. Amazing Stories and Cautionary Tales. New York, Wiley (1996)
10. Goatham, R.: Why Do Projects Fail? Retrieved from http://www.calleam.com/WTPF/?p=3501, 11 Mar 2015
11. Gottschalk, P., Karlsen, J.: A comparison of leadership roles in internal IT projects versus outsourcing projects. Indust. Manag. Data Syst. **105**(9), 1137–1150 (2005)
12. Huang, S., Chang, I., Li, S., Lin, M.: Assessing risk in ERP projects: identify and prioritize the factors. Indust. Manag. Data Syst. **104**(8/9), 681–690 (2004)
13. Ibrahim, R., Ayazi, E., Nasrmalek, S., Nakhat, S.: An investigation of critical failure factors. In: Information Technology Projects. Dissertation, Kuala Lumpur, Malaysia: Advanced Informatics School (AIS)—Universiti Teknologi Malaysia (UTM) (2013)
14. Kaur, B., Aggrawal, H.: Critical failure factors in information system: an exploratory review. J. Global Res. Comput. Sci. **4**(1), 76–82 (2013)
15. McConnell, S.: Software Estimation: Demystifying the Black Art. Micro-soft Press, Redmond, Wa (2006)
16. Putnam-Majaraian, T and Putnam D.:The most common reasons why software projects fail, Retrieved from http://www.infoq.com/articles/software-failure-reaasons, 1 Dec 2016
17. Shang, S., Seddon, P.B.: Assessing and managing the benefits of enterprise systems: the business manager's perspective. Informat. Syst. J. **12**(4), 271–299 (2002)
18. Standish Group 2015 Chaos Report. http://www.Infoq.com/articles/standish-chaos-2015
19. Willard, B. K.: Project Success: Looking Beyond Traditional Project Metrics. Retrieved from http://www.maxwideman.com/guests/metrics/abstract.htm. 8 July 2015
20. Young, R., Jordan, E.: Top management support: Mantra or necessity? Int. J. Project Manage. **26**(7), 713–725 (2008)

Analysing Distribution of Copyrighted Files by Tracking BitTorrent Network

Chintan B. Sidpara, Darshana H. patel, Kunal U. Khimani, Darshan Upadhyay and Avani R. Vasant

Abstract Now a days Internet usage continually increase at a rapid speed; and with that internet-based violation also. BitTorrent is the most popular open Internet application for content sharing. There is free availability of copyrighted content through p2p network. So; it continually grows up the popularity of BitTorrent gateways. This protocol is one of the highest internet bandwidth consumers. Recent tracking summary shows that content being shared are mostly illegal. Thus aim of this paper is to find copyrighted files from torrent networks and to find distributor of that files. So to find illegal copies, I propose crawling algorithm to collect efficient listing portals. And characterizing sharing of illegal copies of content in torrent networks. Crawling algorithm mainly having three engine Scanner Engine, Monitor Engine and Analyzer Engine.

Keywords Piracy detection · BitTorrent · P2P · Torrent monitoring
Content distribution

1 Introduction

File distribution through BitTorrent networks is generating most of the Internet freight. Mainly this traffic is related to share pirated media and software. The large number of internet users who regularly get these content. These bandwidth mainly consumed by these usage increased rapidly between 2010 and 2013 [1, 2]. Globally, 432 million unique internet users were explicitly looking for infringing content during January 2013 [1].

C. B. Sidpara (✉) · D. H. patel · K. U. Khimani · D. Upadhyay · A. R. Vasant
Department of Information Technology, V.V.P. Engineering College, Rakjot,
Gujarat, India
e-mail: chintan.sidpara@gmail.com

X.-S. Yang et al. (eds.), *Smart Trends in Systems, Security and Sustainability*, Lecture Notes in Networks and Systems 18,
https://doi.org/10.1007/978-981-10-6916-1_13

BitTorrent is the most popular p2p content sharing protocol globally. Around 72% of internet bandwidth is consumed by the p2p protocol only. This kind of user searches torrent files for specific content. Torrent files open in a μTorrent then it connects with swarms of other users or down loaders.

Main three key regions North America, Europe, and Asia-Pacific are the absolute users of BitTorrent. They used almost 6,692 petabytes in 2012 only. It is increased by 244.9% from 2010 [1, 2]. BitTorrent users in January 2013 counted 178 million, it is an increment of 23.6% from November 2011, and page views also increase by 7.4 billion, around 30.6% from November 2011 [1, 2].

It is very crucial to identify the distribution of copyrighted content with these vast private BitTorrent networks. However, it is the remarkable loss to the publishing industries like media, books and software. Among these users many of them think that they are doing piracy that is unlawful, but they still continue doing this. But the thing is that with the limited resources and different law enforcement departments, it is not possible for them to crack every member of this huge BitTorrent network.

Lots of surveys have been carried out like [3, 4, 6] in recent past to lookup content spreading by p2p protocol. Among them very few target on the flooding of pirated copies. Take an example, few of them knows that they access original copy to create digital form of that. Or else distribute these copies in BitTorrent network. So it is unidentified which user create pirated copy, and who is the reciprocal to available these in BitTorrent.

Preservation system is compulsory to stop spreading illegal copies through file sharing protocol. So the owners of these contents are very much interested to develop strategy or mechanism to minimize piracy. For that, require the development body or knowledge base processes that find and stop creation and distribution of illegal copies in file sharing communities.

So, it is compulsory to establish an automated program to monitor these increasingly expanded torrent activities. The rest of this paper is structured as follows. The basic of BitTorrent Technology is presented in Sect. 2. Section 3 gives an overview of Scanner engine and Data Collection System. Torrent file separation is examine in Sect. 4.

2 BitTorrent (BT) Basics

BitTorrent is a P2P file sharing protocol, which is developed to allow efficiently distribute large number of data or files and it is very efficient to share large files, like videos, movies, software and MP3s with minimum hardware and transmission bandwidth. BitTorrent uses a peer to peer strategy in which every user's computer contributes. Its idea is to redistribute the cost of upload to downloaders: when several people are downloading the same file at the same time, they upload pieces of the file to each other.

A BitTorrent network is made up of four types of entities:

Mainly Tracker is a server that manages the distribution of files. The role of tracker is to acts as an information switching center from that peers obtains required information about other peers to which they can connect [7].

Torrent file is one kind of file that has metadata about the file to be shared. It consists of the address of the tracker that coordinates communication between peers [7].

Peer is a computer that contributes in a download. For combining all peers (including seeders) sharing a torrent, are called a swarm.

Seeder is one kind of peer that has a whole copy of the file and offers it for download.

2.1 *Creating and Sharing Torrent File*

To distribute a file using BitTorrent, the owner of this file crates a torrent file and that has metadata about the file to be distributed and the URL of the tracker. Normally file will be shared in threads, the torrent file also have thread length and hash code of every piece, so another peers can identify the threads. Once the torrent file is generated, it is certified with a tracker and the file owner has to make the torrent file available to other Internet users by placing it on a torrent portal or any other place. The computer with the main copy of the file is usually referred to as the initial seeder.

3 Overview of Data Collection System

Basically Data Collection System has two modules, torrent file scanner, and torrent file monitoring. Torrent scanner will be responsible to finding of torrent files on the portals specified by the user. Subsequently, the listing of search result will be monitored by monitoring engine and information will be stored into database. This stored information will be passed on to Analyzer engine. Torrent Scanner and Torrent Analyzer will be described in more detail below.

Keyword Searching:

Open portals provide BitTorrents users with a favorable platform to discuss. To distribute files, BitTorrent users usually provide some information and remarks together with the torrent files. This information may be useful analyzing torrent content. Different contents may, however, appear in the links of a forum. Scanner system provide its users to associate a search session with a list of keywords, against which matching will be performed. Listing of matching results will be stored locally.

Monitoring System:

Software system provides a convenient auto-scheduling mechanism for fetching torrent related information. Monitoring system will work on torrent lists generated by scanner engine. Monitoring system will collect information uploaded by user for further processing. Monitoring system will store collected information into database. Monitoring System configured to search and fetch only updated topics, and to analyze torrent files, and have not been examine priory. The auto-scheduling and history-shrewd setting makes it possible to monitor 24 * 7, efficient and effective (Table 1).

3.1 Monitoring System Architecture

Monitoring system designed and developed for examining BitTorrent networks. As shown in Table 2, the architecture contains two main elements: Scanner and Monitor. The Scanner acts as a manager of architecture. Main work of this part is to provide the operator and configure the system and observe the collected results. Monitor responsible for monitoring BitTorrent network and getting results according to requests received from the Scanner. Monitor further divided into different components based on its responsibility to monitoring different group of elements.

Table 1 Parameter list for each torrent file

Index	Parameter name
1	Torrent file name
2	Posting date
3	File hash
4	File size
5	Release date
6	Number of seeds
7	Number of leeches
8	Download link
9	Uploader
10	Movie name
11	Category
12	Number of trackers
13	Tracker link
14	Hostname
15	Tracker status
16	Uploader's reputation score

Table 2 Digitalization processes

Acronym	Source	Estimated time
CAM	Recorded at movie theater	Aprox. 1 week
TS	Recorded at movie theater with exclusive audio source	Aprox. 1 week
TC	Directly copied from theaters media	Aprox. 1 week
PPVRip	Contents exhibited to hotel clients	Aprox. 8 weeks
SCR	Copy distributed to critics and special users	Unpredictable
DVDScr	DVD distributed to special users	Aprox. 8 weeks
R5	Non-edited DVD, launched only on region 5	Aprox. 4 weeks
DVDRip	DVD distributed to general public	Aprox. 10 weeks

First phase of Monitor engine will identify the first seeders of swarms and capture torrent earliest as they are published in BT network. Second phase continuously monitors the web page of collected torrent list in order to collect parameter list show in Table 1.

Monitoring System will collect information for different parameters and based on these parameters analyser engine will work.

Algorithm 1 presents a general view of the monitoring system.

```
Input: time, product for i ← 0 to time do
   list [torrent] ←CollectRecentTorrents (product);
   for j ← 0 to list [torrent].size () do
       Torrent ← list [j];
       Parse_seeds (torrent);
       Parse_leechers (torrent);
       Parse_uploader (torrent);
       DownloadTorrentDetailpage
       (torrent); ReadFile (torrent);
       Parse_Tracker (torrent);
       Parse_Downloadlink
       (); Parse_hash ();
       Parse_ReleaseDate ();
       Parse_PostingDate ();
   end
end
```

Algorithm 1: Monitoring process

4 Torrent Analyzer

Illegal Copies of Movies:

In analysis engine we will analyze torrent information to find illegal torrent files, and distributor of illegal files. Digitalization groups are mainly responsible for generating pirated copies of media through illegal methods [3]. Experimental surveys of BitTorrent communities express that experienced users do not see a torrent file as authentic if it does not have the digitalization group identification. The main use of a pseudonym is to generate prominence, and these groups compete with each other. So, this pseudonym is noticed by both user's producers and consumers. The users choice about whether to download or not particular copy. It is also altered by the type of digitalization process described in the torrent file.

Types of digitalization methods represented in Table 1. Every method is classified by: (1) an acronym; (2) a source (that is, the content or file that provide as basis for generation of the illegal copy); and (3) the time expected to create a pirated copy such a way that it can be founded after the initial premiere of the movie.

Digitalization process details will be used to find illegal copy of content by the analyzer engine.

4.1 Identification of Illegal Copy of Movie

This process will work as follow.

Phase 1: It will find digitalization process acronym applied on movie file by pattern matching algorithm.
Phase 2: Group movies by acronym (as given in Table 1)
Phase 3: Calculate time span between releases of movie and uploading date of movie
Phase 4: Compare timespan (time gap between release of movie and uploaded time) of every content with estimated release time of each group (grouped by acronym)
Phase 5: If timespan is lesser than estimated time, them it is highly probable that content is illegal or pirated.

Based on these phases we can find illegal copies of content.

5 Evaluation of Algorithm

Evaluation of the identification of distribution process has not been done so far.

Analyzer engine will use publisher and producer data collected by monitor engine. These data will be analyzed automatically by our system to figure out the relation between publisher and producer. Further improvements in algorithm will be done after analysis phase.

6 Conclusion

This is a work in progress paper, where we have proposed a torrent monitoring algorithm and illegal copies of content detection algorithm. A detailed introduction of monitoring infrastructure was given where we explored the functionalities of the monitor engine. We have used digitalization process parameter and timespan (time between movie released and uploaded) parameter for identifying illegal copies of movie related torrent.

Further work on the software system would include relationship between producer and publisher of torrent files and would characterize distribution process of illegal copy of content.

References

1. Price, D.: Netnames piracy analysis: sizing the piracy universel. http://www.copyrightalliance.org/sites/default/files/2013-netnames-piracy.pdf. Accessed Oct 2013
2. Envisional: An estimate of infringing use of the internet 2011. http://documents.envisional.com/docs/Envisional-InternetUsage-Jan2011.pdf. Accessed July 2011
3. Zhang C., Dhungel, P., Wu, D., Ross, K.: Unraveling the bittorrent ecosystem. IEEE Trans. Parallel Distrib. Syst. **22**(7), 1164–1177 (2010)
4. Zhang, C., Dhungel, P., Wu, D., Liu, Z., Ross, K.: Bittorrent dark-nets. In: INFOCOM 2010, IEEE International Conference on Computer Communications, pp. 1460–1468 (2010)
5. Le Blond, S., Legout, A., Lefessant, F., Dabbous, W., Kaafar, M.A.: Spying the world from your laptop: identifying and profiling content providers and big downloaders in bittorrent. In: LEET 2010, USENIX Workshop on Large-Scale Exploits and Emergent Threats, pp. 4–4 (2010)
6. Cuevas, R., Kryczka, M., Cuevas, A., Kaune, S., Guerrero, C., Rejaie, R.: Is content publishing in bittorrent altruistic or profit-driven? In: Co-NEXT 2010, International Conference on Emerging Networking Experiments and Technologies, pp. 1–12 (2010)
7. Schollmeier, R.: A definition of peer-to-peer networking for the classification of peer-to-peer architectures and applications, p2p. In: First International Conference on Peer-to-Peer Computing (P2P'01), p.0101 (2001)
8. Sinn, D.: Downloaders shrug off court threat, South China Morning Post, 27 Jan 2006

Ant Colony Optimization Based Load Frequency Control of Multi-area Interconnected Thermal Power System with Governor Dead-Band Nonlinearity

Gia Nhu Nguyen, K. Jagatheesan, Amira S. Ashour, B. Anand and Nilanjan Dey

Abstract The interconnected thermal power system consists of several areas. Various parameters should be provided to reach power systems' firm operation. The current work proposed an optimization algorithm, namely Ant colony optimization (ACO) to optimize the Proportional-Integral-Derivative (PID) controller for the load frequency control of two-area interconnected non-reheat thermal power system with Governor dead band nonlinearity. The ACO in used to determine optimal controller's parameters, where an objective function, namely Integral Time Absolute Error is conducted. A comparative study for the ACO performance to the Craziness based Particle swarm optimization (CPSO) is studied to examine the proposed approach performance in the interconnecting thermal power system. The result established the ACO optimized PID controller response superiority of the compared to the CPSO optimized controller.

Keywords Ant colony optimization · Governor dead band · Automatic generation control · Interconnected power system · Nonlinearity · Load frequency control Optimization · PID

G. N. Nguyen (✉)
Duy Tan University, Danang, Vietnam
e-mail: nguyengianhu@duytan.edu.vn

K. Jagatheesan
Department of EEE, Mahendra Institute of Engineering and Technology, Namakkal, Tamilnadu, India

A. S. Ashour
Faculty of Engineering, Department of Electronics and Electrical Communications Engineering, Tanta University, Tanta, Egypt

B. Anand
Department of EEE, Hindusthan College of Engineering and Technology, Coimbatore, Tamilnadu, India

N. Dey
Department of Information Technology,
Techno India College of Technology, Kolkata, India

X.-S. Yang et al. (eds.), *Smart Trends in Systems, Security and Sustainability*, Lecture Notes in Networks and Systems 18,
https://doi.org/10.1007/978-981-10-6916-1_14

1 Introduction

The Load Frequency Control (LFC) in the power system (PS) control is extremely significant and has been given a notable concern. The LFC is executed to resolve the PS problems that befall due to the system blackouts, external disturbance and small load perturbations [1–21]. The frequency/voltage control and the reactive power are two different types of control issues in the PS. The frequency-/active power-control are generally referred to as load frequency control. Similarly, voltage and real power control are referred to as automatic voltage regulator. Sudden load disturbance affects the system stability and produces a large deviation in the system frequency and tie-line power flow between the control areas. Several control approaches have been implemented to overwhelm these disadvantages in LFC. For interconnected power system Proportional-Integral (PI) controller, Demiroren et al. introduced the decentralized control design method for the LFC [1]. Azar presented an integral controller in the AGC of multi interconnected PS [3]. The integral gain values have been optimized using the parameters-plane technique considering the effects of Generation Rate Constraint (GRC). Typically, an interconnected PS entails large power systems that are separated into dissimilar connected control areas. Chon [4] proposed an AGC monitoring for the frequency of system and tie-line flows. Das et al. [5] applied a Variable Structure Control (VSC) strategy depending on sliding manner model in the AGC of 2 area interconnected reheat thermal PS. Francis et al. [8] realized the PI+ controller for solving LFC problem in PS.

Interconnected PS is employed to raise consistency. Generally, consistent thermal-thermal or hydrothermal type systems are considered. In [11], the LFC dynamic performance for hydrothermal power system was discussed via implementing optimal control theory. Continuous discrete mode optimization technique was proposed in [12] for turning on the integral controller gain value in AGC of two-area hydrothermal PS. Pan and Liaw analyzed the LFC control of the single area PS [13]. Authors introduced adaptive controller and effectiveness of the proposed technique by considering GRC effect and wide range of the system parameter changes. Typically, conventional tuning methods are unsuitable for tuning the controller gain value in some operating criterion due to some factors, such as the size, power system complexity, non-linear characteristics of the load requirements and continuously varied operating point. All these factors lead to the necessity for introducing the Artificial Intelligence (AI) based ideas. Several AI based control and optimization techniques are proposed with the AGC/LFC of single/multi-area PS. There is several controllers' gain tuning techniques that reported in literatures. Such techniques include the Artificial Neural Network (ANN) [1], Bacterial Foraging Optimization Algorithm (BFOA) [2], Fuzzy Logic Controller (FLC) [3], classical controller, and the Stochastic Particle Swarm Optimization (SPSO) [9]. Numerous LFC/AGC controllers of multi-area PS include the Integral Controller (I), PI [10, 11], PID controller [10], and Integral Double Derivative (IDD) controller [14]. The AGC problem of the two-area PS implemented by fuzzy-PID controller was discussed [16]. From the preceeding literatures, the PS performance is based on the used controller as well as controller gain optimization technique. Hence, the proposed algorithm is implemented in the power system for proper controller design. In addition, it improved the system performance during a sudden and continuous load demand considering 1% SLP in area 1 and GDB non linearity effect.

2 Two Area Power System Model

Generally, the nonlinear equation can be employed to illustrate the system dynamic behavior ability to be linear nearby an operating point through the small load changes. The ACO based PID control under examination entails two-area interconnected non-reheat thermal PS with governor non linearity illustrated in Fig. 1 [17].

For analysis and design of the automatic LFC, this system is extensively used, where U_1 and U_2 in p.u MW/Hz represent the frequency bias factor; R_1 and R_2 in p.u. Hz refer to self-regulation parameter for a governor; Tg_1 and Tg_2 in seconds refer to speed governor time constant; Tt_1 and Tt_2 in seconds refer to the time constant of steam chest; Tp_1 and Tp_2 refer to the PS time constant; Kp_1 and Kp_2 represent the load frequency constants; T_{12} and T_{21} denote the synchronization coefficient in MW/rad; K_p refer to proportional controller gain, while K_i represent the integral controller gain, K_d refers to the derivative controller gain; control input from ACO-PID controller is represented by U1 and U2, and finally in area 1 and 2, the frequency deviations is represented by delF1 and delF2; respectively. In the current work, a governor dead band non-linearity effect is employed and the GDB

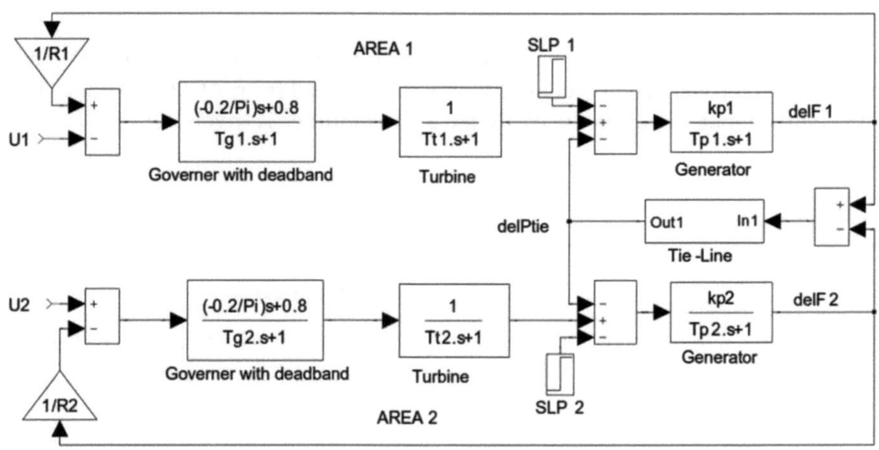

Fig. 1 Two area interconnected thermal PS transfer function model with governor dead band non-linearity

non linearity produces sinusoidal continuous oscillations for the natural period of 2 s. The governor transfer function with the dead band (Gg) GDB is represented by [17]:

$$G_g = \frac{-\frac{0.2}{\Pi}S + 0.8}{T_g S + 1} \tag{1}$$

It contains two equal areas. The ACO-PID controller input/output control the input (u_1, u_2) and ACE (ACE_1 and ACE_2). The ACEs for 2-area interconnected PS are:

$$ACE_1 = \Delta P_{tie1} + B_1 \Delta f_1 \tag{2}$$

$$ACE_2 = \Delta P_{tie2} + B_2 \Delta f_2 \tag{3}$$

where, $\Delta f_1, \Delta f_2$ in Hz represent the system frequency deviations.

2.1 PID Controller

The deviations in frequency control and tie-line power flow between the control areas are reduced by providing the PID controllers in each area of the PS. Figure 2 demonstrates the PID controller transfer function (TF) [10, 11].

The controller input is the error signal (ACE1 and ACE2). The TF of PID controller is:

$$G(s) = K_p + \frac{K_i}{S} + K_d S \tag{4}$$

Three different modes are available including P-mode, I-mode and the D-mode in the PID controller. The PS control inputs are obtained from ACO-PID controller. The control input signal produced by the PID controller is given by:

Fig. 2 PID controller TF

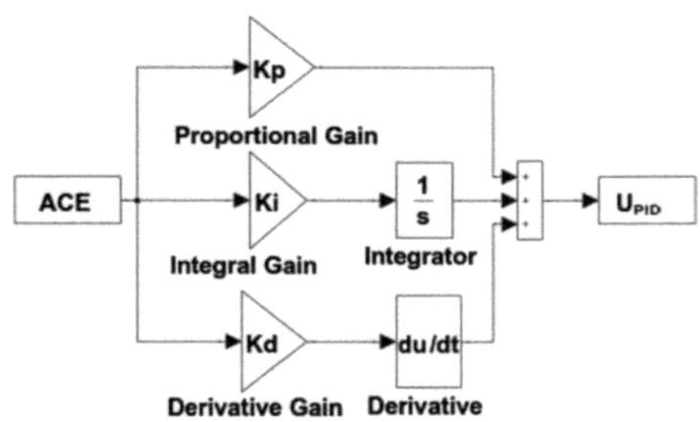

$$u_1 = K_{p1}(ACE_1 + \frac{1}{ST_{i1}} ACE_1 + ST_{d1} ACE_1) \tag{5}$$

$$u_2 = K_{p2}(ACE_2 + \frac{1}{ST_{i2}} ACE_2 + ST_{d2} ACE_2) \tag{6}$$

The foremost aim of the LFC work is to set the ACE value to zero to achieve better control performance. During this investigation, the objective function, namely Integral Time Absolute Error (ITAE) is used [11], which is given by:

$$J = \int_0^{\infty} t |\{\Delta f_i + \Delta P_{tiei-j}\}| dt \tag{7}$$

where, in tie line power, ΔP_{tiei-j} represents the incremental change and t refer to the simulation time.

3 Ant Colony Optimization (ACO) Approach

The real ants search behavior inspired the ACO technique [10]. The major three different phases of the ACO are the initialization, evaluation of the ant solution and updating pheromone concentration. Generally, the AI techniques require proper cost function for optimizing the PID controller parameters, namely K_p, T_i and T_d. The flow of the optimization procedure of PID controller is as follows: (1) Start the optimization process, (2) Initialize the number of ants, pheromone and probability, (3) Run the process model, (4) Evaluate the cost function, (5) Update the pheromone and probability, (6) Calculate the optimal value of PID controller parameters, (7) Check if maximum iteration is reached or not, (8) If No, then run the process again, and (9) If Yes, then stop the process. The proper selection of the ants-, nodes-, iteration-, variables- numbers and pheromone quantity provided good results. In the current work, the factors are nominated as follows: pheromone (τ) = 0.6, number of ants = 50, 0.95 evaporation rate (ρ) and 100 number of iterations.

4 Results Analysis

The proposed system TF 1 is established using MATLAB software. Table 1 reported the tuned controller parameters.

The developed model is simulated using an ACO algorithm as 1% SLP in area 1 with the simulation time of 180 s. The maximum-, minimum-, average- and standard- values of the settling times are reported (Table 2). The settling time

Table 1 Tuned PID parameters with proposed algorithm

Optimization technique	Controller gain of area 1			Controller gain of area 2		
	Kp_1	Ki_1	Kd_1	Kp_2	Ki_2	Kd_2
ACO-PID	0.28	0.87	0.26	0.23	0.72	0.73
CPSO-PI-J1 [20]	−0.5762	0.1962	–	−0.5762	0.1962	–
CPSO-PI-J2 [20]	−0.4000	0.3000	–	−0.4000	0.3000	–

Table 2 Settling times comparisons of CPSO-PI-J1, CPSO-PI-J2 and Proposed ACO-PID

Response	Settling time (s)		
	ACO PID	CPSO-PI-J1 [20]	CPSO-PI-J2 [20]
delF1	5.97	9.65	11.64
delF2	4.25	10.98	12.42
delPtie	7.51	13.6	17.55

comparisons of frequency deviations in area 1 and 2 and tie-line power flow between control areas are given in Table 2.

The proposed performance of optimization based PID is associated to the CPSO technique based controller. Figures 3 and 4 depicted the dynamic performance comparisons of the investigated system to the proposed approach. For comparison purpose, the graphs are illustrated only for a few seconds (0–30 s).

Figure 3 illustrates the comparison between the proposed ACO-PID and the CPSO optimized controller for frequency deviation in area 1. It is obvious that the CPSO-PI-J1 and CPSO-PI-J2 controllers provided more damping oscillations with more over and undershoot value and they take more time to settle with their nominal value compared to the proposed ACO-PID controller (delF1: 5.97 < 9.65 < 11.64; delF2: 4.25 < 11.64 < 12.42). Figure 4 illustrates the responses' comparison of the tie-line power flow between the control areas and ACE of area 1 and 2 with proposed controller and optimization technique.

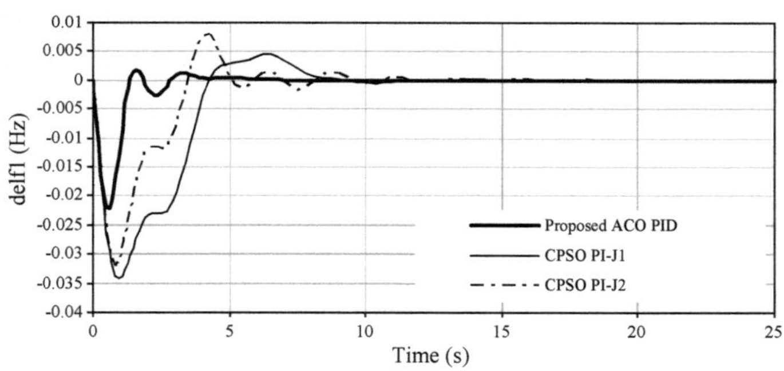

Fig. 3 Frequency deviations in area 1 (delF1)

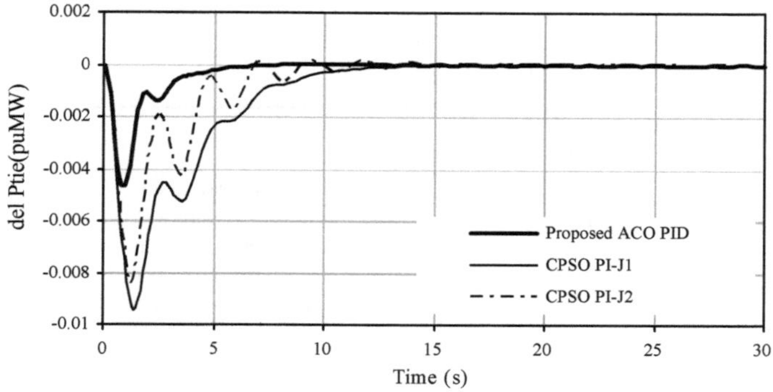

Fig. 4 Tie-line power deviations (delPtie)

It showed clearly that the proposed approach reduced effectively the damping oscillations, the maximum undershoot and overshoot in their response compared to CPSO optimized controllers.

5 Sensitivity and Robustness Analysis

For varied parameters of system within −50% to +50% range, the sensitivity analysis is demonstrated to clarify the proposed approach strength in the step of 25%. The time constants of turbine governor, and generator are varied from −50% to + 50% to check the effectiveness in area 1 of a system with 1% step load perturbation. The system parameters, frequency deviations in areas 1 and 2, tie-line power flow variations between areas 1 and 2 and ACE for different cases are shown in Figs. 5 through 10.

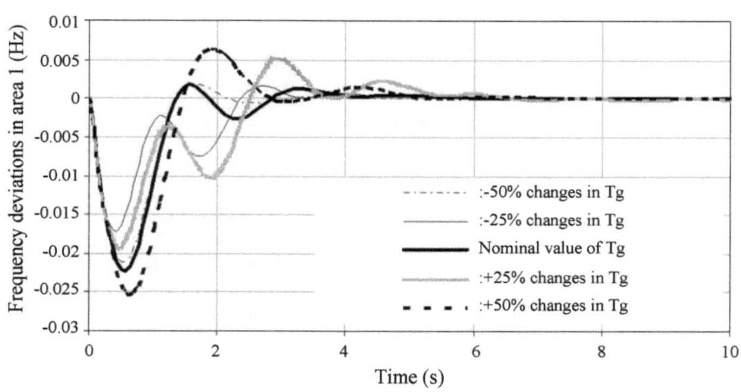

Fig. 5 Frequency deviations in area 1 for different variations in Tg

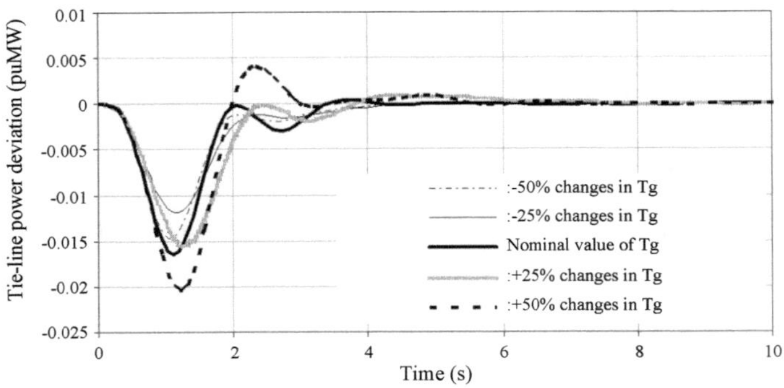

Fig. 6 Tie-line power deviations for different variations in Tg

Figures 5 and 6 present the system response comparisons at the wide range of governor time constant tug from −50% to +50% in steps of 25%. (−50% = 0.1 s, −25% = 0.15 s, Nominal = 0.2 s, +25% = 0.25 s, +50% = 0.3 s).

Figures 7 and 8 show the system response comparisons for a wide range of turbine time constant Tt from −50% to +50% in steps of 25%, where −50% = 0.15 s, −25% = 0.225 s, Nominal = 0.3 s, +25% = 0.375 s, and +50% = 0.45 s.

Figures 9 and 10 demonstrate the system response comparisons at a wide range of generator time constant Tp from −50% to +50% in steps of 25%. (−50% = 10 s, −25% = 15 s, Nominal = 20 s, +25% = 25 s, +50% = 30 s). Generally, Figs. 5 through 10 established that the proposed controller provided superior control performance with ±50% time constant variations and 1% step load perturbation. Consequently, the proposed approach superior performance is proved by comparing its results to that obtained by the optimization CRAZYPSO technique [18]. Additionally, the sensitivity and robustness analysis was done by varying turbine/governor time constant and inertia constant of −50% to + 50% range from its normal value in step of 25%.

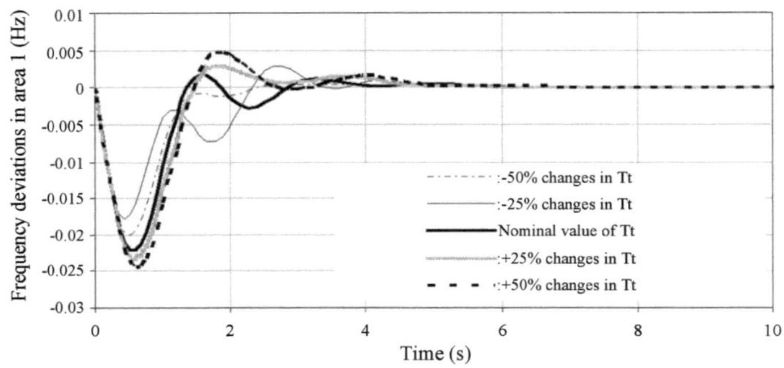

Fig. 7 Frequency deviations in area 1 for different variations in Tt

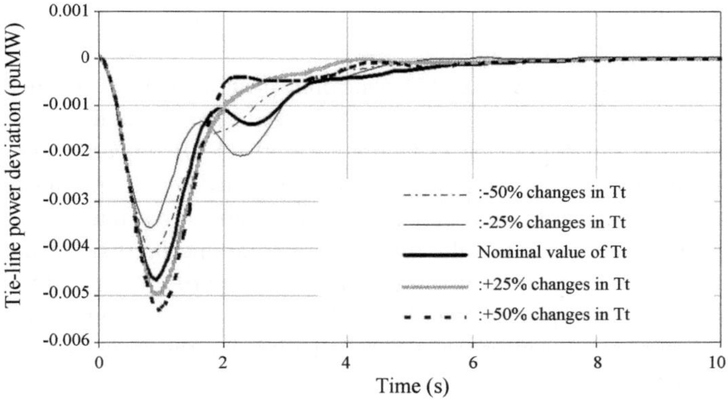

Fig. 8 Tie-line power deviations for different variations in Tt

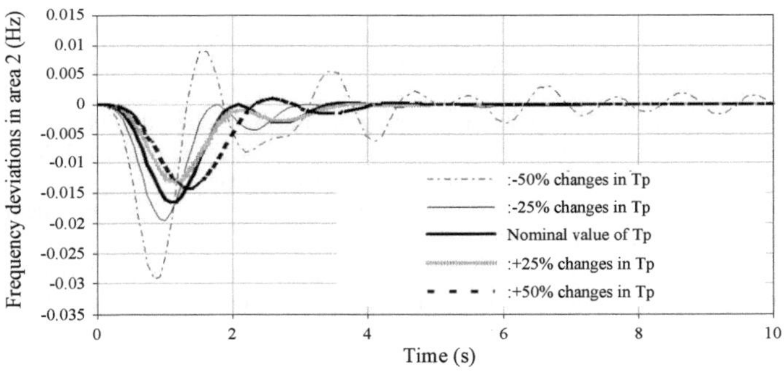

Fig. 9 Frequency deviations in area 2 for different variations in Tp

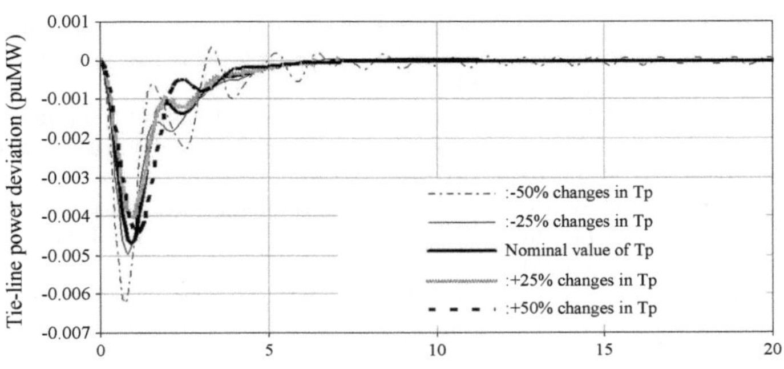

Fig. 10 Tie-line power deviations for different variations in Tp

6 Conclusion

The present work introduced the PID controller design for LFC of interconnected PS with governor dead non-linearity. The PID controller optimal gain values have been obtained by employing ACO technique with ITAE objective function. The proposed optimization technique superiority established by comparing results with CPSO. On the other hand, the proposed approach sensitivity analysis is calculated by varying the time constant of the turbine, governor and inertia constant of PS in the −50% to +50% range from its normal value in step of 25%. Finally, the results and sensitivity analysis proved that the ACO-PID controllers were more effective. Furthermore, the ACO performance is insensitive for the broad range of power system parameters.

References

1. Demiroren, A., Zeynelgil, A.L., Sengor, N.S.: The application of ANN technique to load frequency control for three-area power systems. IEEE Porto Power Tech Conference on 10–13 Sept 2001, Portugal, pp. 1–5
2. Ali, E.S., Abd-Elazim, S.M.: Bacteria foraging optimization algorithm based load frequency controller for interconnected power system. Electric Power Energy Syst. **33**, 633–638 (2011)
3. Azar, A.T.: Fuzzy Systems. IN-TECH 2010, Vienna, Austria
4. Chon, N.: Some aspects of tie-line biased control on interconnected power systems. Am. Inst. Elect. Eng. Trans. **75**, 1415–1436 (1957)
5. Das, S., Kothari, M.L., Kothari, D.P., Nanda, J.: Variable structure control strategy to automatic generation control of interconnected reheat thermal system. IEE Proc-D. **138**, 579–585 (1991)
6. Dey, N., Samanta, S., Chakraborty, S., Das, A., Chaudhuri, S.S., Suri, J.S.: Firefly algorithm for optimization of scaling factors during embedding of manifold medical information: an application in ophthalmology imaging. J. Med. Imaging Health Inform. **4**, 384–394 (2015)
7. Dey, N., Samanta, S., Yang, X.S., Chaudhuri, S.S., Das, A.: Optimization of scaling factors in electrocardiogram signal watermarking using cuckoo search. Int. J. Bio-Inspired Comput. (IJBIC) **5**, 315–326 (2014)
8. Francis, R., Chidambaram, I.A.: Optimized PI+ load-frequency controller using BWNN approach for an interconnected reheat power system with RFB and hydrogen electrolyser units. Electric Power Energy Syst. **67**, 381–392 (2015)
9. Jagatheesan, K., Anand, B., Ebrahim, M.A.: Stochastic particle swarm optimization for tuning of PID controller in load frequency control of single area reheat thermal power system. Int. J. Electr. Power Eng. **8**, 33–40 (2014)
10. Jagatheesan, K., Anand, B., Dey, N.: Automatic generation control of thermal-thermal-hydro power systems with PID controller using ant colony optimization. Int. J. Serv. Sci. Manag. Eng. Technol. **6**, 18–34 (2015)
11. Jagatheesan, K., Jeyanthi, S., Anand, B.: Conventional load frequency control of an interconnected multi-area reheat thermal power systems using HVDC link. Int. J. Sci. Eng. Res. **5**, 88–92 (2014)
12. Ebrahim, M.A., Mostafa, H.E., Gawish, S.A., Bendary, F.M.: Design of decentralized load frequency based-PID controller using stochastic particle swarm optimization technique. In: International Conference on Electric Power And Energy Conversion System, pp. 1–6 (2009)

13. Pan, C.T., Liaw, C.M.: An adaptive controller for power system load-frequency control. IEEE Trans. Power Syst. **4**, 122–128 (1989)
14. Jagatheesan, K., Anand, B., Dey, N: Evolutionary algorithm based LFC of single area thermal power system with different steam configurations and non-linearity. In: Proceedings of The third International Conference on Information Systems Design and Intelligent Applications 2016 (INDIA-2016), pp. 185–194 (2016)
15. Samanta, S., Acharjee, S., Mukherjee, A., Das, D., Dey, N.: Ant weight lifting algorithm for image segmentation. In: 2013 IEEE International Conference on Computational Intelligence and Computing Research (ICCIC), pp. 1–5 (2013)
16. Sahu, R.K., Panda, S., Padhan, S.: A hybrid firefly algorithm and pattern search technique for automatic generation control of multi area power systems. Electric Power Energy Syst. **64**, 9–23 (2015)
17. Gozde, H., Taplamacioglu, M.C.: Automatic generation control application with craziness based particle swarm optimization in a thermal power system. Electr. Power Energy Syst. **33**, 8–16 (2011)
18. Gozde, H., Taplamacioglu, M.C., Kocaarslan, I.: Comparative performance analysis of Artificial Bee Colony algorithm in automatic generation control for interconnected reheat thermal power system. Electric Power Energy Syst. **42**, 167–178 (2012)
19. Jagatheesan, K., Anand, B., Baskaran, K., Dey, N., Ashour, A.S., Bala, V.E.: Effect of non-linearity and boiler dynamics in automatic generation control of multi-area thermal power system with proportional-integral-derivative and ant colony optimization technique. Recent advances in nonlinear dynamics and synchronisation with selected applications, Springer (Germany), in the Series "Studies in Computational Intelligence" (2016)
20. Jagatheesan, K., Anand, B., Dey, N., Ashour, A.S.: Artificial intelligence in performance analysis of load frequency control in thermal-wind-hydro power systems. Int. J. Adv. Comput. Sci. Appl. **6**(7) (2015)
21. Jagatheesan, K., Anand, B., Samanta, S., Dey, N., Santhi, V., Ashour, A.S., Balas, V.E.: Application of flower pollination algorithm in load frequency control of multi-area interconnected power system with non-linearity. Neural Comput. Appl. (NCAA) (2016)

A Qualitative Review of Two Evolutionary Algorithms Inspired by Heuristic Population Based Search Methods: GA & PSO

Kuldeep S. Raghuwanshi

Abstract PSO is relatively recent Evolutionary computational technique which is inspired by swarming behavior of biological populations. PSO and GA are resembles in sense that both are heuristic population based search methods. In other words both starts from random variable and reach to a final desired solutions without any user input. GA has been utilizing in many research and in many form because its easiness in implementation and ability to solve highly complex non-linear engineering problems but it is costly. This paper claims that PSO is more effective then GA while it is very recent but in sense of implementation, in solving complex problems it is more efficient.

Keywords Genetic algorithm · Particle swarm optimization
Evolutionary algorithm · Heuristic search

1 Introduction

PSO brings a great revolution in Evolution computation techniques [1], it was invented by Kennedy Eberhart in 1990s during the choreography of graceful motion of flying swarms of birds as a part of Sociocognitive study investigating the "collective intelligence" of biological populations [2]. As per study of great white pelicans it has found that birds flying in formation use up to fifth less energy than those flying alone. As per Reynolds's rule for flocking swarm theory works on three parameters as Separation, Alignment and Cohesion. PSO terminology includes Population, particle, evolution span, fitness function and population manager. It works on imitate, evaluate & compare principles. Swarming theory has introduced

K. S. Raghuwanshi (✉)
Sensus Edusoft Pvt. Ltd., Jaipur, India
e-mail: ksraghuwanshi@ieee.org

X.-S. Yang et al. (eds.), *Smart Trends in Systems, Security and Sustainability*, Lecture Notes in Networks and Systems 18,
https://doi.org/10.1007/978-981-10-6916-1_15

in various system design applications like aerospace systems utilizing it for formation flying of aircraft and space craft [3]. PSO acts on the behavior and searches for the best solution in vector space for search as per environment of problem. Every single solution is treated as particle, which has cost being calculated by a function to be minimize. Each particle has a velocity which stimulates the behavior of particle in search vector space. PSO proven its performance in various application in comparison with any other evolutionary computation techniques due to easy of implementation and less no. of parameters to be calculated in it [4]. In PSO initially random positions along with random velocities are assigned. After each step particle compare its solution with two best values, the first is its own cognitive best solution in vector search space which have lowest cost (Highest fitness function) called P, and other is best solution in swarm by any other particle which is called is Pg, now as per this solution each particle is adjust it velocity and position as per the following equation,

$$v' = v0 + c1 \cdot R1 \cdot (P - x) + c2 \cdot R2 \cdot (Pg - x0)$$
$$x' = x0 + v'$$

where v1 is current velocity and v′ will be new velocity after computation, x1 is the current position and x′ will be the new position of particle after adjusting it position in accordance with pB and gB. R1 and r2 are randomly distributed numbers in interval of (0, 1) and c1 & c2 are accelerating coefficients which influence the cognitive and social behavior of particle.

Another Evolutionary computation technique is Genetic Algorithm (GA), it was come into existence in the mid of 1970s by Johan Holland and his colleagues and his students at University of Michigan [5]. It inspired by the biological population reproduction, evaluation and on basis of genetic behavior observed in the populations. The basic principal for GA is 'Survival for Fittest'. In the search process of GA to generate individual solutions that are adapt the environment. Due to this after many generations (iterations) desirable solutions at each generations (iterations) will come in existence and remains in composition of the population, which comes less efficient in desirable characteristics. Aspects of evolution in GA are Survival for fittest, Genetic Combination and Mutation as discussed above, it uses Evolutionary techniques to optimize the parameters. GA has basic terminology like Population, Generation, Fitness function, Evaluation span, Life span and Population manger. GA is most popular Evolutionary Computation techniques since a long time because it most suited to solve complex optimization problems because it is capable to handle discrete as well as continuous variables, functions without using gradient information.

This paper claims that PSO is more efficient and effective than GA in finding optimum solution for global problems, Computation has performed on both GA as well as PSO and found that PSO has better computational efficiency than the GA. due to its less cost and easy computation as well as its parameters are used in computation are well suited than GA's parameters.

2 PSO

In this study PSO algorithm is analyzed with the principle of PSO and also presented a discrete version of algorithm. It is to be noted that GA is discrete i.e. it encounters the variable in 0's and 1's in this way it becomes easier to handle discrete variables by GA.

The basic PSO algorithm consist of three steps as Initializing the particle's position and velocity by random value, velocity update as per the cognitive and social velocity, and finally update in position in vector space. In this a every particle refers to point in vector space that changes its position in every iteration as per the velocity. In this implementation i represent the particle and k represent the time at instant. Initially the position x0 and velocity v0 of particles are randomly assigned using Xm and Xmn of design value as per following equations a and b. as per equations a and b, R is a randomly distributed variable in range of 0 and 1. Through this process swarm particle get the random position in swarm at initial level.

$$\mathrm{x0} = \mathrm{Xm} + R(\mathrm{Xm} - \mathrm{Xmn})$$

and

$$\mathrm{v0} = \mathrm{Xmn} + R(\mathrm{Xm} - \mathrm{Xmn})/\Delta\mathrm{t} = \mathrm{position}/\mathrm{time}$$

After this velocity of all particle will be updated at k + 1 time instant using the fitness value and the desired point (result). The fitness function value for each particle represents that which particle has the best global value in swarm, Pg(k), it aslo determine the best cognitive position over the time, Pi during all iterations. Now velocity will be updated as per these two above mentioned parameters along with Current motion, cognitive memory and swarm's best value by a particle, with two factors like cognitive influence factor c1 and swarm influence factor c2,

Now velocity of particle after k time at k + 1 will be,

$$\mathrm{Vk} + 1 = \mathrm{Vk} + \mathrm{c1} \cdot \mathrm{R}\{(\mathrm{P} - \mathrm{Xk})/\Delta\mathrm{t}\} + \mathrm{c2} \cdot \mathrm{R}\{(\mathrm{P} - \mathrm{Xk})/\Delta\mathrm{t}\}$$

Basic PSO [6] uses 2 for c1 and c2 both but in this we have used 1.4 for both c1 and c2 which provide the best convergence rate for all tests. Updating in position is last step in all moves. The position of each particle is updated as per following equation,

$$\mathrm{Xk} + 1 = \mathrm{Xk} + \mathrm{Vk} + 1\Delta\mathrm{t}$$

All three steps will be repeated until a desire convergence criterion is achieved. In this we use a stopping criteria (S) is that the maximum changes in best fitness should be equal to specified tolerance for specified numbers of moves.

In comparison with GA PSO can take any value based on the position of particular particle and calculated velocity in vector space. This may create a problem so to avoid that it is defined that whenever design variables (solutions) will breaks the limit of Xm and Xmn they should brought back into their boundaries like birds do in their swarm in directions.

Venter's [4] research suggests the simple but effective way to implement discrete design variable with PSO this strategy is used in this study which reflects the optimize performance of PSO. Size of Swarm in PSO and Chromosomes size in GA should be use similar in all test.

3 GA

As over a long time span GA is popular in among researchers and it is also using in many applications or in complex problem solving of all domains, but still it works on three operators as Selection, Crossover and Mutation [6]. It uses Evolutionary technique to parameter optimization. Evolutionary loop is as followed to achieve optimization in GA, Fig. 1.

As depicted in figure GA optimize its performance in every generations with set fitness function or desire solution. It also perform task on its parameter to achieve optimum performance as shown in Fig. 2.

All the three main operations in of GA can be implemented in many ways but all follows above mentioned generation loop for optimization in desire solutions.

Fig. 1 Several generation's evolutionary loop

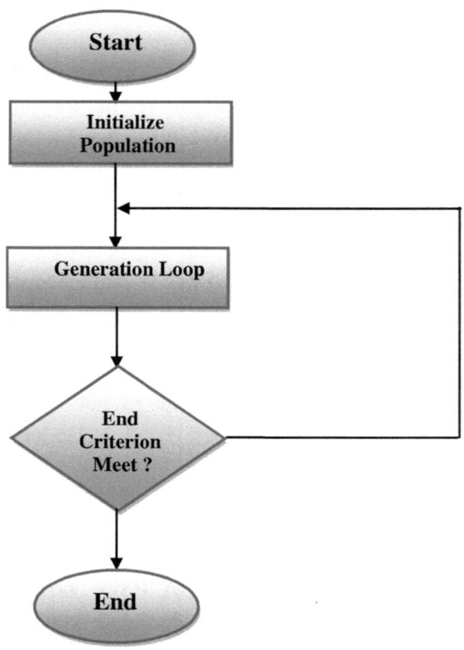

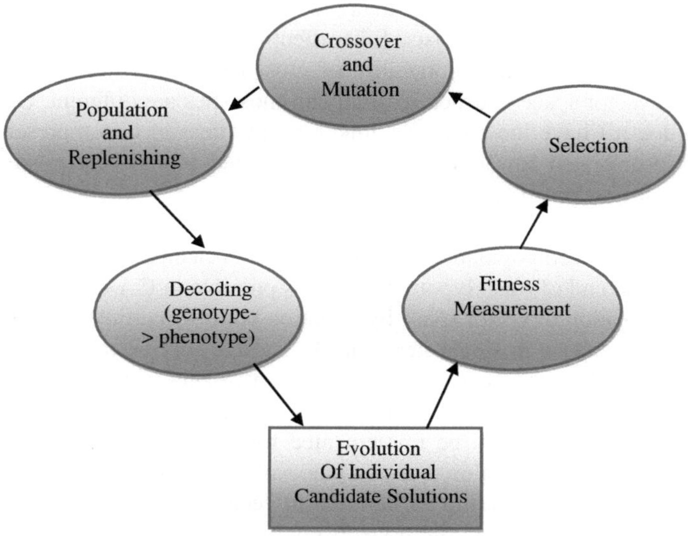

Fig. 2 Generation loop

GA represents each design variable with string of 0's and 1's that refers to chromosomes. It is to be noted that GA encodes the design parameters for discrete as well as continuous in one problem. This encoding accomplish within the upper and lower boundary i.e. there will be no violation in functional constraints of GA. In GA evaluation generally starts with a random population, in this evaluation process of each generation fitness for individual in population is encountered and mutated to form a new generation as per desired variable. As the desired fitness function or criterion is achieved than iteration will automatically stops i.e. algorithm stops. On getting fitness function and genetic presentation GA starts with random population and improve its generation with many applications of crossover, mutation and selection operators until a desired fitness level is achieved.

4 PSO and GA

PSO is recent Evolutionary computation technique, which is based on cognitive as well as social behavior of biological population, than GA. PSO is also resembles with GA in some like both technique based on information sharing among their population members using probabilistic and deterministic approach. On the contrary GA is more stabilize and old also after many research it has many versions. Like PSO [7], GA begins it process from a randomly generated populations that doesn't require any input from user.

GA implements all operators in many ways as per problem domain but PSO doesn't but still similarity is exists. Like effect of crossover is significant due to randomized populations the probability of crossover varies during run of chromosomes in vector space if structure of all chromosomes are different. On the other side PSO doesn't have the crossover function but effect of it is represented when particle is accelerated towards its previous best and global best position or the neighborhood best position.

Effect of mutation is less during starting of run and maximum at the end, because of initially randomized population which ensures less chances of changing in bits as maximum at the end of run. Hypothetically it is possible in GA to any chromosome to reach any position of vector space by mutation but PSO particle cannot reach any point at problem space in single iteration. As particle survive in every iteration so after enough iteration they can g oat any point in problem space.

The effect of selection in GA helps in full filling the principle of "survival for fittest". Selection in GA can be implemented by many ways like roulette wheel selection, tournament selection but PSO doesn't use selection parameter so all particles of swarm exists during all iteration in vector space, so the possibilities of optimum result is high in few titrations as compared to GA.

In GA crossover occurs in randomly selected parents. While the evaluation of chromosome depends on the exchanging of genes with quite randomly selected. While particle is influenced by its cognitive behavior and neighborhood if position is best among others. In some cases PSO is also influenced by its topological neighbors and particle geometry remains the same throughout the run in vector space. So these are not influenced from random strangers.

5 Conclusion

A numerous ideas have come during study of this comparison of GA and PSO. In this paper analysis is focused on the operators of GA and PSO to make search better and how the using of one operator can other technique might be improved. While PSO is new heuristic search method of Evolutionary computation technique but it is more efficient and accurate as per our study shows that PSO outperforms the GA with lager computational efficiency.

References

1. Eberhart, R.C., Shi, Y.: Comparison Between Genetic Algorithms and Particle Swarm Optimization, Indiana University
2. Kennedy, J., Eberhart, R.: Particle swarm optimization. In: Proceedings of the IEEE International Conference on Neural Networks, pp. 1942–1945, Perth, Australia (1995)
3. Weimerskirch, H., et al.: Energy saving in flight formation. Nature **413**, 697–698 (2001)

4. Particle swarm model selection. J. Mach. Learn. Res. **10**, 405–440 (2009) Submitted 1/07; Revised 8/08; Published 2/09
5. Goldberg, D.: Genetic Algorithms in Search, Optimization and Machine Learning, pp. 1–25. Addison-Wesley, Reading, MA (1989)
6. Hassan, R., Cohanim, B., de weck, O.: A Comparison of Particle Swarm Optimization and the Genetic Algorithm, AIAA 2005-1897
7. Venter, G., Sobieski, J.: Particle Swarm Optimization, AIAA 2002-1235, 43rd AIAA/ASME/ASCE/ AHS/ASC

Configuration of IRS Tool on Linux OS Using Low Cost Low Power Computer

S. R. Kattamuri, V. Kakulapati and N. Muthyala

Abstract Infringement detection system and infringement avoidance systems (IDS/IAS) are crucial elements of computer network security. By using a freeware IDS tool like Snort, the outcomes of research are presented. Research has been made on a Snort IDS on a flexible, inexpensive reasonably powered machine known as Raspberry Pi2 (Model B), with a precise aim of finding their capability, competence and worth in the computer network atmosphere, where the price is an influencing aspect. SOHO and learning organization computer networks are some examples.

Keywords Infringement detection system · IDS · Raspberry Pi2 model B OS · Operating system

1 Introduction

Intrusion detection systems (IDSs) are hardware (appliance) or software (operating system) based systems that can detect abnormal or malicious activities or attacks using specific rules or configuration scripts [1]. From a computer network viewpoint, IDSs are commonly either network-based (NIDS) and host based (HIDS). NIDS are network based intrusion detection systems that detect network traffic among network environments [2]. HIDS are host-based IDS that are configured on individual host systems to guard the single targets [3, 4].

S. R. Kattamuri (✉) · V. Kakulapati · N. Muthyala
Sreenidhi Institute Science and Technology, Yamnampet,
Ghatkesar 501301, Telangana, India
e-mail: ksmurthy@sreenidhi.edu.in

V. Kakulapati
e-mail: Vijayalakshmik@sreenidhi.edu.in

N. Muthyala
e-mail: nagarajum@sreenidhi.edu.in

X.-S. Yang et al. (eds.), *Smart Trends in Systems, Security and Sustainability*, Lecture Notes in Networks and Systems 18,
https://doi.org/10.1007/978-981-10-6916-1_16

NIDS can either be configured in-line or passive monitoring modes. NIDS installed in inline mode monitor and intercepts traffic passes through it. NIDS in passive mode monitors traffic sent to it either by an agent computer or another network device. IDSs are an important component in network security as they act as a first line of defense (along with firewalls) on computer networks to detect and mitigate attacks and store logs for further analysis. However, IDSs in passive mode cannot prevent networks from being attacked [5]. Raspberry Pi is a multi-purpose and low-cost ARM processor-based miniature device that can be used for intrusion detection in computer network environments such as SOHO (small office/home), educational institutions or in developing countries. This paper outlines the results of an experiment in operating Snort IDS on a Raspberry Pi2 device, (referred to as Pi-IDS; configured in NIDS passive mode), in test-bed environment with a view open source IDSs—particularly to simulate network attacks and then evaluating the performance.

The structure of the document is arranged in the following way: Part 2 gives a concise presentation of literature and an overview of the two key components used in the experiment—Raspberry Pi, Snort IDS. Section 3 introduces how to configure snort on raspberry pi and the experimental environment architecture. Section 4 explains a different mode of operation. Section 5 provides the study of gathered information and sniffer mode operation along with conversation. Section 6 presents the conclusion and future work.

2 Literature Review

A review of literature is provided in this part, including an overview of Raspberry Pi, Snort IDS.

2.1 Raspberry Pi

Raspberry Pi is a card-sized small computer that works on either central power or on battery power. The operating system of Raspberry is Raspbian, and there are other flavors to it like ARM Linux which functions on it. This mini machine can be employed in temperature examining devices, arcade machines, robotics. It can be used for *MATLAB* applications, among others, and arrives in different models with different interfaces for different requirements. Raspberry Pi2 Model B (the latest version of the hardware) has 1 gigabyte (GB) of random access memory (RAM), a 900 MHz quad-core ARM processor, four universal serial bus (USB) interfaces, an Ethernet port, a mini USB for power supply and high definition multimedia interface (HDMI) for display. The OS is flashed onto a micro Secure Digital (SD) card [6]. The running device is accessed directly using a USB keyboard, mouse and display or via a LAN port by creating a secure shell (SSH) session distantly.

2.2 *Snort IRS*

Snort is widespread IRS installed in the network security atmosphere and it is to have a vast information collection of signs for malignant actions [7]. It verifies for exact information in the network flow and intimates every example of a specific sign [7–9]. The data coming from the network goes into the *Packet Decoder* of Snort IRS. This data is prepared and sent to the *Pre-processor* and modified to suit the needs of a detection engine, which analyses the info for the existence of any intrusion activity. Normal flow is bypassed whereas doubtful flow is recording into logger mode. Then, the *Output Module* accepts the logs and generates the final output [10–12].

3 The Configuration of Snort-IRS

The configuration of snort-IRS is different in Linux OS when we compare it in with Windows OS. Here we are deploying Snort-IDS on Linux OS (Ex: Ubuntu Mate) which is compatible with Raspberry Pi2.

4 Modes of Operations

In order to work with Snort-IDS we have three modes of operations.

1. Sniffer method
2. Sachet logger method
3. Set of connection infringement recognition method (SCIRM).

4.1 *Sniffer Method*

Headed for publishing the TCP/IP sachet header information in the direction of the monitor (i.e. sniffer method).

$$./\text{snort} - \text{v}$$

This runs the grunt and demonstrates different headers like the Internet Protocol and TCP/UDP/ICMP headers and shows application data in transit.

./snort – vd

Display the packet data along with header data. It provides data link layer headers for elaborative show.

./snort –vde

We can write this command line switches either separately or in a combined way.

./snort –d –v – e is equivalent to ./snort –vde.

4.2 Sachet Logger Method

Used for storing content to the disk, for that we required providing a sorting register and grunting moves to sachet logger method:

./snort – dev – l./log

That means we have to log register during the present working register. When there is no directory, Snort throws an error. In packet logger mode, Snort gathers every information it checks and provides that info as a hierarchy depending on the Internet Protocol address of one of the congregations in the datagram. When –l switch is used, it is observed that it utilizes remote machine address where it stores data or local host address. If home network to be used, we need to mention which is domicile network:

./snort – dev – l./log – h192.168.1.0/24

The rule exhibits the information connection and TCP/IP header information in addition to request information into the log register and desire to monitor the sachets comparative in the direction of the 192.168.1.0 C set of connections. Every arriving packet will be logged interested in subdirectories of the record index, where index name depends lying on the remote host address (non-192.168.1) congregation.

Remember when the source and target congregations together present taking place the same domicile network, logs are stored in the directory where the greater port number of two is given as the name or the source address as the name if both have same port numbers. It would be good if you log in binary form when one is on the high speed network or when one need the logs to be compact. The logs will be in tcp dump format and to a single binary file when you choose binary mode:

./snort – l./log – b

The control line format modifies here. As binary mode logs all details into a single file, we are not required to mention any home network, and we are not required to mention the output directory structure. Apart from it, in binary mode, we are not required to use –d or –e switches anymore or not required running in verbose mode as the whole packet is recorded and not just sections. We are required to use only –l to specify the logging directory and –b to mention that we want to run in binary mode than default ASCII mode.

After the data has been recording to the binary file, one can search the packet flow by any tcp protocol dump binary format supporting sniffer like tcp protocol dump or Ethereal. Playback mode is also available in Snort where we can read the packets back by –r switch. Sachets from any tcp dump arranged categorizer be capable of developed through sniff in any of its run methods. For instance, if you desired to dart a dual log categorizer in the course of sniff in the sniffer method to deposit the sachets to the monitor, you can attempt amazing like this:

./snort – vd – r packet.log

Snort's packet logger mode and IRS mode is used for manipulating the data in the file. The BPF interface from the command line will also serve the same purpose.

We can make snort search only particular type of flow like ICMP from the log file, we can do so by only stating the BPF riddle at the control procession, then the grunt will observe ICMP sachet rush forward in the categorizer.

./snort – dvr sachet log icmp

4.3 *Set of Connection Infringement Recognition Method (SCIRM)*

Through Set of connection infringement recognition method, we are not required to evidence each sachet received through the line:

./snort – dev – l./log – h 192.168.1.0/24 – c Snort.conf

Snort.conf is the given name of the sniff arrangement file. Sachet's exploit is depends on the design regulations revealed in Snort.conf categorizer. The default yield register is /var/log/snort if no yield listing name is specified. Another important point regarding the preceding control procession is that if sniff is to be employed in extensive expression employ then –v control is supposed to absent off the control procession for speed.

The exhibit is of time-consuming place to, and sachets can be failed while writing to the display. We can even omit –e switch as data link headers are that much.

$$./snort\ -\ d\ -\ h\ 192.168.1.0/24\ -\ l./log - cSnort.conf$$

Above control ensures sniff to dart in its most fundamental SCIRM structure, logging sachets that generate regulations specified in the Snort.conf in basic ASCII to compact disk using a hierarchical register formation (identical to sachet logger method).

SCIRM Method Yield Options

We can configure Snort output in several ways in SCIRM method. The full alert mechanism with decoded ASCII format is the default recording system. This mechanism displays the alert message along with full packet headers. Various alert output modes are even offered at the command line apart from 2 logging services.

Comparatively, alert modes are difficult. Alert modes offered seven modes: none, cmg, the console, the syslog, the socket, fast, and full. Among these seven, six can be right to use with-A control procession control. Those are:.

Option	Description
-A fast	Alert method, which is fast. Displays alert with the alert message, timestamp, destination, and source ports/IPs in a simple format
-A full	Alert mode, which is full and it is the default alert mode if none is specified
-A unsock	Alerts sent to Unix socket where other programs here
-A none	The alert mechanism turned off
-A console	Alerts of "high-speed-method" alerts sent to the console
-A cmg	"cmg style" alerts are generated

With –b option of the command line, packets are recorded in binary file or default ASCII format. With –N option, we can make packet logging inactive.

From configuration file, we can find output modes, see Section.

Remember all output options mentioned in configuration file are overridden with command line logging. With this, configuration issues debugging is fast.

$$\text{Syslog alerts can be sent by using } - \text{ s option.}$$

LOG_ALERT and LOG_AUTHPRIV are the default facilities for syslog alerting. With this output plug-in directives in the Snort.conf, we can configure other facilities for output of syslog. We can verify below parts for much information on presenting the output of syslog.

For instance, we can use below the command line for logging to default (decoded ASCII) facility and alerts can be sent in syslog:

$$./\mathrm{snort}\ -\ \mathrm{c\,Snort.conf}\ -\ \mathrm{l./log}\ -\ \mathrm{h\,192.168.1.0/24}\ -\ \mathrm{s}$$

One more instance, we can use below command line to record the alerts in /var/log/snort default facility and alerts can be sent too rapid alert categorizer:

$$./\mathrm{snort}\ -\ \mathrm{c\,Snort.conf}\ -\ \mathrm{A\,fast}\ -\ \mathrm{h\,192.168.1.0/24}$$

5 Findings

Five important factors were considered for comparison: central processing unit utilization, random access memory utilization, sachet loss rate, log script swiftness and the number of detected attacks. The command "top" was used for collecting data about Central processing unit and random access memory usages. Packet loss rate and log writing speed for Snort IRS were easily collected from the console after stopping the sniff service.

5.1 Snort IRS Alert Information

SYN Flood attack alert showed the IP address and MAC address of the source and destination (see Figs. 1 and 2). The alert information also contains type of packets.

The following figure (Fig. 3) shows the detected ARP attacks with timestamps.

SYN Port Scanning alert (Fig. 4) was also detected by Snort IDS which displayed port types being scanned and priority numbers configured using Snort rules.

Fig. 1 DoS alert from Snort IDS

Fig. 2 DoS alert from Snort IDS

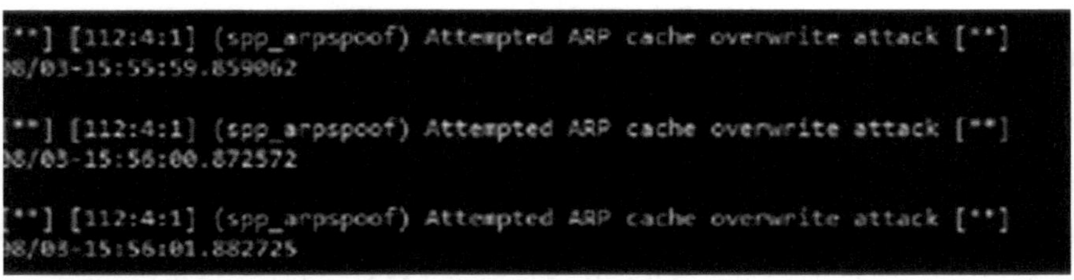

Fig. 3 ARP spoofing alert from Snort IDS

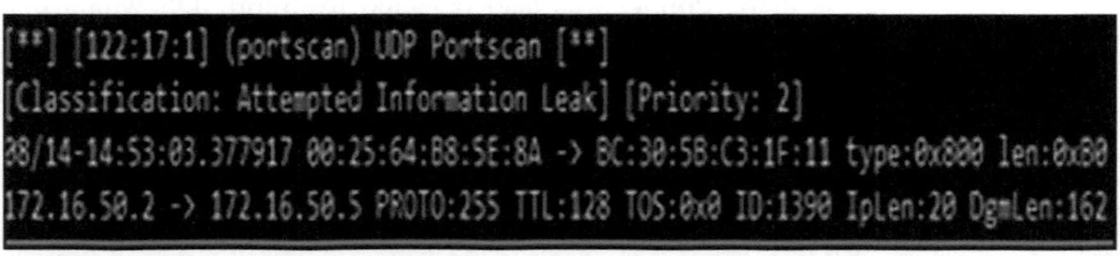

Fig. 4 Port scanning alert from Snort IDS

5.2 Working of Snort-IDS in Sniffer Mode

After, running snort in sniffer mode i.e. by using command "snort-v" we get output as in Fig. 5.

Figure 6 gives output of command "snort-vd".

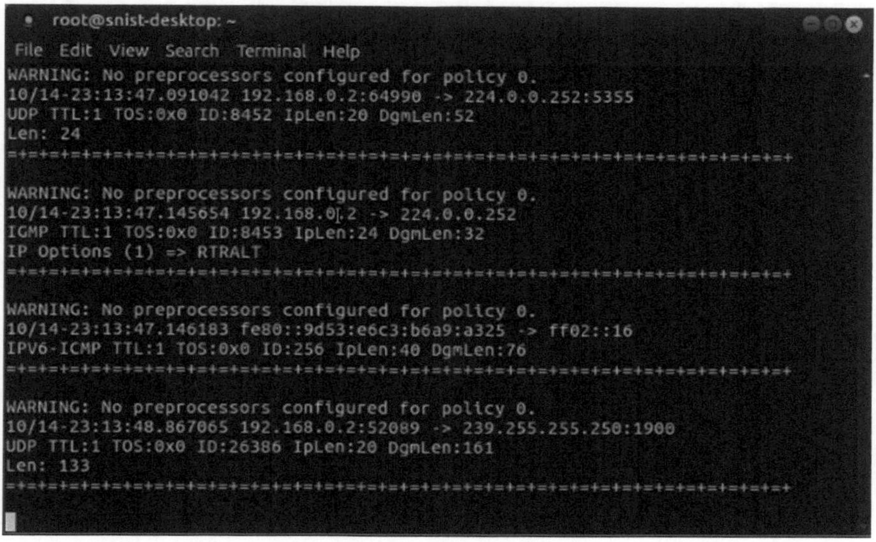

Fig. 5 Snort-IDS in sniffer mode

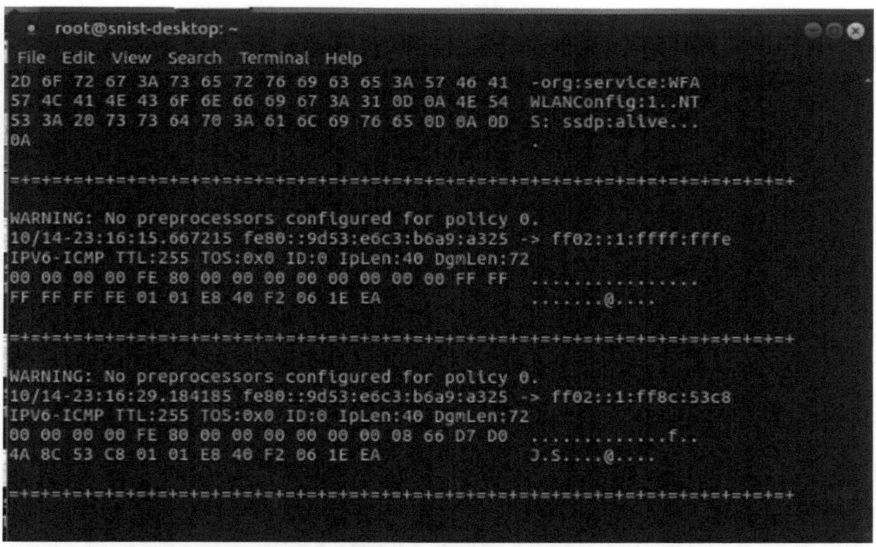

Fig. 6 Snort-IDS in sniffer mode with "snort -vd"

Figure 7 gives output of command "snort -vde".

Figure 8 gives you the information about packets received and analyzed in each and every command output. Sachets information will be given as follows.

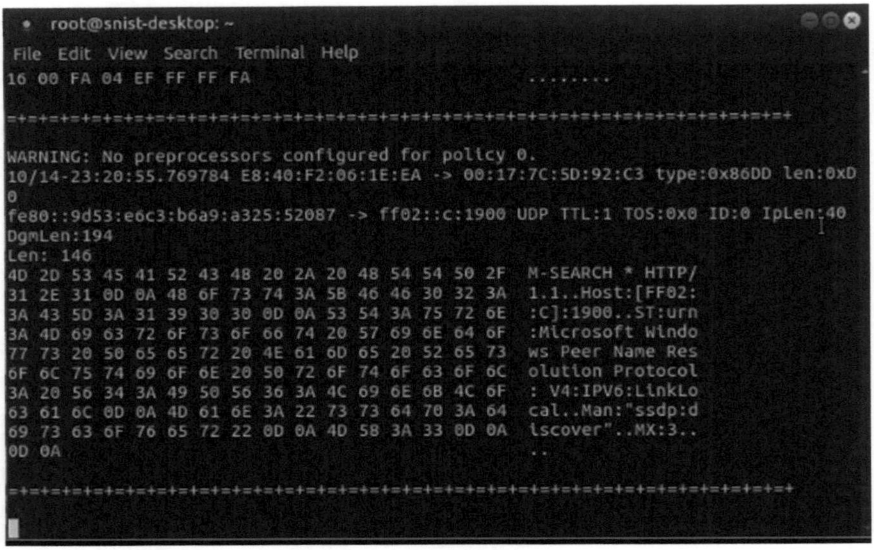

Fig. 7 Snort-IDS in sniffer mode with "snort -vde"

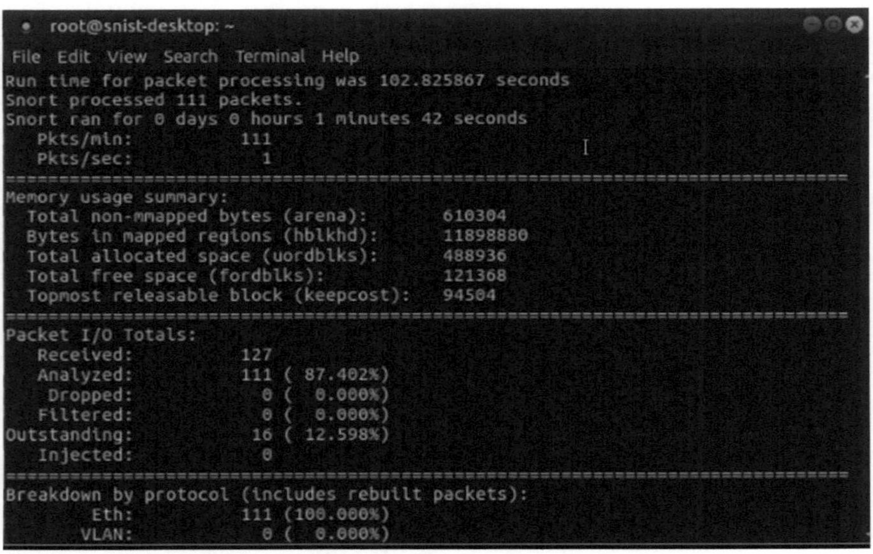

Fig. 8 Packets received and analyzed details

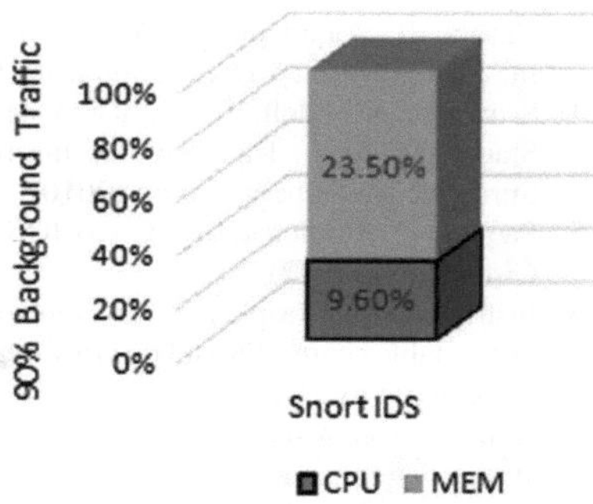

Fig. 9 Port scanning attack with 90% background traffic

5.3 CPU and Usage of Memory

The Port Scanning attack was measured with background traffic at 90% (Fig. 9). When this attack was performed, with Snort IDS, the CPU and RAM usages rose to 9.60 and 23.50%.

5.4 IP Addresses Storage

Incoming addresses are all stored in log file that stores if any event occurs.

6 Conclusion

As a part of our study, we assessed the Snort IDS performance and running on a Raspberry Pi2. As per outcomes, the performance of Snort IDS is superior on a Raspberry Pi2. In the future, Splunk applications will be configured to analyse logs from Snort IDS. Similarly, a Security Onion box will be integrated into the Local Area Network and the records generated by the IDS will be analyzed by using Snort by or Enterprise Log and Search Archive (ELSA). Additionally, many structured network attacks will be performed to test the performance of the Snort IDS by using penetration testing tools such as Kali Linux.

References

1. Mehra, P.: A brief study and comparison of snort open source network intrusion detection system. Int. J. Adv. Res. Comput. Commun. Eng. **1**(6), 383–385 (2012)
2. Depren, O., Topallar, M., Anarim, E., Ciliz, M.: An intelligent intrusion detection system (IDS) for anomaly and misuse detection in computer networks. Expert Syst. Appl. **29**(4), 713–722 (2005)

3. Cusack, B., Alqahtani, M.: Acquisition of evidence from network intrusion detection systems. In: 11th Australian Digital Forensics Conference, pp. 36-43. Perth, Australia (2013)
4. Scarfone, K., Mell, P.: Intrusion detection and prevention systems. In: Stavroulakis, P., Stamp, M. (eds.) Handbook of Information and Communication Security 1, pp. 177–192. Springer, Heidelberg, Berlin (2010)
5. Paquet, C.: Implementing Cisco IOS Network Security (IINS 640-554). Indianapolis, Ind., Cisco Press (2012)
6. Instructables: Raspberry Pi Firewall and Intrusion Detection System. http://www.instructables.com/id/RaspberryPi-Firewall-and-Intrusion-Detection-Syst/step13/SNORT/. Accessed 24 Oct 2015
7. Hoque, M.S., Mukit, M.A., Naser, A., Bikas, M.: An implementation of intrusion detection system using genetic algorithm. Int. J. Netw. Security Appl. **4**(2), 109–120 (2012)
8. Parkansky, K.: How to configure a snort IDS intrusion detection system on debian linux. Aboutdebian (2013). http://www.aboutdebian.com/snort.htm. Accessed 24 Oct 2015
9. Aldeid: Suricata-vs-snort-aldeid (2015). http://www.aldeid.com/wiki/Suricata-vs-snort. Accessed 24 Oct 2015
10. Rødfoss, J.T.: Comparison of Open Source Network Intrusion Detection Systems. University of Oslo, Oslo, Norway (2011)
11. Rani, S., Singh, V.: SNORT: an open source network security tool for intrusion detection in campus network environment. Int. J. Comput. Technol. Electr. Eng. (IJCTEE) **2**(1), 1–3 (2012)
12. Snort: Snort blog: snort 2.9.7.3 (2015). http://blog.snort.org/2015/05/snort-2973-is-nowavailable.html. Accessed 24 Oct 2015
13. Pir, R.M.: Intrusion detection techniques and open source intrusion detection (IDS) tools. Int. J. Eng. Develop. Res. **2**(3), 3290–3295 (2014)

Intelligent Twitter Spam Detection: A Hybrid Approach

Varad Vishwarupe, Mangesh Bedekar, Milind Pande and Anil Hiwale

Abstract Over the years there has been a large upheaval in the social networking arena. Twitter being one of the most widely-used social networks in the world has always been a key target for intruders. Privacy concerns, stealing of important information and leakage of key credentials to spammers has been on the rise. In this paper, we have developed an Intelligent Twitter Spam Detection System which gives the precise details about spam profiles by identifying and detecting twitter spam. The system is a Hybrid approach as opposed to single-tier, single-classifier approaches which takes into account some unique feature sets before analyzing the tweets and also checks the links with Google Safe Browsing API for added security. This in turn leads to better tweet classification and improved as well as intelligent twitter spam detection.

Keywords Twitter · Spam · Machine learning · Google safe browsing
Hybrid classifiers

V. Vishwarupe (✉) · A. Hiwale
Department of Information Technology, MIT College of Engineering, Pune, India
e-mail: varad44@gmail.com

A. Hiwale
e-mail: anil.hiwale@mitcoe.edu.in

M. Bedekar
Department of Computer Engineering, MAEER's MIT, Pune, India
e-mail: mangesh.bedekar@mitpune.edu.in

M. Pande
MIT School of Telecom & Management Studies, Pune, India
e-mail: director@mitsot.com

X.-S. Yang et al. (eds.), *Smart Trends in Systems, Security and Sustainability*, Lecture Notes in Networks and Systems 18,
https://doi.org/10.1007/978-981-10-6916-1_17

1 Introduction and Related Work

Twitter is a free Micro-blogging service that allows users to post messages, called tweets, up to 140 characters in length. Its value is in its ease of sharing and accessing user-generated content, including opinions, news, and trending topics. Thus, Twitter provides an opportunity to generate large traffic and revenue, especially since it has hundreds of millions of users.

However, these opportunities make Twitter a prime target of spammers. It is easy for humans to distinguish spammers from actual users, but the existence of spammers wastes user time and attention, puts users at risk in accessing malicious and dangerous content, and devalues Twitter's services and the overall online social network. Some of the related work in the domain of spam detection is highlighted below:

The features extracted from each Twitter account for purpose of spam detection can be categorized into:

(i) User-based features
(ii) Content-based features.

User-based features are based on a user's relationships e.g. those whom a user follows (referred to as friends) [1–3], and those who follow a user (referred to as followers) or user behaviors e.g. the time periods and the frequencies when a user tweets.

For content-based features, we use some obvious features. The average length of a tweet, words, syntax etc. [4–8].

2 Problem Statement and Solution Approach

In Intelligent Twitter Spammer Detection we introduce features which exploit the behavioral-entropy, profile characteristics, spam analysis for spammer's detection in tweets. We take a supervised approach to the problem, but leverage existing hashtags in the Twitter data for building training data. Twitter is one such popular network where the short message communication (called tweets) has enticed a large number of users. Spammer tweets pose either as advertisements, scams and help perpetrate phishing attacks or the spread of malware through the embedded URLs [9]. In this system, we fetch twitters tweets for a particular hashtag. Each hashtag may have 1000s of comments and new comments are added every minute, in order to handle so many tweets we are using twitter4j API and perform preprocessing by removing quotes, hash symbols and spam analysis through URL, Number of Unique Mentions (NuMn), Unsolicited Mentions (UlMn), Duplicate Domain Names (DuDn) techniques and Google safe browsing API. The aforementioned features are described in [10] which are quite relevant to our proposed model for classification of tweets. Hence using these unique feature sets and additional elements conceptualized by us such as AFINN dictionary mapping for some words, account type, etc. a better classifier is developed. This classifier hybrid uses Support Vector Machine (SVM) and Naïve Bayes (NB) for initial classification based on a 8-point feature set explained in 4.4. The malicious URLs if any are checked with Google Safe Browsing API. If a match is found, the tweet is more likely to be a SPAM tweet. A decision is made not only based on the URL found in the tweet but also takes into consideration other factors that might contribute to its classification.

3 System Modelling and Architecture

The above system architecture gives an overview of the Intelligent Twitter Spam Detection System which consists of unique feature vectors having features such as UIMn, NuMn, DuDn and comparison with Google Safe Browsing lists for detection and cross identification of URLs in Spam Tweets (Fig. 1).

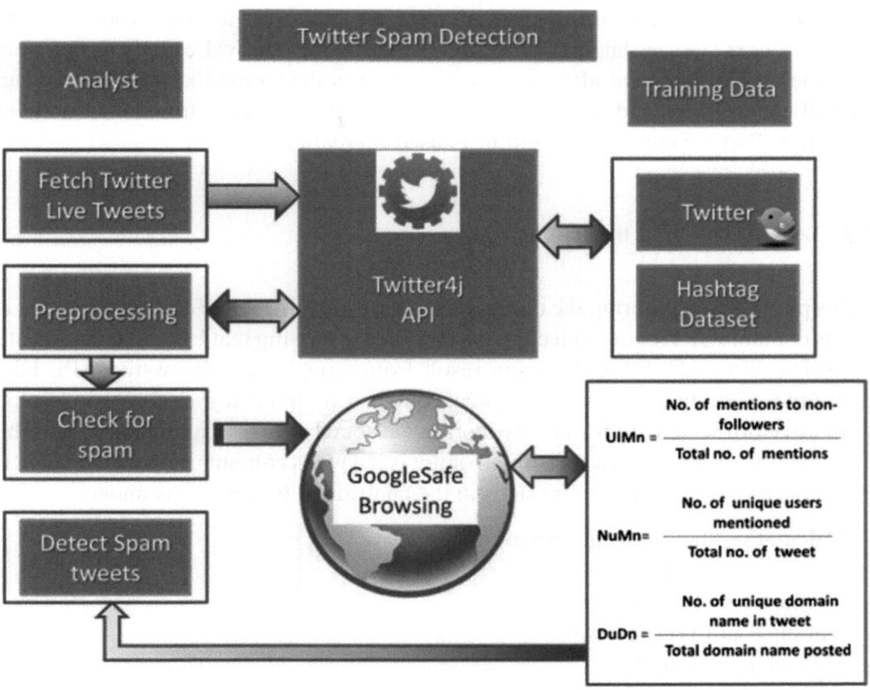

Fig. 1 System architecture

4 Algorithm and Tweet Analysis

4.1 Integrate the System with Twitter

The system will integrate with twitter and able to read the tweets for particular hash tags.

4.2 Data Set for Hashtags

Based on the trending topics on twitter (We used the trending topics in India during a period of 3 weeks in 2016).

4.3 Pre-processing

The first pre-processing technique is remove @ which means it scans the whole document of input dataset and after comparing it with @ it deletes @ from every available comment with @. The next step of pre-processing is remove URL where the whole input document gets scanned and compared with http:\\... and the comments having URL are deleted. Further we move on to stop word removal being the next step in data pre-processing. Stop word removal exactly means that from the whole statement after scanning it removes the words like and, is, the, etc. and only keeps noun and adjective. Tokenization and Normalization are carried out thereafter. Porter Stemmer Algorithm is used thereafter.

4.4 Analysis of Tweets

In this phase we are filtering the tweets on the basis of some criterion Checking that is tweets contains URL or not, if it contains then we are passing that URL to Google Safe Browsing API and getting the spam result from Google Safe Browsing API. Furthermore we are checking that is the tweet contains any spam keyword. We have a list spam keyword which are banned by worldwide social networking organization. The Basic criteria to declare a tweet is spam or not is if a tweet contains both a SPAM URL as well as SPAM WORDS. The steps in the analysis of tweets are as under:

a. Calculating the duplicate tweet count.
b. No of Unique Mentions.
c. Duplicate domain names.
d. Variance in tweet intervals.
e. Unsolicited mentions.
f. AFINN dictionary for finding word and their sense.

g. Finding negative annotations in the sentence and reverse the weight.
h. Detecting spam tweets.
i. Detecting spam accounts.
j. Blocking the spam accounts.
k. Finally, positive, negative or neutral count for that particular #hashTag will be calculated.

4.5 Feature Sets

a. Duplicate tweet count.
b. No. of Unique Mentions.
c. Duplicate domain names.
d. Variance in tweet intervals.
e. Unsolicited mentions.
f. Time interval of user being online.
g. Annotations.
h. Links to malicious URLs.
i. Account type: Verified/Unverified.

On the basis of the above constraints we will block/suspend the account.

5 Implementation and Results

In the implementation part of Intelligent Twitter Spam Detection using a Hybrid Approach, we used Twitter4J API, Google Safe Browsing Toolkit, A combination of classifiers including NB and SVM and unique feature sets that provide an intelligent spam detection solution. The system consists of 6 tabs which were designed in NetBeans using JavaSwing for the front end part. The Twitter Spammer tab consists of buttons that Fetch Live Tweets and Recent Twitter Feed related to the trending hashtags. The tweets are then stored on a local dataset which are then loaded for further analysis. Thereafter pre-processing takes place and the tweets are sent to the next stage for implementation of classification algorithms using the Hybrid Approach. A decision is made based on the code that runs on the filtered tweets using the aforementioned feature vectors and the tweets are classified as SPAM. Furthermore there is a provision for showing Trends-Wise Analysis for a particular hashtag in the next stage after which the final stage shows the suspended twitter accounts which were labelled as SPAM by the system. The analysis for this research was conducted by mining 10,782 tweets comprising of 72 hashtags that were trending on twitter. The system was able to classify 2,466 tweets as SPAM while the others were found legitimate. After cross-checking with Google Safe Browsing it was found that 2,153 tweets did contain a malicious URL that re-directed the user to suspicious websites. This affirmed that the accuracy of the said system stands at **87.30%** with this multi-tier approach which is greater than systems which use a single-classifier and non-hybrid approaches. The snapshots of the system at various stages are as follows (Figs. 2, 3 and 4):

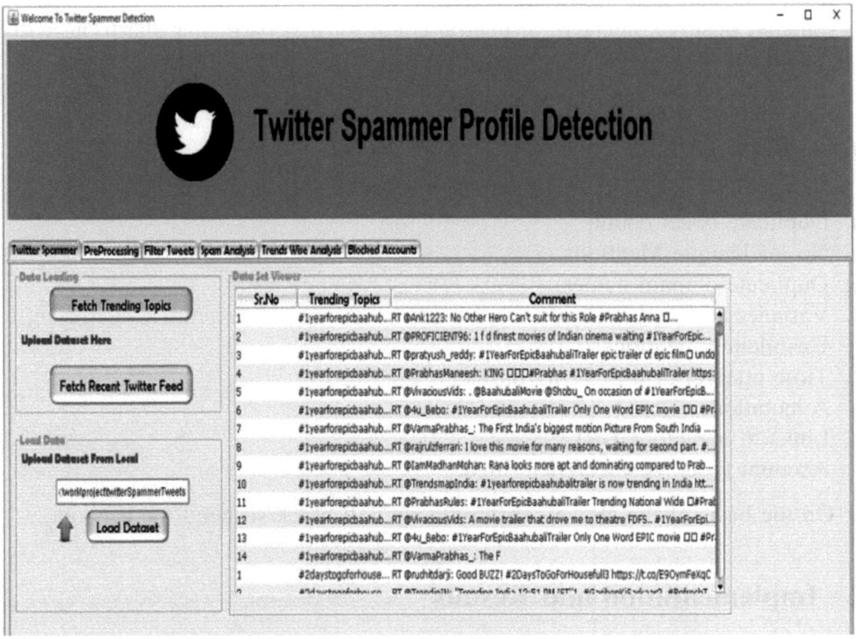

Fig. 2 Phase 1: fetching of tweets and trends

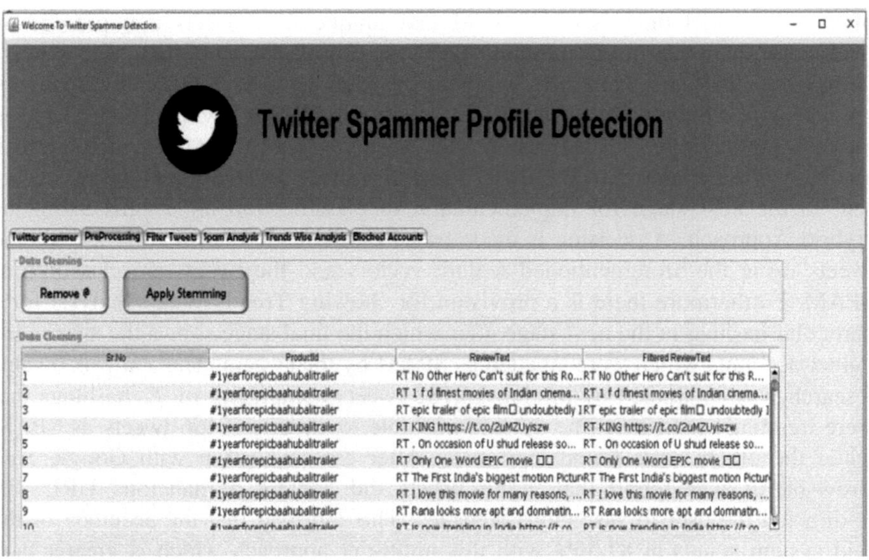

Fig. 3 Phase 2: preprocessing

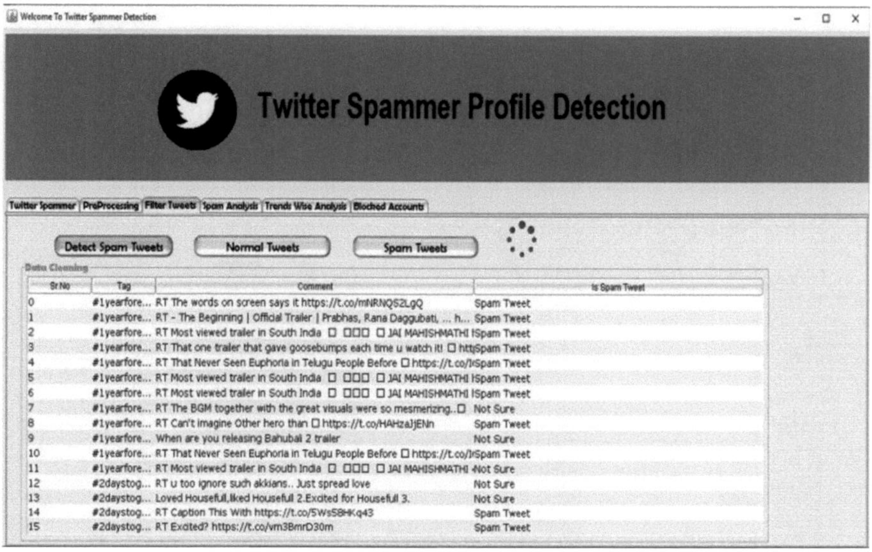

Fig. 4 Phase 3: analysis and detection of SPAM tweets

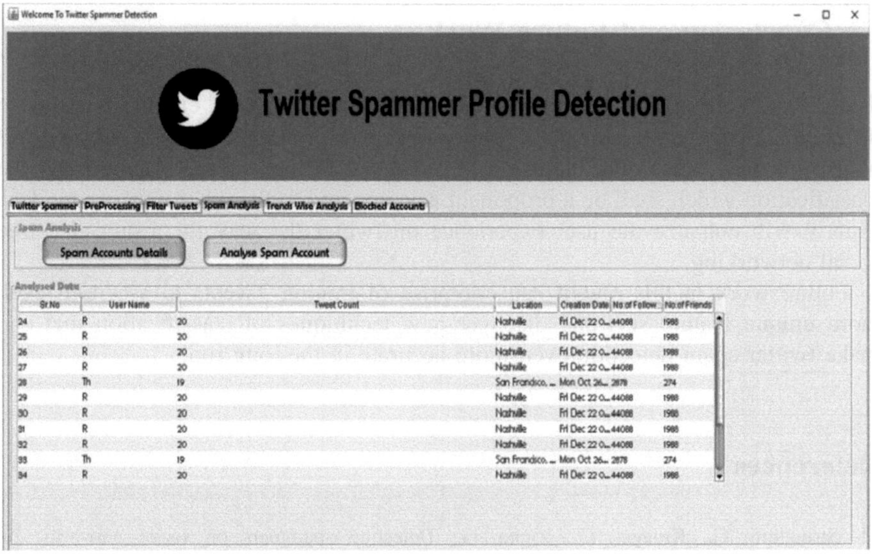

Fig. 5 Phase 4: SPAM account details

After this, the details of SPAM account with the Tweets and Location are displayed (Figs. 5 and 6).

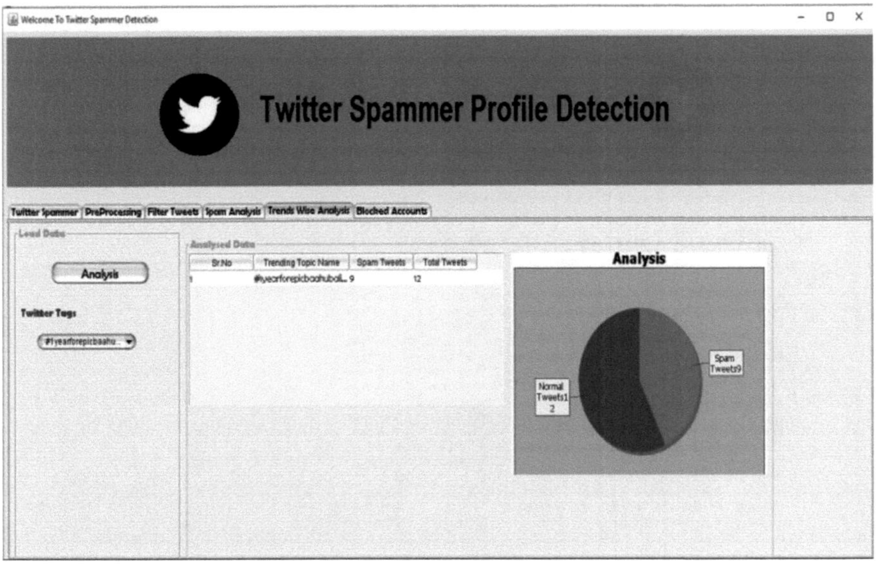

Fig. 6 Phase 5: Trend-wise analysis for a particular hashtag

6 Conclusion and Future Work

Twitter spammer profile detection makes the efficient use of classifiers so as to differentiate between legitimate accounts and spam profiles. Our approach is novel in its own way due to the inclusion of enhanced feature vectors for building the classification which shall be a proponent as compared to the other existing models. Thus it will enhance the user experience on twitter the way for a nuisance free social networking.

Future work in this regard can comprise of mining Tweets to ascertain even more unique feature sets and discover new techniques of classification that can make twitter spam detection even more accurate in the long run.

References

1. Stringhini, G., Kruegel, C., Vigna, G.: Detecting spammers on social networks. In: Proceedings of the 26th Annual Computer Security Applications Conference, pp. 1–9. ACM (2010)
2. Sala, A., Cao, L., Wilson, C., Zablit, R., Zheng, H., Zhao, B.: Measurement-calibrated graph models for social network experiments. In: Proceedings of the 19th International Conference on World Wide Web (WWW'10), pp. 861–870. ACM (2010)

3. Ahmed, F., Muhammad, A.: An MCL-based approach for spam profile detection in online social networks. In: 2012 IEEE 11th International Conference on Trust, Security and Privacy in Computing and Communications, pp. 1–7
4. Halim, Z., Gul, M., ul Hassan, N., Baig, R., Rehman, S., Naz, F.: Malicious users' circle detection in social network based on spatio-temporal co-occurrence. In: 2011 International Conference on Computer Networks and Information Technology (ICCNIT), pp. 35–39, July 2011
5. Perez, C., Lemercier, M., Birregah, B., Corpel, A.: SPOT 1.0: scoring suspicious profiles on Twitter. In: 2011 International Conference on Advances in Social Networks Analysis and Mining (ASONAM), pp. 377–381. IEEE (2011)
6. Tian, X., Guang, Y., Peng-yu, L.: Spammer detection on Sina micro-blog. In: 2014 (21st) International Conference on Management Science and Engineering, pp. 1–6, 17–19 August 2014
7. Qunyan, Z., et al.: Duplicate detection for identifying social spam in microblogs. In: 2013 IEEE International Congress on Big Data (BigData Congress) (2013)
8. Amleshwaram, A.A., et al.: CATS: characterizing automation of Twitter spammers. In: 2013 Fifth International Conference on Communication Systems and Networks, New York. IEEE (2013)
9. Chen, C., et al.: Battling the internet water army: detection of hidden paid posters. In: Proceedings of the 2013 IEEE/ACM International Conference on Advances in Social Networks Analysis and Mining. ACM (2013)
10. Joshi, S., et al.: Detection of spam tweets by using machine learning. Int. J. Adv. Res. Comput. Sci. Softw. Eng. **7**(4). ISSN: 2277 128X

Use of Learning Style Based Approach in Instructional Delivery

Ravindra Vaidya and Manish Joshi

Abstract Technology Enabled Learning (TEL) has started journey from off-line, non-interactive content available on storage media, and now current destination of that journey is personalised e-learning. Instructional Delivery is an important phase in e-learning environment. In our Personalised e-learning model, we have used Learning Style as deciding factor in Instructional Delivery mechanism. We tested our model on 111 learners. Our result shows that Learning style based learning object selection and their delivery elevates learning which in turn improves learner understanding in that subject.

Keywords Learning style · Intelligent tutoring system · Personalized e-learning Teaching-learning process · Taxonomy of e-learning

1 Introduction

Technology enabled Teaching-Learning (TL) process has transformed the conventional way of imparting education. The transformation helped to overcome the limitation of the conventional approach such as distance, language, rigidity, lack of personalization etc. This multi fold transformation in TL process using several electronic tools and appropriate methodologies is broadly referred as e-learning.

Effective instructional delivery in on-line learning enhances learners' experience [12]. Learning Style is one of the factors used and tested to improve learning efficiency of the learner [4]. Learning Style refers to the way learner understands the subject. Initially, psychiatrist and psychoanalyst C.G. Jung proposed this theory. In 1940, Myer-Briggs Type Indicator (MBTI) test which is based on Jung's theory,

R. Vaidya (✉)
MES's Institute of Management and Career Courses (IMCC), Pune, India
e-mail: rpv.imcc@mespune.in

M. Joshi
North Maharashtra University, Jalgaon, India
e-mail: joshmanish@gmail.com

X.-S. Yang et al. (eds.), *Smart Trends in Systems, Security and Sustainability*, Lecture Notes in Networks and Systems 18,
https://doi.org/10.1007/978-981-10-6916-1_18

became very popular and widely used. If teacher knows learners' learning style, then he/she can change teaching style(instructional delivery) to improve learners' learning experience. In open and distance education system separate instructional delivery is possible to every learner using technology. "Can we incorporate learning style in the instructional delivery to improve learning of the students?", was the million dollar question that we have pursued and satisfactorily answered.

Today there is a complete paradigm shift from technology enabled single terminal group learning to e-learning based personalized self learning system. Mulwa et al. [7] discussed about Technology Enhanced Learning (TEL) with specific reference to Adaptive Educational Hypermedia System (AEHS). Adaptivity can be achieved by personalization in AEHS by measuring Learning Style (LS) of every learner. Mulwa et al. discussed various models of Learning Styles and emphasized the importance of incorporating LS in various AEHS. They claimed that such blending of educational psychology and technology helped to increase efficiency in learning experience and achieved better learning outcomes.

In this research, we highlight the transformation in e-learning with respect to learning approach. We categorize research work in e-learning using Teaching Learning mode and learning approach aspects. Our focus is on personalized/adaptive e-learning approach and Furthermore we analysed nine different research experiments. The analysis shows following observations.

- Many researchers build Learner Model based on the LS, cognitive traits, Learning behaviour etc. after content delivery.
- Very few adapted building of LM before content delivery and evolve after it.
- Many researchers had adopted course level content delivery.
- Very few have partially used LO level learning content delivery.

This analysis shows that there is paradigm shift to personalised e-learning which enhances learning experience. This also shows that very few researchers have partially implemented LO level personalization, so there is wide scope for personalization at LO level. Further sections describes our model and explains experimental work conducted based on the model. At the end we describes analysis of of experimental works and conclusion.

In this paper we discusses the outcomes of our two research experiments based on our personalized e-learning architecture. This paper is subdivided in three sections. In Proposed model and experiments section, our personalized e-learning architecture is discussed and explained experimental methodology in detail. Experimental results are analysed and presented in Result Analysis section. The paper ends with conclusion section.

2 Proposed Model

Based on the observations, we decided to propose new Personalized e-learning architecture as shown in Fig. 1. This architecture suggest personalization at Learning

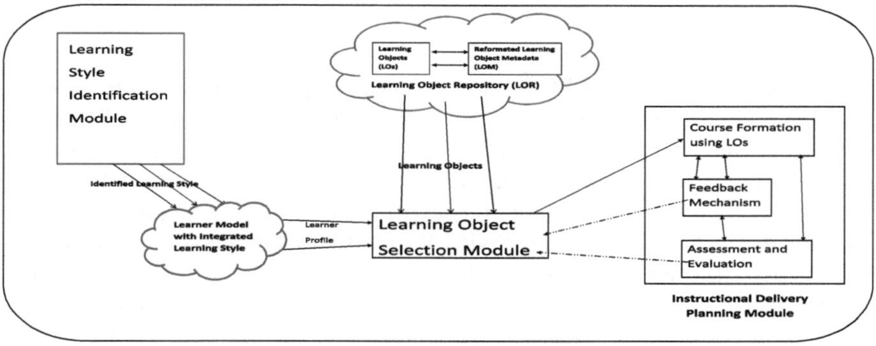

Fig. 1 Personalized e-learning architecture

Object (LO) level. This proposed model is subdivided in three module namely Learning Style Identification module, Learning Object Selection module and Instructional Delivery Planning module.

1. **Learning Style Identification Module (LSIM)**: In this module, we used Felder-Silverman Learning Style Model (FSLSM). The process of identification of LS in FSLSM was done through ILS questionnaire. This questionnaire helps us to identify various dimensions of LS of the learner.
2. **Learning Object Selection Module (LOSM)**: The selection of LOs from Learning Object Repository (LOR) has been done in accordance with LM. LOR contains LOs and its metadata called LOM. LOM contains various characteristics and attributes of LOs in the form of elements. Every elements has name and the value.
3. **Instructional Delivery Planning Module (IDPM)**: This module delivers the LOs selected in LOSM. This delivery has been done through Learning environment.

In this section we suggested LO level personalization based new personalized e-learning architecture. Next section discusses implementation strategy and experimental work in detail.

3 Experimental Work

Our experimental work is categorised in two parts. Each part of the experiment was conducted to test different but inter-related hypothesis.

3.1 Automatic Classification

The first part of the experiment was done to check empirical role of LS in Teaching Learning process. There are two streams of thought on the usefulness of Learning Style. Both sides have presented and supported their claims. We need to test the usefulness of LS by providing different Learning Objects to students with varying LS. The hypothesis that is tested in this experiments is "Learners prefer Learning Object(s) that suit to their Learning Style". We used automatic classification model—decision tree classifier to test our hypothesis. A decision tree is a predictive machine-learning model that decides the target value (dependent variable) of a new sample based on various attribute values of the available data. The next subsection describes decision tree classifier and decision tree induction algorithm and why we use it in testing our hypothesis?

3.1.1 Decision Tree Classifier

According to Han and Kamber [3] any prior knowledge or parameter setting is not required to construct decision tree classifier. Relatively learning speed is faster and accuracy is higher in decision tree classifier. Classification rules generated by decision tree classifier are simple and easy to understand. Hence we decided to use decision tree classifier for our experimentation and used J48 algorithm. J48 algorithm is an implementation of C4.5 algorithm suggested by Quinlan [9, 10]. J48 algorithm follows greedy approach and tree is constructed in top-down recursive divide and conquer manner. The J48 Decision tree classifier uses information gain. The algorithm for inducing decision tree from the training sample.

For experimentation we have used data mining tool called weka version 3.6.13. This tool has facility for various data mining techniques like classification, clustering, association etc. We use J48 decision tree classifier algorithm. We are experimenting with three dimensions of LS proposed by Felder-Silverman Learning Style Model (FSLSM) namely Active/Reflective, Visual/Verbal,and Sequential/Global. Apparently, we have developed eight distinct LOs that corresponds to combination of these three dimensions as **Active(0)/Reflective(1)**, ı**Visual(0)/Verbal(1)** and **Sequential(0)/Global(1)**.

In order to carry out the experiments to test the hypothesis, it is subdivided in three phases namely Sample selection, Preparation and selection of Learning Objects and actual experimentations. The complete process is explained in next sub-section.

3.1.2 Sample Selection

We have used random sampling technique and ensured that the selected learners must have following criteria.

- Selected Learners (sample) must have computer background.
- Selected Learners must have same level of knowledge about LOs' domain.

3.1.3 Preparation and Selection of Learning Objects

All learners are computer literate. We selected Learning objects from various domains viz. science, arts, commerce etc. LOs on common topic were selected, which require primary level knowledge to understand the subject, so every learner should understand it easily. Every learner has basic knowledge about the subjects, also they (learners) have same level of knowledge, hence we selected LOs on common topics of that subjects.

As all participants are from Computer background, LOs were selected mostly from non-computer domain. We ensured that these LOs are from varied domains of knowledge and most of our users did not have prior knowledge of these topics. These LOs are tagged with the corresponding LS as proposed by us [5].

3.1.4 Experimentation

The main objective of the experimentation was to collect data from the participants and obtain automated classification rules to investigate whether our proposition is true or false? Data was collected from learners at various locations by conducting the experiment. Experiment was conducted in the group of 20 students.

The procedural steps followed for each group of students are enlisted below

1. Each Learner has to complete 2-choice Index of Learning Style (ILS) questionnaire which was used to identify learners' Learning Style.
2. There are total 8 LOs, each tagged with unique LS, but we did not disclosed these styles to the learner.
3. We presented each LO one after another to the learners and instructed them to comprehend it. Depending on the type of content of LO, learners watch/listen, carried out activities in order to understand the subject presented in LO.
4. At the end of each LO, learners were asked to fill-up feedback of questionnaire which ask learner to write her/his preferred LO
5. This questionnaire has two part
   (a) Based on the understanding of the LOs, learner was asked to rank these LOs on the scale of 1(least preferred) to 5(most preferred).
   (b) The learner was also asked to mention the most preferred LO with the reason at the end of the experiment.

3.1.5 Observations

A J48 classification algorithm on Weka platform revealed that participants LS and the tag of the most preferred LO is matching for most of the participants. The data shown in the Table 1 depicts observations.

Some of the classification rules in the form of decision Tree are as shown in Fig. 2.

Table 1 Data output from J48 classification algorithm

LS of learner	Total number of learners	LS of preferred LO	Number of learners	Percentage of matching (%)
LS000	40	LS000	25	80
LS001	30	LS110	15	50
LS010	12	LS010	06	50
LS011	05	LS011	04	80
LS100	30	LS100	25	83
LS101	26	LS110	15	58
LS110	04	LS101	03	75
LS111	06	LS111	03	50

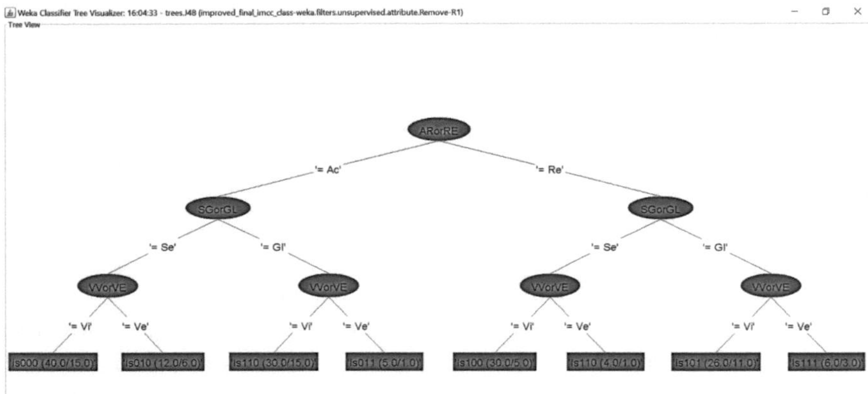

Fig. 2 Classification tree of learners

From decision tree, we observe that out of 30 Re-Vi-Se learners, 25 prefers LO with same tagging. This corresponds to the rule

```
if (ACorRE = "Re" and SEorGL = "Se" and VIorVE = "Vi"
then LO with tagging LS100 get selected
```

In general more than 50% learners prefer matching LOs except Ac-Vi-Gl(LS001) learners and in particular, more than 75% of learners with learning style Ac-Vi-Se, Ac-Ve-Gl, Re-Vi-Se and Re-Vi-Gl prefers matching LOs. 50% of Ac-Vi-Gl(LS001) learner prefers LOs with Re-Ve-Se tagging.

Our observation justifies correctness of the hypothesis that Learners prefer Learning Object(s) that suit to their Learning Style.

Our next experiment explores the possibility when if delivery of matching LO shall have more impact on elevation of understanding as compared to delivery of non-matching LO. The hypothesis and method to justify its correctness presented in next session.

3.2 Learning Style Driven Instructional Delivery

The automatic classification methods used in previous experiment revealed that Learning Style of most of the learners and the LS of the most preferred LO selected by these learners is matching. Hence, we decided to implement our Personalized e-learning architectures as shown in Fig. 1 explained in previous section. In continuation with first experiment, this experiment was conducted to test multiple hypothesis:

1. Active/Reflective learners improves learning experience which in turns improves learners' understanding, after absorbing Active/Reflective learning objects.
2. Visual/Verbal learners improves learning experience which in turns improves learners' understanding, after absorbing Visual/Verbal learning objects.
3. Sequential/Global learners improves learning experience which in turns improves learners' understanding, after absorbing Sequential/Global learning objects.

As researcher is from computer field, we decided to select subject from computer domain. Hence we chose Data Structure as subject for this experiment. As Learning Objects to be developed and/or to be selected, we decided to use stack as subtopic for this experiments.

3.2.1 Selection of Participants

We have to test impact of Learning Style based LOs selection and delivery on learners understanding. For this experiment also we have used random sampling techniques with some criteria listed below

- Selected learners are from computer background.
- Selected learners without any prior knowledge of Data Structure.
- Selected learners are from under graduate category.

We selected participants who do not have prior knowledge of subject i.e. Data Structure. We selected 176 students studying in First Year of B.Sc.(CS), B.Sc.(IT) and BCA. Data Structure is the part of their second year syllabus.

3.2.2 Experimentation Methodology

This experiment was conducted in three different phases. This section is subdivided in 3 subsections which are three phases of this experiment.

3.2.3 Learning Style Identification Phase

In this phase we asked student to fill Felder-Soloman Index of Learning Style (ILS) questionnaire [1]. For experimentation we developed a web-based interface where students

1. created login by filling up necessary information.
2. after successful login, students attempted ILS questionnaire, which comprises of 33 two choice questions.
3. upon completion and submission of ILS questionnaire, the system identifies Learning Style of student and consequently appropriate Learner Model is created.

Out of 176 learners, 146 learners successfully completed first phase of experiment. These 146 learners has been distributed in Learning Style Dimension wise.

3.2.4 Learning Object Selection Phase

The selection of LOs was done in accordance with Learner Model (LM) generated in previous phase. LOs for each subtopic of DS, with different attributes are stored in Learning Object Repository (LOR). We also developed some activity based learning objects for this experiment. Each LO in LOR has been tagged as per our reformatted Learning Object Metadata (LOM). Learner Model of each learner contains values for each dimension Active(0)/Reflective(1), Visual(0)/Verbal(1), Sequential(0)/Global(1).

As proposed in our Personalized e-learning architecture, we selected LOs at each subtopic level. This process of selection was done using manual match-making of LOM and LM.

E.g. For "Operation on Stack" topic under "stack", we had different LOs with different tagging like Ac-Vi-Se(000), Re-Vi-Gl(101) etc. This tagging was done based on content type, activity and delivery mechanism. We did manual match-making and selected appropriate LOs for respective LM.

3.2.5 Instruction Delivery Phase

In Intelligent Tutoring System (ITS), tutoring module decides tutoring strategies based on student module and domain module. Instruction delivery phase decides delivery of LOs selected in Learning Object Selection phase, based on Learners' model. Instruction delivery mechanism ensures accurate delivery of LOs in appropriate manner. We used Learning Content Management System—MOODLE as an agent of delivery.

Course formation is the first step in this delivery. For every group, selected LOs were delivered in appropriate way. Courses created delivered to learner who shows respective learning style. Participants were added as student user in each course

Table 2 Learner distribution—course-wise

Course name	Number of learners	Course name	Number of learners
LS000	16	LS100	15
LS001	18	LS101	21
LS010	23	LS110	20
LS011	15	LS111	18

according to their Learning Style identified in Learning Style Identification phase. All 146 learners distributed among these eight different courses. Tabular information of distribution is given in Table 2.

3.2.6 Evaluation

In each course, before delivery of LOs, Pre-learning test was conducted. The purpose of the test is to investigate learners' prior knowledge about the subject.Out of 146 learners, 111 learners participated this phase of experiment. Every learner completed the course and feedback in the form of Post-learning test was taken. Grading in Pre-learning test and Post-learning test are used as performance indicator in this experiment.

3.2.7 Observations

We analysed grades obtained by learners in Pre-learning Test and Post-learning test. In this analysis performance improvement has been investigated. All learners have been provided with LOs. Hence, it is apparent that the performance of each student will improve as compared to the performance of Pre-learning test grades. In order to cross-check this hypothesis, we propose to use Performance Indicator (PI) which is calculated as

$$PI = G_{pos} - G_{pre} \tag{1}$$

where G_{pos} is grade received in Post-learning test while G_{pre} is grade received in Pre-Learning test and PI is performance Indicator.

Learner with positive PI is termed as Improved Learner. Following table shows data of Improved learner when matching LOs has been delivered to each learner.

Combining of learning style dimension effect on the percentage of Improved learner. The data in Table 3 shows that combined LS dimensions increases percentage of Improved learner except LS010 and LS101. In other combined LS dimension, percentage of Improved learners increases significantly.

Table 3 Performance improvement chart—combined LS dimension wise

Combined LS dimension	Number of learners with matching LOs	No. of improved learner	Per. (%)
LS000	8	6	75
LS001	10	6	60
LS010	6	2	33.33
LS011	3	2	75
LS100	9	8	88.88
LS101	10	3	30
LS110	2	2	100
LS111	3	2	75

4 Conclusion

The aim of this research is to evaluate the effect of Learning Style in personalized e-learning system. Many researchers [2, 6, 11] are in favour of using learning style in personalization. On the other hand, doubts about the concept of Learning Style itself are raised and its usage in learning process is questioned [8]. Amidst these 'for and against' claims, our research revealed some important findings.

Students from five different institutions of four different cities (from rural and urban area) were selected for experimental study. Although the developed course was new for all the students few students might have been already aware of few topics of the course. Hence, all students were evaluated before delivering course contents (LOs) to them. This pre-learning test score is later used as a base to appraise the performance of students. Instead of using absolute post-learning test score, we used relative difference in performance to ensure that the experimental results are not biased.

It was observed that lower at 53% of Global learner to higher 71% Sequential learner who have been provided with LOs matching to their learning style improvised their performance. Even if we use combined LS dimension (LS000, LS001, LS011, LS100, LS110, LS111) then percentage of Improved learner increases i.e. lower 60% to Higher 100%. This shows that if we use combined LS dimension then result is improved. Although, this emphasizes the need of instructional delivery of LO as per the LS of user, we also came across some interesting observations.

In short, our research builds a bridge between two strong opposite opinions on use and effectiveness of Learning Style. We have demonstrated how a learning environment for learning style based instructional delivery can be set. We are certain that it will be used in future for on-line e-learning personalized systems.

Acknowledgements The data references/selected Learning objects from various domains viz. science, arts, commerce etc. LOs on common topic were selected, which require primary level knowledge to understand the subject. The information has also been collected from different authentic sources such as textbooks, journals and sorted data sources.

References

1. Index of Learning Styles Questionnaire
2. Graf, S., Liu, T.-C., Kinshuk: Analysis of learners' navigational behaviour and their learning styles in an online course. J. Comput. Assist. Learn. **26**(2), 116–131 (2010)
3. Han, J., Kamber, M.: Data mining: concepts and techniques. In: The Morgan Kaufmann Series in Data Management Systems, 2nd ed. Elsevier, Morgan Kaufmann, Amsterdam, Boston, San Francisco, CA (2006)
4. Hazra, A.K., Patnaik, P., Suar, D.: Influence of learning styles and academic achievement on performance and preference in different learning modes: an adaptive hypermedia approach. In: 2012 4th International Conference on Intelligent Human Computer Interaction (IHCI), pp. 1–7. IEEE (2012)
5. Joshi, M., Vaidya, R.: A review of paradigm shift from conventional to personalized e-learning. In: International Conference on Advances in Computing, Communications and Informatics (ICACCI), pp. 1263–1269. IEEE (2013)
6. Joshi, M., Vaidya, R., Lingras, P.: Automatic determination of learning styles. In: Proceedings of the 2nd International Conference on Education and Management Technology, vol. 13. IACSIT Press, Singapore (2011)
7. Mulwa, C., Lawless, S., Sharp, M., Arnedillo-Sanchez, I., Wade, V.: Adaptive educational hypermedia systems in technology enhanced learning: a literature review. In: Proceedings of the 2010 ACM Conference on Information Technology Education, p. 7384 (2010)
8. Pashler, H., McDaniel, M., Rohrer, D., Bjork, R.: Learning styles concepts and evidence. Psychol. Sci. Public Int. **9**(3), 105–119 (2008)
9. Quinlan, J.R.: C4.5: Programs for Machine Learning. Morgan Kaufmann Publishers (1993)
10. Quinlan, J.R.: Induction of decision trees. Mach. Learn. **1**(1), 81–106 (1986)
11. Sonamthiang, S., Cercone, N., Naruedomkul, K.: Discovering hierarchical patterns of students' learning behavior in intelligent tutoring systems, pp. 485–485. IEEE (2007)
12. Tellis, P.S., Narayan, P.S., Dhamapurkar, S., Sasikumar, M., Mudur, S.P.: "Vasistha"-An instruction delivery framework for online learning. In: Proceedings of ELELTECH India, National Seminar on e-Learning and e-Learning Technologies, Hyderabad (2001)

An Intelligent Hardware Calibration Scheme for Real Time VoIP Applications

Chinu Singla, Gurjot Kaur, Jasleen Kaur, Nitish Mahajan, Shubhani Aggarwal and Sakshi Kaushal

Abstract VoIP (voice over IP) is a methodology for the transmission of data, voice, video, messaging and chat services over Internet Protocol. The quality of VoIP is heavily dependent on the type of hardware used. Hardware calibration is a mechanism for selecting a suitable hardware for the required system characteristics. Sometimes, it becomes quite difficult to select the possible hardware with optimal cost from a class of processors which provide the same services. The aim of this paper is to use an Intelligent hardware calibration scheme for VoIP based real-time applications by considering different parameters like CPU utilization, cost, RAM utilization, jitter and delay. The selection of the processor by using an Intelligent scheme saves the execution time and availability of resources when it takes calls per second as input and gives output as whichever hardware is recommended for call manager to handle a different number of calls at an optimized cost. In this paper, Fuzzy Analytic Hierarchy Process (FAHP) based scheme is proposed which allows the user to make the best decision from a set of available choices to compare a set of processors that would help service providers to optimally provide different types of VoIP based services.

C. Singla (✉) · G. Kaur · J. Kaur · N. Mahajan · S. Aggarwal · S. Kaushal
Computer Science and Engineering, University Institute of Engineering and Technology, Panjab University, Chandigarh, India
e-mail: cheenusingla10@gmail.com

G. Kaur
e-mail: gkaur0403@gmail.com

J. Kaur
e-mail: kaurjasleen873@gmail.com

N. Mahajan
e-mail: nitish7mahajan@gmail.com

S. Aggarwal
e-mail: shubhaniaggarwal529@gmail.com

S. Kaushal
e-mail: sakshi@pu.ac.in

X.-S. Yang et al. (eds.), *Smart Trends in Systems, Security and Sustainability*, Lecture Notes in Networks and Systems 18,
https://doi.org/10.1007/978-981-10-6916-1_19

Keywords Hardware calibration · Fuzzy analytic hierarchy process
Voice over IP · SIP · FreeSWITCH · RTP

1 Introduction

Voice over IP (VoIP) is a technology that carries and transforms the voice communication over IP network like Internet. Instead of using circuit switched network, we use packet switched network where audio is transmitting in the form of digital packets [1]. VoIP is encapsulating digital voice in IP packets. For voice and data traffic, traffic engineering is an important aspect. Once the session is initiated, RTP packets can be exchanged between the initiator and the receiver end [2]. RTCP is a protocol that controls end to end information about the session of RTP packets [3]. SIP is an application layer protocol used for initiating, establishing, modifying and terminating the session [4]. It is one of the most popular protocols that are designed for establishing the VoIP calls. It combines with other IETF protocols like RTP, RTCP, and Gateway Control Protocol (GCP) to establish the entire multimedia architecture [5]. It is very similar to HTTP, the web protocol, or SMTP. SIP is responsible for finding the location on the basis of URL of the end users and can be done with the help of DNS server [6].

User Agent Client (UAC) sends requests to SIP server and SIP server sends back responses. The SIP server acts as a proxy server and send requests to some another server on the favor of a client [7]. The SIP server acts as a registrar and it also checks that UAC is authorized or not as shown in Fig. 1.

FreeSWITCH is an open source IP based platform [8]. Its design is based on the central core which is stable and has modules for specific functionality [9]. A session is initiated by the connection between FreeSWITCH and a SIP protocol.

Hardware calibration is done using specialized software and hardware and it enables fast and easy calibration in a shorter period of time [10, 11]. We will design an Intelligent scheme by considering parameters such as delay, jitter, cost, RAM utilization and CPU utilization using simulation model [12]. Thus, it is necessary to

Fig. 1 SIP server

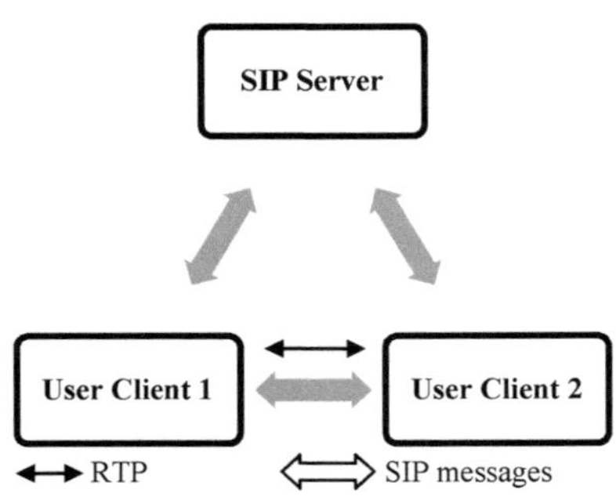

choose an optimal processor that which fulfills the demands of the users in less CPU utilization and in an optimal cost.

Fuzzy Analytic Hierarchy Process (FAHP) based decision-making model is used for selecting a processor from a huge variety of processors [13]. In order to make an appropriate decision, multiple parameters need to be determined and perform the comparison between them according to their importance. Conventional AHP needs pair-wise comparison for the selection of arbitrary values that involve uncertainty [14]. Instead of conventional AHP, FAHP is more appropriate and suitable method when there is uncertainty in pair-wise comparison of values [15]. Thus in this paper, we have used fuzzy AHP which is an extension of conventional AHP for hardware calibration to provide different types of VoIP based services.

The rest of this paper is organized as follows. Section 2 presents the basic concepts of Analytic Hierarchy Process (AHP) method, Fuzzy set theory and also describes the concept and existing work related to these methods in decision-making process for selection of different processors. Proposed Technique and simulation results are presented in Sect. 3 and Conclusion is presented in Sect. 4.

2 Background

This Section demonstrates the basic concepts of Analytic Hierarchy Process (AHP) method and Fuzzy set theory for selection of different processors.

2.1 The Analytic Hierarchy Process (AHP) Method

AHP is a theoretical approach to measure through comparisons based on priority scales aiding complex decision-making [16]. These scales are constructed corresponding to the dominance of a particular element in making a judgment. In a multiple criteria decision problem, AHP narrows down a bigger complicated problem into well constructed smaller elements.

Firstly, we define the unstructured problem by clearly stating its objectives. Next, we disintegrate the problem into the hierarchical structure. We now perform a pair-wise comparison of the resulting criteria. AHP assigns a weight for each of these evaluation criteria as per the result of the comparisons made. Then, we use the Eigen value method to deduce the relative scores. While we perform some pairwise comparisons, it may result in some inconsistencies. The next step hence is to provide an effective solution to check for these inconsistencies. Finally, we rank the elements by checking the final scores and then aggregate those to obtain the overall performance of all the elements combined.

2.2 *Fuzzy Set Theory*

Fuzzy set theory provides a mathematical tool to address the uncertainty problems [17]. A fuzzy set is defined by a membership function which states the degree related to a particular element. Its value is specified from 0 to 1. Value 1 corresponds to a complete member and value 0 corresponds to a non-member. The in between range characterizes fuzzy members, which relate to the fuzzy set partially [18].

In next section, we have used FAHP for decision making which is the combination of AHP and fuzzy set theory. Although this method has been used in many papers but none of the researchers have proposed any Intelligent scheme for hardware calibration in VoIP.

3 The Proposed Model

In this section, we have presented an Intelligent Fuzzy AHP based scheme for hardware calibration by using simulation model.

3.1 *Simulation Model*

We have used Star Trinity as simulation framework for simulating different VoIP calls. Star Trinity is an emulator used for load testing and performance of SIP servers [12]. It also monitors the VoIP quality of live IP network servers and tells about delay, jitter, packet loss etc. It is able to monitor and simulate thousands of simultaneously outgoing and incoming SIP calls and build real-time reports. In our simulation study, we have considered four PCs such that two PCs behave as call originators and two PCs are used to receive the incoming calls (receiver). Total 2000 users were registered on each. An input to Call Manger (SIP Server) can be given in form of Busy Hour Call Attempt (BHCA), which can be further converted to Calls per Second (CPS). We have carried out this simulation for one hour and analyze the results for 5, 10, 30 and 40 CPS. Simulations results are discussed in next section.

3.2 *Fuzzy Analytic Hierarchy Process (FAHP)*

While making decisions, sometimes it becomes difficult to fuzzify the ambiguous conditions. AHP is not capable of transforming the uncertainty into crisp values. The fuzzy logic, on the other hand, is unable to measure the degree of consistency

Table 1 9-Point scale and its definition

Satty's scale	Definition
1	Equally important
3	Moderately more important
5	Strongly more important
7	Very Strongly more important
9	Extremely more important
2, 4, 6, 8	Intermediate values

Table 2 Fuzzy pair wise comparison scale and its definition

Fuzzy pair-wise scale	Definition
0.95	Extremely important
0.85	Very strongly important
0.75	Strongly important
0.65	Important
0.55	Slightly important
0.5	Equally important

in judgments. By merging the two techniques, we can overcome these problems. Hence, Fuzzy Analytic Hierarchy Process (FAHP) is the solution obtained by combining the advantages of two above mentioned techniques that help in making the best decision [19]. FAHP process in defined in the following steps [20]:

Step 1 Develop pair-wise comparison matrix (shown in Table 1) of the different criteria through 9 point scale Satty [16].

Step 2 Check consistency in the comparison matrix.
If the comparison matrix is consistent it satisfies:

$$s_{11} = 0.5, s_{ij} + s_{ij} = 1, \frac{1}{s_{ij}} - 1 = \left(\frac{1}{s_{ik}} - 1\right) * \left(\frac{1}{s_{ki}} - 1\right) \tag{1}$$

Step 3 The scores of the pairwise comparison is transformed into fuzzy variables by evaluating fuzzy positive matrix. The fuzzy variables have values ranging from 0 to 1. Resulting fuzzy pairwise comparison scale is shown in Table 2.

Step 4 Evaluate the fuzzy weights of decision elements.

Step 5 Aggregate all the opinions by taking the Geometric mean.

Step 6 By calculating the Consistency Index (CI), obtain the final ranking. Consistency Ratio (CR) is given by:

$$CR = CI/RI \tag{2}$$

Here, RI is Random Consistency Index. Its values are shown in Table 3. Matrix is consistent if $CR < 0.1$.

Then, we compared the different processors according to predetermined criteria like cost, CPU utilization, RAM utilization, delay and jitter which are calculated by fuzzy AHP. We performed the simulation on Star trinity for different number of Calls per Second (CPS) and we neglected the packet loss parameter due to a small number of CPS.

To set up the hierarchical model, we listed three hierarchies as shown in Fig. 2.

Different evaluation criteria's priority is assumed as shown in Table 4 by using fuzzy pair-wise comparison scale of Table 2. The precedence value of the criteria's can be different in different conditions.

By using the FAHP method, we calculate the priority weights, CI, RI and CR as shown in Table 5.

CI is calculated as 0.056, RI = 1.12 because n = 5

Thus, CR = CI/RI = 0.056/1012 = 0.05.

Table 3 Values of RI

Matrix size	1	2	3	4	5	6	7	8	9	10
RI	0.0	0.0	0.58	0.9	1.12	1.24	1.32	1.41	1.45	1.49

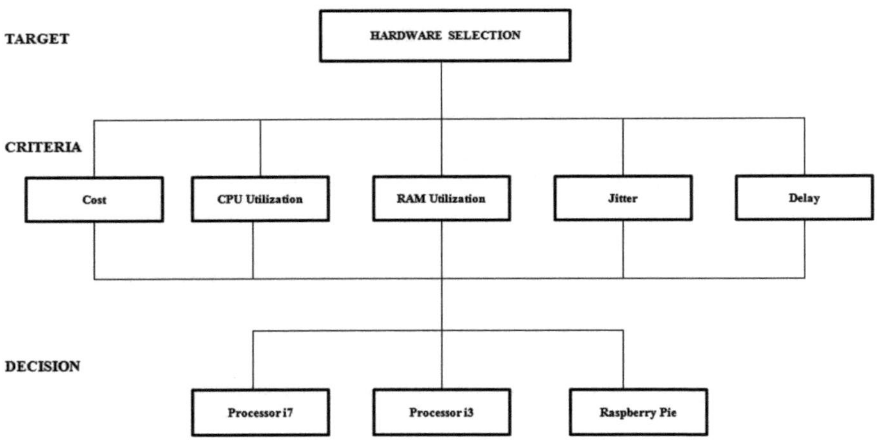

Fig. 2 Hierarchical structure for hardware selection

Table 4 Fuzzy comparison matrix at different criteria level

Criteria	Cost	RAM utilization	Jitter	CPU utilization	Delay
CPU utilization	0.5	0.7	0.8	0.5	0.9
RAM utilization	0.3	0.5	0.6	0.3	0.7
Jitter	0.2	0.4	0.5	0.1	0.4
Cost	0.5	0.7	0.9	0.5	0.7
Delay	0.1	0.3	0.6	0.3	0.5

Table 5 Results obtained from FAHP

Criteria	Weights
Cost	0.371
RAM utilization	0.061
Jitter	0.153
CPU utilization	0.348
Delay	0.067

Table 6 Evaluation matrix for hardware selection for 40 calls per second (CPS)

Processors	Cost	RAM utilization	Jitter	CPU utilization	Delay
i7 processor	H	L	L	L	M
i3 processor	M	M	M	M	M
Raspberry Pie	L	M	L	H	L

Table 7 Final results

Processors	Cost	RAM utilization	Jitter	CPU utilization	Delay	Result
i7 processor	0.7	0.0	0.0	0.0	0.3	
i3 processor	0.3	0.3	0.3	0.3	0.3	
Raspberry Pie	0.0	0.3	0.0	0.7	0.0	
Weights	0.371	0.061	0.153	0.348	0.067	
i7 processor	0.2597	0.0	0.0	0.0	0.0201	0.2798
i3 processor	0.1113	0.0183	0.0459	0.1044	0.0201	0.3
Raspberry Pie	0.0	0.0183	0.0	0.2436	0.0	0.2619

From Table 5, we conclude that CPU utilization and cost are the most important criteria. Also, the weights are consistent as consistency ratio is $0.05 < 0.1$. Thus, we can say that the weights are consistent under FAHP method.

Let the evaluation ranking be set as High (H), Medium (M) and Low as described in Table 6.

Hence, from Table 7, we obtain that i3 processor is the most cost efficient processor having less CPU utilization as compared to Raspberry Pie, when we simulate maximum 40 Calls per Second (CPS). With our proposed scheme, we can simulate a different number of calls per second efficiently. For testing any kind of processor (high end/low end), cloud services can also be hired and test call load on it by using Star Trinity software as a load generator for desired results. This can save physical resources of VoIP service providers and hence would help them to provide proficient services optimally by reducing their operational costs.

4 Conclusion

This paper proposes a Fuzzy Analytic Hierarchy Process (FAHP) based technique to select an ideal hardware from a different set of processors by determining the weights of five different criteria and by obtaining the final ranking of alternative processors. Results indicate that cost and CPU utilization are the important criteria among other criteria's. For simulation study, we have used Star Trinity emulator for load testing and performance of Call Manger. Results show that our proposed scheme is able to calibrate hardware requirements given any Call load in form of CPS or BHCA. This would help VoIP service providers to provide efficient services optimally by reducing their operational costs and hence users can experience better QoS.

References

1. Yu, J., Al Ajarmeh, I.: Design and traffic engineering of VoIP for enterprise and carrier networks. Int. J. Adv. Telecommun. **1** (2008)
2. Dantu, R., Fahmy, S., Schulzrinne, H., Cangussu, J.: Issues and challenges in securing VoIP. Comput. Secur. **28**(8) 743–753 (2009)
3. Ott, J., Wenger, S., Sato, N., Burmeister, C., Rey, J.: Extended RTP profile for real-time transport control protocol (RTCP)-based feedback (RTP/AVPF) (2006)
4. Mäkelä, M.: Signaling in session initiation protocol (2003)
5. Schulzrinne, H., Wedlund, E.: Application-layer mobility using SIP. Mob. Comput. Commun. Rev. **4**(3), 47–57 (2000)
6. An sari, A.M., Nehal, M.F., Qadeer, M.A.: SIP-based interactive voice (2013)
7. Rosenberg, J., Schulzrinne, H., Camarillo, G., Johnston, A., Peterson, J., Sparks, R., Schooler, E.: SIP: session initiation protocol (2002)
8. FreeSwitch: http://www.freeswitch.org. Accessed April 2013
9. Minessale, A., Schreiber, D.: FreeSWITCH 1.0.6. Birmingham, UK: Packt Publishing (2010)
10. Dougherty, B., White, J., Thompson, C., Schmidt, D.C.: Automating hardware and software evolution analysis. In: 16th Annual IEEE International Conference and Workshop on Engineering of Computer Based Systems, ECBS 2009. IEEE (2009)
11. Queenette, U.I., Jerome, I.O.: Selection criteria for computer software and hardware, a case study of six university libraries in Nigeria. Chin. Libr. Int. Electron. J. **32** (2011)
12. www.StarTrinity.com
13. Singla, C., Kaushal, S.: Cloud path selection using fuzzy analytic hierarchy process for offloading in mobile cloud computing. In: 2015 2nd International Conference on Recent Advances in Engineering and Computational Sciences (RAECS), pp. 1–5 (2015)
14. Yu, C.S.: A GP-AHP method for solving group decision-making fuzzy AHP problems. Comput. Oper. Res. **29**, 1969–2001 (2002)
15. Ji, D.C., Song, B.F., Yu, T.X.: The method of decision-making based on FAHP and its application. Fire Control Command Control **11**, 011 (2007)
16. Saaty, T.L.: The analytic hierarchy process (1980)
17. Zadeh, L.A.: Fuzzy sets. Inf. Control **8**, 338–353 (1965)

18. Gupta, M.M., Sanchez, E.: Fuzzy automata and decision processes. Elsevier North-Holland, New York (1977)
19. Xin, C.H.E.N.: Appliction of fuzzy analytic hierarchy process for optimum selection in decision-making. Comput. Eng. Design **10**, 068 (2004)
20. Lee, A.H., Chen, W.C., Chang, C.J.: A fuzzy AHP and BSC approach for evaluating performance of IT department in the manufacturing industry in Taiwan. Expert Syst. Appl. **34**, 96–107 (2008)

Automatic Sub Classification of Benign Breast Tumor

Aparna Bhale and Manish Joshi

Abstract This paper describes about automatic classification of benign lesions. Our re-search work mostly aims at development of a data driven effective classification system to determine type of benign breast tumors. The existing system in medical field identifies the type of benign breast tumor based on histopathological report obtained after a painful surgical process. Our target is to develop a system that would work without histopathological attributes to avoid pains to the patient. Our focus was to eliminate the role of histopathological attributes in the detection of benign tumor type. So we tried to identify correlation of histopathological features with mammographic image features and patient history features in order to explore if histopathological features can be replaced by corresponding correlated features. With replaced attributes we gain training accuracy for J48 as 79.78% and with SVM 81.91%. We obtained testing accuracy for J48 and SVM as 100% and 90.90% respectively.

Keywords Mammogram · Histopathology/clinical attributes · Feature extraction · Correlation · Classification

1 Introduction

One technique for breast cancer identification is provided by using mammograms for suspect patients descriptions further utilized by physicians to recognize potential abnormal areas thorough optical examination. When digital mammograms are existing, computer-aided based investigation may help the surgeon in having an

A. Bhale (✉) · M. Joshi
School of Computer Sciences, North Maharashtra University, Jalgaon, India
e-mail: aparnakulkarnibhale@gmail.com

M. Joshi
e-mail: joshmanish@gmail.com

X.-S. Yang et al. (eds.), *Smart Trends in Systems, Security and Sustainability*, Lecture Notes in Networks and Systems 18,
https://doi.org/10.1007/978-981-10-6916-1_20

additional precise result. The commonly used system for breast cancer revealing is mammography. A mammogram is an x-ray of a breast that is utilized to identify irregularities in a breast, mainly breast cancer. With the help of mammographic images, medical practitioners (experts) can easily determine whether the deposited tissues (tumor) are malignant (cancerous) or not. But if a tumor is benign, then to determine the exact class of the benign lesions a doctor recommends further investigations. Mostly Histopathological Analysis reveals the class of the tumor. Most of the Computer Aided Detection (CAD) research with respect to mammograms is conned to detection of malignancy. We did not come across any attempts of sub classification of benign tumors, which is essential for Medical experts in order to advise patients to help in preventing transformation of benignity into malignancy. A record of few of the current research work is mentioned in Sect. 2. We have made an attempt to correlate various features of mammogram with histopathology report of certain class. We have used histopathology reports in association with mammograms. Such multi modal information data is used to train the automatic classification system. The basic idea is to use data driven techniques to automatically classify tumors. Determination of appropriate features (mammographic features, histopathological features, patients medical history features other related features) i.e. feature selection and extraction is done to generate training and testing data set. We have worked on live mammogram images obtained from various hospitals. We have mined texture characteristics with Gabor filter and statistical characteristics by gray level co-occurrence matrix. Techniques used for feature extraction are discussed in Sect. 3. Section 3 also elaborates benign breast tumor classifier in detail.

Since the widely held of benign lesions are not correlated with an increased threat for subsequent breast cancer, needless surgical measures should be avoided. It is significant for pathologists, radiologists and oncologists to identify benign lesions, both to distinguish them from in situ and invasive breast cancer and to measure a patients risk of growing breast cancer, so that the most suitable treatment modality for each case can be established. Abnormal growth of cells that leads to the development of tumor is known as cancer. One of the main reasons of casualty among women is Breast cancer. Advances in screening and treatments have improved the survival rate in recent years. In breast, malignant and benign are abnormal growth of tumor cells. Section 4 gives focus on PINK classifier. This classifies the benign sub tumors. This classifier is working without histopathology attributes so its a pain reducing pink classifier. Result analysis is done in Sect. 5. This section includes results of correlation method, replacement of attributes. Accuracy of both the classifiers using SVM and J48 are compared. Pruned tree elaborate results of classifiers. The paper is organized as follows. Related work is discussed in Sect. 2. Feature extraction and details of benign breast tumor classifier are elaborated in Sect. 3. Section 4 describes the working of PINK classifier whereas results and observations are put forward in Sect. 5. Conclusions are in Sect. 6.

2 Related Work

This section explains the earlier work available in the literature. The utilization of statistical features is speedily increasing and one can see numerous research documents depicting the use of statistical features. We have gone through the research papers by Gurevich and Koryabkina [6], Mohanty et al. [13], Jasmine et al. [9] on statistical features and papers by Lakshmi and Manoharan [11], Kilic et al. [10], Pradeep et al. [16], XU et al. [23], Nagi et al. [15] for various preprocessing techniques and techniques used to extract the features. In our research we have utilized clinical features also. We studied papers by Thongkam et al. [21], Rosenberg et al. [19], Titus-Ernsto et al. [22], Friedenreich et al. [5] in which clinical attributes like age, age of menarche, number of kids, dietary habits are discussed. Many researchers have focused on these attributes and also studied its correlation with breast cancer such as Mohanty et al. [13], Titus-Ernsto et al. [22], Hartmann et al. [7], Robsahm and Tretli [18]. In order to augment the precision of all classifiers, we decided to work with pre processing feature selection methods. We have studied diverse feature selection methods like Cfssubset Evaluator, Chi square, and Gain ratio. We observed that Cfssubset evaluator gives us improved results. By using Cfssubset, we obtained 5 attributes, namely standard Deviation, Kurtosis, Homogeneity, Proliferation and Structure. Buciu and Gacsadi [2], Raghavan [17], Mohan Kumar and Balakrishnan [12], Chang et al. [3], El-Naqa et al. [4] and Shi et al. [20], Zaïane et al. [24], Hejazi and Ho [8] presented the classification methods and their observations. We found that in most of the cases binary classification is done using SVM. Attributes used for the classification belongs to either one or two categories such as clinical data or statistical data. But we have used unique combination of mammographic, Histopathological and clinical attributes for automatic classification of benign tumour.

3 Experimental Details

In this section we elaborate the stepwise experimental process followed to obtain classification of benign breast tumor without using histopathology attributes which we have termed as PINK. Initially we have extracted features and then developed two classifiers namely Benign breast tumor classifier and PINK classifier. In this section we have elaborated the phases of feature extraction and benign breast tumor classifier's working.

3.1 Feature Extraction

Once an image is pre processed to obtain the enhanced image, it is fed as input to the next processing phase, which can be feature extraction or image classification. The enhanced images are processed to extract relevant discriminating features for the next processing phase, which can be the image classification phase. The significance of the extracted features determines the performance of the overall system. Benign breast mammographic dataset is not available. To meet out our requirement of dataset, we have developed our own dataset. Unique feature of this data set is that it contains not only mammogram images but also histopathology reports and clinical details of benign cases. We have developed a database, which includes mammograms, patients history and corresponding histopathology reports that can be used as a standard data set in future by researchers worldwide.

Pre-processing is completed using morphological reconstruction operation. Gradient magnitude of an image is pre-processed to remove the noise. Watershed transformation applied to re ne gradient magnitude image. Marker Control Watershed Segmentation MCWS is used for deduction of over segmentation occurred by watershed. We have obtained Region of Interest (ROI) after segmentation by MCWS technique. Gabor filter is applied on ROI to extract features. ROI locates a lesion. Large number of statistical, textural features can be extracted. Appropriate features are selected that could significantly determine presence of a breast lesion and architectural distortions. By Gabor filter and GLCM (gray-level co-occurrence matrix), we extracted a few textural features from mammogram, which are the quantitative measures of texture that are Contrast, energy, entropy, homogeneity. Many researchers studied various aspects, which show the relation of breast cancer to age [14], age of menarche, breast cancer history [7], dietary habits [18], and recall of early menstrual history [14]. We have collected this information from patients. We have extracted important histopathology features like cellular change, sign, structure and other features from histopathology report for classification. Same features importance is authenticated by pathologist.

So in all we have extracted features from three main categories namely mammogram, Histopathology reports and clinical history.

3.2 Benign Breast Tumor Classifier

Benign breast tumor classifier uses final set of attributes comprises of mammographical, clinical and histopathological attributes. Patient history, mammogram and histopath contribute the feature that will be used for classification. In order to extract most significant features we conducted various experiments. It shows that mammographic features, if augmented with appropriate set of clinical features, can improve the overall classification result [1]. We obtained 5 attributes out of 23. With reduced number of attributes, we reduced complexity and processing time for classification.

3.3 *Approach Used for Benign Breast Tumor Classifier*

In this classifier mammographic, clinical, histopathological attributes are used. Following steps elaborate the approach of classification in brief:

1. Identify tumor is malignant or benign.
2. Once we observe that tumor is benign, we further classify it into 3 types, namely: Fibrocystic change/disease (FCC/FCD), Fibroadenoma, Phyllodes.
3. We further classify it into subtypes of above mentioned tumors as: (A) FCC with epitheliosis, (B) FCC with mastitis, (C) FCC breast with nuclear atypia, (D) Bilateral FCC, (E) FCC with epitheliosis Papillomatosis, (F) Bilateral fibroadenoma. We used our live dataset for the experiments. Figure 1 is a pictorial representation of benign breast tumor classifier.

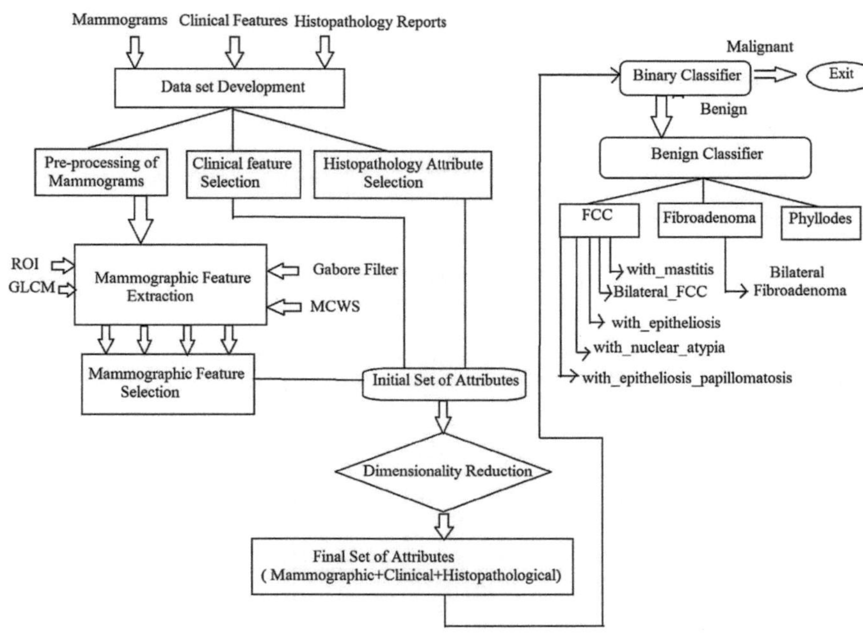

Fig. 1 Benign breast tumor classifier

4 Painless Intelligent BeNign Tumor Classifier (PINK)

We Proposed PINK classifier that eliminates the histopathological attributes by using correlation attribute module. This module yields revised set of attributes, which includes correlated attributes for histopathological attributes (Fig. 2).

4.1 What Is PINK Classifier?

We all discern that pink is the colour generally related with love, beauty, charm, politeness, sensitivity, tenderness, sweetness, childhood, and femininity. The pink ribbon is an international sign of breast cancer realization. Pink ribbons and the

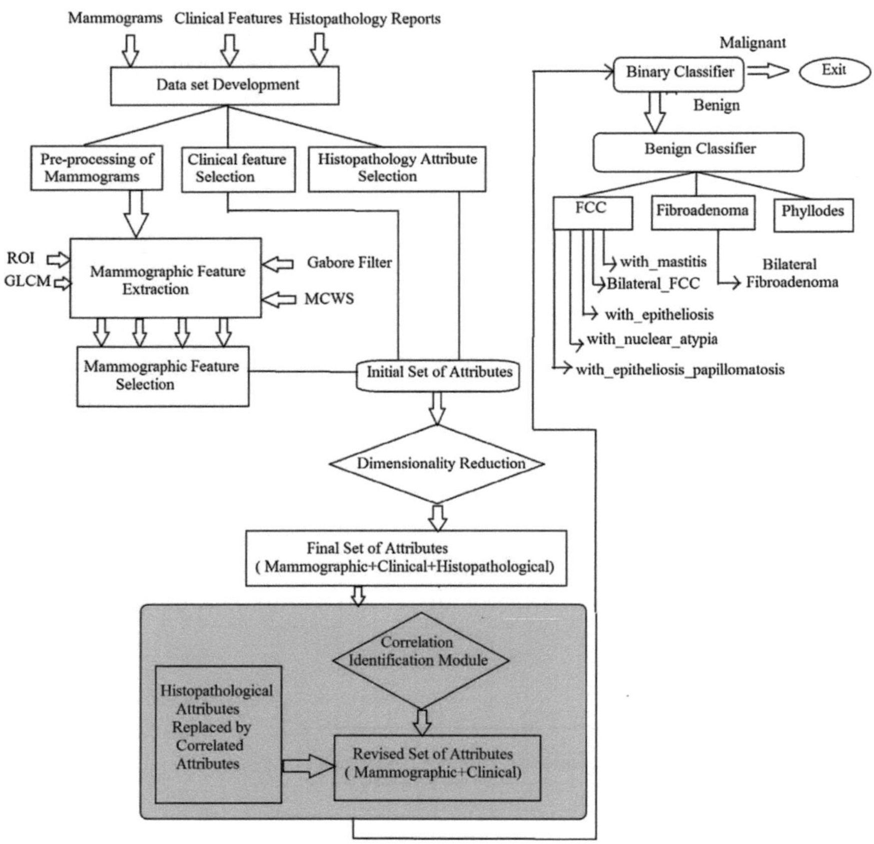

Fig. 2 PINK classifier

colour pink in common, recognize the sup-porter with the breast cancer trademark and convey moral support for women with breast cancer. Pink ribbons are most normally observed during National Breast Cancer Awareness Month that is October. So, we labelled our unique benign breast tumor classifier as Pink Classifier.

4.2 Unique Feature

To get histopathological attributes/findings patient's biopsy is needed, which is a painful procedure. To reduce the sufferings and pain of patients we obtained the correlated attributes for histopathology attributes by chi square method.

4.3 Correlation

In statistics, reliance or alliance is any statistical association, whether primary or not, between two random variables or two sets of data. Correlation is any of a broad class of statistical relationships linking dependence, even if in common usage it most often refers to the amount to which two variables have a linear connection with each other. Correlations are useful because they can designate an analytical relationship that can be exploited in practice. Often a number of quantitative variables are calculated on every member of a sample. If we regard as a couple of such variables, it is commonly of interest to set up if there is a association between the two to see if they are correlated.

4.4 Replacement of Attributes

To reduce the pain and sufferings of patient we want to replace histopathology attributes. So we carried out some experimental and statistical operations. We found that histopathological attributes can be replaced by mammographical attributes and clinical attributes. We performed chi square analysis to obtain correlation of histopathology attributes.

4.5 Chi Square Method

Chi-squared analysis is frequently constructed from a sum of squared errors, or through the sample discrepancy. Test statistics that follow a chi-squared distribution arise from an assumption of independent normally distributed data, which is suitable in numerous cases due to the central limit theorem. The chi-squared test is utilized to decide whether there is a significant variation between the expected frequencies and the observed frequencies in one or more categories.

5 Results and Observations

This section includes results of Benign breast tumor classifier and results of PINK classifier using correlation method, replacement of attributes. Accuracy of both the classifier using SVM and J48 are compared. Pruned tree elaborate results of classifiers.

5.1 Observations of Benign Breast Tumour Classifier

In this benign breast tumor classifier live data set's few instances out of 640 instances used as a training data set. We tested the classifier accuracy on unseen data. The SVM and J48 classifiers are applied on weka based java programming environment. Table 1 displays the results of experiments with reduced attributes. we trained classifier by J48 with an accuracy of 98.93% and testing precision observed is 81.81%. SVM classifier gives 85.10% training accuracy and testing accuracy as 90.90%.

5.2 Observations of PINK Classifier

Here are few examples of the experimental calculations of chi square test:

Energy with Proliferation: Energy attribute is extracted from mammogram. We discretized energy into 3 bins. A clinical attribute proliferation holds two attributes yes and no. We found that chi square value for these two attribute is 0:640. This 0:640 > 0:211($x^2$0:90). So null hypothesis is rejected and we conclude that these two attributes are correlated (Table 2).

Table 1 Testing accuracy of benign classifier with reduced attributes

Dataset	Features	Classifier	Activity	Accuracy (%)
Live Dataset	Reduced attributes (5)	J48	Training	98.93
Live Dataset	Reduced attributes (5)	SVM	Training	85.10
Live Dataset	Reduced attributes (5)	J48	Testing	81.81
Live Dataset	Reduced attributes (5)	SVM	Testing	90.90

Table 2 Chi square of energy with proliferation

Energy/Proliferation	Yes	No	Expt. freq. for yes	Expt. freq. for no	Chi square
E1	3	3	2.80851	3.25532	0.64057
E1	10	16	12.1702	14.1064	
E3	31	32	29.4894	34.1809	
Deg. Freedom = 2					

Table 3 Chi square of age with structure

Age/Structure	s	d	w	e	c	r	en	e:s	e:d	e:w	e:e	e:c	e:r	e:en
A1	6	1	12	1	4	1	15	5.53	1.74	10.63	1.27	2.97	2.55	15.31
A2	7	3	13	2	3	5	21	7.46	2.29	14.36	1.72	4.02	3.44	20.68
Degree of freedom = 6														

Age with Structure: Age attribute is from clinical history. We discretized age into 2 bins.

We grouped it as if Age <30 then 0 and Age >=30 then 1. Chi square = 0.773 Table 3 shows attributes as: (i) s: stroma is abundant myxoid cellular. (ii) d: dense, (iii) w: well circumscribed, (iv) e: enlarged terminal duct, (v) c: congested blood vessels, (vi) r: round oval cell, (vii) en: encapsulated, (viii) e:s—represents expected frequency for stroma, (ix) e:d—represents expected frequency for dense, (x) e:w—represents expected frequency for well circumscribed, (xi) e:e—represents expected frequency for enlarged terminal duct, (xii) e:c—represents expected frequency for congested blood vessels, (xiii) e:r—represents expected frequency for round oval cell, (xiv) e:en—represents expected frequency for encapsulated, (xv) cs—Chi square. (xvi) df—Degree of freedom. We found that chi square value for these two attribute is 0:773: This 0:773 > 0:67 ($x^2$0:995). So null hypothesis is rejected and we conclude that these two attributes are correlated.

Homogeneity with Proliferation: Homogeneity attribute is extracted from mammogram. We discretized Homogeneity into 3 bins. A histopathological attribute proliferation holds two attributes yes and no. We found that chi square value for these two attribute is 0.272. This 0.272 > 0.211 ($x^2$0:90). So null hypothesis is rejected and we conclude that these two attributes are correlated. Results of replacement attributes (Fig. 3 and Table 4).

Pruned tree: Pruning is a method in machine learning that decreases the dimension of decision trees by eradicating segments of the tree that offer small power to categorize instances. Pruning trim downs the intricacy of the final classifier, and hence improves analytical accuracy.

From Fig. 3 we observed that Standard Deviation is grouped into two bins as (<=54:99) and (>54:99). Further it has two nodes of homogeneity attributes. When homogeneity is (<=0:4) it has age as a node. Again age is grouped into two bins as 0 and 1. 0 indicates age <=30 and 1 indicates age >30. We observed that when age

Fig. 3 Pruned tree by J48 for replaced attributes

```
------------------

Std <= 54.995051
|   Homogeneity <= 0.472993
|   |   Age = 0: Bilateral_fibroadenoma (3.0/1.0)
|   |   Age = 1: Bilateral_FCC (4.0/1.0)
|   Homogeneity > 0.472993
|   |   Kurtosis <= 53.584671: FCC (7.0)
|   |   Kurtosis > 53.584671
|   |   |   Kurtosis <= 62.790357: FCC_with_epitheliosis_papillomatosis (4.0/1.0)
|   |   |   Kurtosis > 62.790357: FCC_with_mastitis (6.0)
Std > 54.995051
|   Kurtosis <= 39.365633
|   |   Kurtosis <= 19.408508
|   |   |   Homogeneity <= 0.511785: Fibroadenoma (14.0/1.0)
|   |   |   Homogeneity > 0.511785
|   |   |   |   Std <= 72.136465: Fibroadenoma (2.0)
|   |   |   |   Std > 72.136465: phyllodes (9.0)
|   |   Kurtosis > 19.408508: FCC_breast_with_nuclear_atypia (7.0/1.0)
|   Kurtosis > 39.365633
|   |   Energy <= 0.8625: Fibroadenoma (12.0)
|   |   Energy > 0.8625
|   |   |   Homogeneity <= 0.490776: FCC (18.0/1.0)
|   |   |   Homogeneity > 0.490776: Fibroadenoma (8.0/1.0)
```

Table 4 Chi square of homogeneity with proliferation

Homo/Proli	Yes	No	Ex. fr. yes	Ex. fr. no	Chi square
H1	2	2	3.351	3.648	0.272
H2	5	5	3.351	3.648	
H3	38	42	38.297	41.702	
DF = 2					

is 0 that is age <=30 bilateral Fibroadenoma type of tumor is identified. Figures (3:0 = 1:0) indicated that total number of instances of bilateral Fibroadenoma are 3 and 1 indicates that number of misplaced instance. At the same time when age >1 that is age >30 we observed Bilateral FCC type of tumor is diagnosed. In all we have 4 Bilateral FCC type of tumors out of that only 1 is misclassified. In this way can interpret this pruned tree obtained from J48 classifier.

Figure 4 represents training and testing accuracy of PINK classifier with replaced attributes.

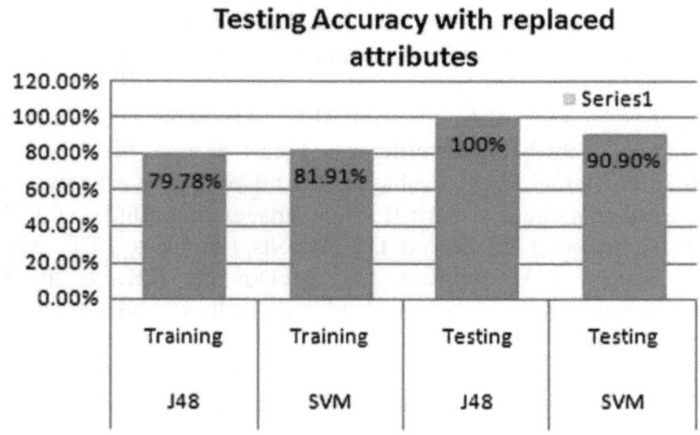

Fig. 4 Testing accuracy with replaced attributes

6 Conclusions

In this paper we discussed in detail about benign breast tumor classifier and PINK classifier. Attributes used for the classification belongs to either one or two categories such as clinical data or statistical data. But we have used unique combination of mammographic, Histopathological and clinical attributes for automatic classification of benign breast tumor. While testing with unseen data we achieved 81.81% with J48 and 90.90% with SVM classifier accuracy of benign breast tumor is obtained. PINK classifier focus to eliminate the role of histopathological attribute for the detection of type of benign breast tumor. We obtained testing accuracy by replacing histopathological attribute Proliferation and structure by mammogram attribute Energy and clinical attribute age. With replaced attributes we gain training accuracy for J48 as 79.78% and with SVM 81.91%. We obtained testing accuracy for J48 and SVM as 100% and 90.90% respectively.

Acknowledegment (Ethical Approval) The authors desires to express gratitude towards Dr. U. V. Takalkar [M.S. (Gen.Surg.) M.E.D.S FUICC (SWITZERLAND) FAIS, MSSAT (USA), Cancer, Chairman, Kodlikeri Memorial Hospital] for their backing and priceless guidance. Images were offered by Kodlikeri memorial Hospital, Aurangabad and Hedgewar Hospital, Aurangabad, Maharashtra, India.

References

1. Bhale, A., Joshi, M., Patil, Y.: Role of clinical attributes in automatic classification of mammograms. In: Emerging ICT for Bridging the Future-Proceedings of the 49th Annual Convention of the Computer Society of India (CSI), vol. 1, pp. 283–292. Springer (2015)
2. Buciu, I., Gacsadi, A.: Directional features for automatic tumor classification of mammogram images. Biomed. Signal Process. Control, **6**(4), 370–378 (2011)
3. Chang, R.-F., Wu, W.-J., Moon, W.K., Chou, Y.-H., Chen, D.-R.: Support vector machines for diagnosis of breast tumors on us images. Acad. Radiol. **10**(2), 189–197 (2003)

4. El-Naqa, I., Yang, Y., Wernick, M.N., Galatsanos, N.P., Nishikawa, R.M.: A support vector machine approach for detection of microcalcifications. IEEE Trans. Med. Imag. **21**(12), 1552–1563 (2002)
5. Friedenreich, C.M., Bryant, H.E., Alexander, F., Hugh, J., Danyluk, J., Page, D.L.: Risk factors for benign proliferative breast disease. Int. J. Epidemiol. **29**(4), 637–644 (2000)
6. Gurevich, I.B., Koryabkina, I.V.: Comparative analysis and classification of features for image models. Pattern Recogn. Image Anal. **16**(3), 265–297 (2006)
7. Hartmann, L.C., Schaid, D.J., Woods, J.E., Crotty, T.P., Myers, J.L., Arnold, P.G., Petty, P.M., Sellers, T.A., Johnson, J.L., McDonnell, S.K., et al.: Efficacy of bilateral prophylactic mastectomy in women with a family history of breast cancer. New Engl. J. Med. **340**(2), 77–84 (1999)
8. Hejazi, M.R., Ho, Y.-S.: Automated detection of tumors in mammograms using two segments for classification. In: Advances in Multimedia Information Processing-PCM 2005, pp. 910–921. Springer (2005)
9. Jasmine, J.S.L., Govardhan, A., Baskaran, S.: Classification of microcalcification in mammograms using nonsubsampled contourlet transform and neural network. Eur. J. Sci. Res. 531–539 (2010)
10. Kilic, N., Gorgel, P., Ucan, O.N., Sertbas, A.: Mammographic mass detection using wavelets as input to neural networks. J. Med. Syst. **34**(6), 1083–1088 (2010)
11. Lakshmi, N.V.S.S.R., Manoharan, C.: A novel hybrid ACO based classifier model for mammogram microcalcifications using combined feature set with SVM. Eur. J. Sci. Res. **53**(2), 239–248 (2011)
12. Mohan Kumar, S., Balakrishnan, G.: ME. Statistical features based classification of microcalcification in digital mammogram using stochastic neighbor embedding. Int. J. Adv. Inf. Sci. Technol. (IJAIST) ISSN 2319:2682
13. Mohanty, A.K., Senapati, M.R., Lenka, S.K.:. Retracted article: An improved data mining technique for classification and detection of breast cancer from mammograms. Neural Comput. Appl. **22**(1), 303–310 (2013)
14. Must, A., Phillips, S.M., Naumova, E.N., Blum, M., Harris, S., Dawson-Hughes, B., Rand, W.M.: Recall of early menstrual history and menarcheal body size: after 30 years, how well do women remember? Am. J. Epidemiol. **155**(7), 672–679 (2002)
15. Nagi, J., Abdul Kareem, S., Nagi, F., Ahmed, S.K.: Automated breast profile segmentation for ROI detection using digital mammograms. In: 2010 IEEE EMBS Conference on Biomedical Engineering and Sciences (IECBES), pp. 87–92. IEEE (2010)
16. Pradeep, N., Girisha, H., Sreepathi, B., Karibasappa, K.: Feature extraction of mammograms. Int. J. Bioinform. Res. **4**(1), 241 (2012)
17. Raghavan, G.: Mammogram analysis: Tumor classification. (2005)
18. Robsahm, T.E., Tretli, S.: Breast cancer incidence in food-vs non-food-producing areas in norway: possible beneficial effects of World War II. Br. J. Cancer **86**(3), 362–366 (2002)
19. Rosenberg, R.D., Yankaskas, B.C., Abraham, L.A., Sickles, E.A., Lehman, C.D., Geller, B. M., Carney, P.A., Kerlikowske, K., Buist, D.S.M., Weaver, D.L., et al.: Performance benchmarks for screening mammography 1. Radiology **241**(1), 55–66 (2006)
20. Shi, X., Da Cheng, H., Hu, L.: Mass detection and classification in breast ultrasound images using fuzzy SVM. In: JCIS (2006)
21. Thongkam, J., Xu, G., Zhang, Y., Huang, F.: Toward breast cancer survivability pre-diction models through improving training space. Expert Syst. Appl. **36**(10), 12200–12209 (2009)
22. Titus-Ernsto, L., Tosteson, A.N.A., Kasales, C., Weiss, J., Goodrich, M., Hatch, E.E., Carney, P.A.: Breast cancer risk factors in relation to breast density (United States). Cancer Causes Control **17**(10), 1281–1290 (2006)
23. Xu, S., Liu, H., Song, E.: Marker-controlled watershed for lesion segmentation in mammograms. J. Digital Imag. **24**(5), 754–763 (2011)
24. Zaïane, O.R., Antonie, M.-L., Coman, A.: Mammography classification by an association rule-based classifier. MDM/KDD 62–69 (2002)

Prevention of Conjunct Black Hole MANET on DSR Protocol by Cryptographic Method

Jyoti Prabha, Dinesh Goyal, Savita Shivani and Amit Sanghi

Abstract As in MANETs the infrastructure is not present they are more vulnerable for attack which encourages the attacker nodes to become part of the network. A lot of efficient protocols have been proposed in this field. All of these routing protocols depend only in the conviction and supportive environment. Conjunct Black hole attack is a network layer attack in which the malicious node makes use of routing protocol to pretend that it has the shortest path for the destination. In our research work we proposed a protocol, using RSA algorithm to eliminate Conjunct Black hole attack, even comparison with the DSR protocol is shows by the effects of the performance metrics in terms of delay and throughput of the network in the AD Hoc network environment.

Keywords Network · Ad hoc · Conjunct · Black hole attack D.S.R · Security · Simulation

J. Prabha (✉) · D. Goyal
Department of Computer Science & Engineering, Suresh Gyan Vihar University, Jaipur, Rajasthan, India
e-mail: jyotiprabhasingh@gmail.com

D. Goyal
e-mail: dr.dinesh.goyal@mygyanvihar.com

S. Shivani
Department of Computer Science & Engineering, Jaipur National University, Jaipur, Rajasthan, India
e-mail: savita.shivani@mygyanvihar.com; savita.shivani@gmail.com

A. Sanghi
Department of Computer Science & Engineering, Marudhar Engineering College, Bikaner, Rajasthan, India
e-mail: dr.amitsinghi@gmail.com

X.-S. Yang et al. (eds.), *Smart Trends in Systems, Security and Sustainability*, Lecture Notes in Networks and Systems 18,
https://doi.org/10.1007/978-981-10-6916-1_21

1 Introduction

Mobile ad hoc network (MANET) is a new emerging type of wireless communication which is made of collection of mobile nodes. In this kind of network, nodes communicate with each other without the help of any infrastructure. Each mobile node acts as both the router as well as the host. Ad hoc networks are either temporary or special purpose network where each device is capable to allocate or reallocate mobile nodes in any direction. Each device must forward traffic that is not intended to it.

There are various challenges which ad hoc network faces to maintain the network without any interrupt. In MANETs every single node can forward packet to the next hop and manage route traffic [1], so the network can work independently or can be connected to a large network if required.

MANET's open nature of communication makes it vulnerable for security threats [2], hence malicious nodes attempt to compromise the communication between the nodes. Thus security is a major challenge in Ad hoc network.

In case of attack, there is a possibility that such network may drop the privacy of communication where the malicious node acts as a normal node and causes the problem of eavesdropping [3]. This type of selective forwarding attack is generally known as the black hole attack. There are various security threats in MANETs which may include eavesdropping, interfacing, Worm hole, Black hole etc.

One of the routing protocols used in ad hoc network is DSR (Dynamic Source Routing Protocol). In MANETs the security of the DSR protocol is compromised by Black Hole attack which is performed in the network layer. In Black Hole attack, a malicious node uses routing protocol to advertise that it has the shortest path to the destination node and eventually drops all the packets [4]. The attacker node is also able to reply to the RREQ (route request) by creating the data packet and carry it to the sender. In the routing protocol, the Black hole node replies before the reception of replying the actual node and so the forged route is created [5].

2 Problem Investigation

2.1 DSR Protocol

Dynamic Source Routing Protocol, a well known reactive protocol, is an On-Demand Routing protocol. This efficient routing protocol is used in wireless ad hoc network whenever there is a requirement to send data packet. It commonly updates the available route and if it finds any new route available then sends the packet through. This efficient DSR protocol is composed in two parts:

2.1.1 Route Discovery and Route Maintenance

- The source node floods the RREQ (route request) packets to the neighbor nodes which contain various information related to the destination.
- The intermediate node either forwards the packet to its neighbor nodes till the destination or SEND the Route Reply (RREP) if it has a new route to destination. The sequence number of intermediate nodes is compared with the destination sequence number until RREQ reaches the destination.
- The destination node sends the route reply (RREP) packet back to the source node in the reverse path.

Route Maintenance

Route maintenance uses two mechanisms that is Route Error (RERP) packet to find whether the packet reaches to destination successfully and Acknowledgment(ACK) which sends back the ACK message to the source and if it finds any problem in the communication link it send the packet back. Finally, when the node receives the route error packet back it removes the hop from the route cache.

2.2 Black Hole Attack

The black hole attack is a network layer attack in MANETs which advertise that it has the shortest path to reach the destination, and it collects the entire packet from the source node and then drops it [6]. The black hole attack can be categorized into:

2.2.1 Single Black Hole

In Single black hole attack, only a single node acts as a fake node and collects all the packets from the source node and drops the packets to achieve the malicious task. If the S node sends request to the destination node D it sends the RREQ packet to the neighbor node and if it has the valid route to reach to destination then it sends the packets through it and if not it forwards the request to the next node until it reaches the destination.

2.2.2 Conjunct Black Hole

In the conjunct black hole attack, more than one node combined mutually and acts as fake node and they generally found in group of active malicious node [7, 8] .

3 Proposed Solution

The proposed solution is based on cryptographic approach hence we are using RSA algorithm which does the computation of the sequence number to eliminate the Black hole attack and for this we are considering the three large Prime number to calculate d and e value. The RREQ is considering as M. At the end side that is the sender side and then RREQ is send to the next node or the neighbor node. Now, if the neighbor node knows the key value it will be able to decrypt the RREQ and will able to generate RREP in response to be forwarded to sender.

The protocol proposed represents the Conjunct Black hole detection and it includes the following steps.

3.1 *Formation of Network*

It is the initial step which involve NS2 tool and initialize with 80 numbers of nodes, the communication between the nodes are done by the Random Waypoint Model and uses IEEE 802.11 Standers for communication the ration range is 230 m and simulation run for 50 s.

3.2 *Key Exchange for Authentication*

For key sharing we have used RSA key agreement algorithm in our protocol and uses various input such as D, S, Timer, Hop Count, and Threshold.

3.3 *Route Formation for Data Sharing*

Route discovery process chooses the best optimal route from the sender to receiver and follows two steps that is *authentication of valid node* and *Verification B.L.*

3.3.1 Authentication of Valid Node

In node authentication uses the two matrices such as time as well as hop count. When the S node send RREQ to the next node then neighbor node check and evaluates the number of Hop required reaching the D and then it reply it by sending RREP. If time matches between the RREQ and RREP that is TTL (Time to Live) then only the responder node is consider to authentic else it listed it in B.L and consider it as unauthentic node.

Another metric for distinguishing the true node and malicious node is H.C (Hop Count). The number of Hops required to reach the destination from the source is consider by the H HC, so if the hop count exceeds in limit the corresponding node is consider as the malicious and added to the black list HC.

3.3.2 Verification B.L

Black list contains the ID of malicious node that has been obtained from the previous attack history. Every node in MANET contains the BL (black list). If the S node request for the route from the neighbor node it verifies the black list and then forward the RREQ that is not present in black list when the node detects a black hole then it adds the ID of the corresponding node.

3.3.3 Threshold–Diff

After that RREP computes the threshold–diff and will be able to send the packet from source to destination. If the difference of sequence value is below to the

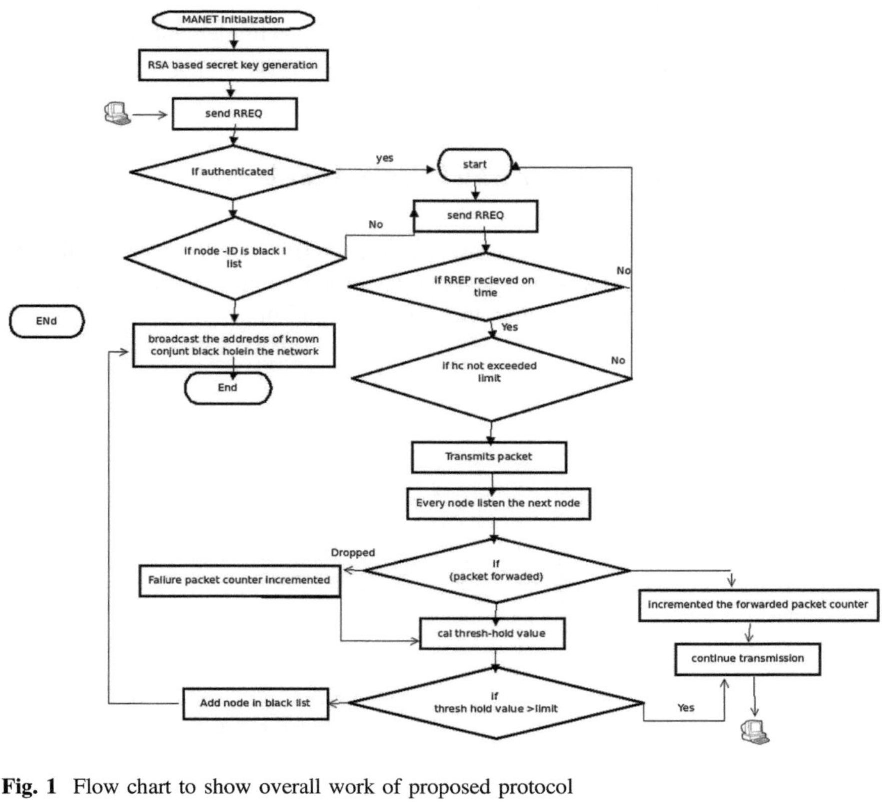

Fig. 1 Flow chart to show overall work of proposed protocol

threshold-value then the node is considered as the legitimate node and if the difference of sequence no is greater than threshold–value then it is consider as the false node as shown in the algorithm proposed and given below as well as the flowchart shown in Fig. 1.

3.4 Transmission of Packets from S to D

After the verification of the true node and conjunct black hole node the packet transmission process initialize, in this process every node follow to the neighbor node. The transmission of the packets increment the PC (Packet Counter) and unsuccessful transmission of packets (Fig. 2).

4 Result

This section illustrates the behavior and the result of the proposed protocol for detection and avoidance of Conjunct Black hole attack. RAS algorithm and DSR protocol is being used for comparison of various matrixes.

The existing algorithm and the proposed protocol are implemented in the NS2 simulation tool and even the results are comparing for various parameters.

4.1 Detection of Conjunct Black Hole Attack

Figure 3 shows the number of successful Conjunct Black hole detection for the existing DSR protocol and it is analyzed that the proposed protocol provides higher black hole detection probability than the existing algorithm.

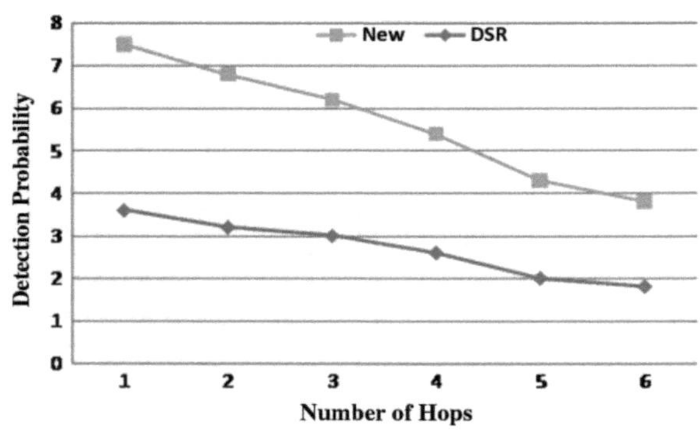

Fig. 2 Detection of conjunct black hole

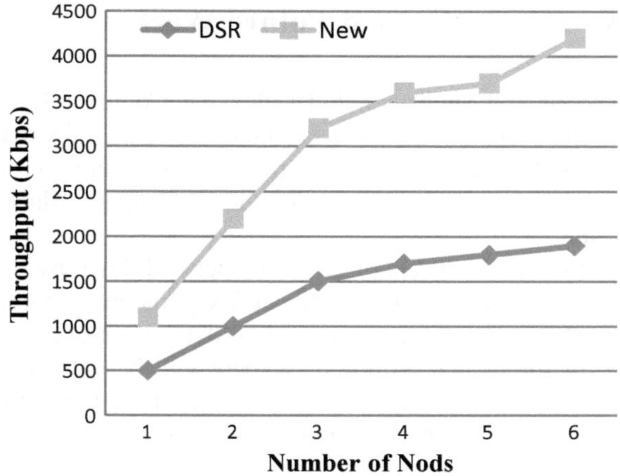

Fig. 3 Comparison of throughput for DSR and new protocol

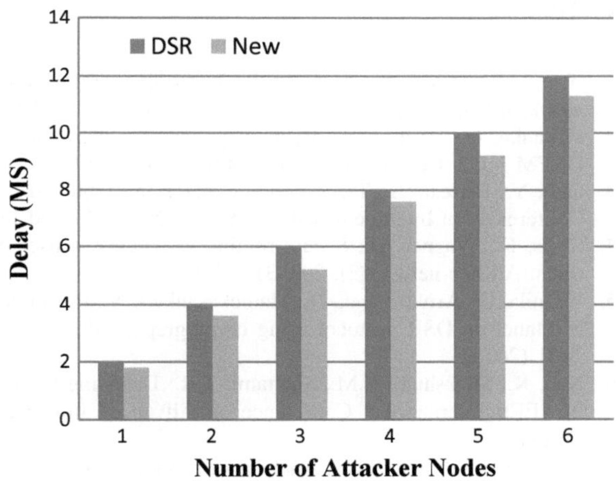

Fig. 4 Comparison of delay for DSR and new protocol

4.2 Throughput

The rate of received packet from the S to D to send the data packet from the source hub is given by throughput. The throughput is ascertained from the condition.

4.3 Delay (End to End)

Comparison of end to end delay with respect to number of attacker node where the proposed protocol provides minimal delay comparing with DSR (Fig. 4).

5 Conclusion and Future Work

The overall study concludes that the Conjunct Black Hole attack is one of the severe security threat attacks in the ad hoc network that occurs due to vulnerabilities of DSA and other routing protocol. Concept of proposed protocol has shown improve result after elimination of the Conjunct Black Hole attack in the simulation result.

Elimination takes place on the network layer by broadcasting the information of attacker node. So the existing DSR pootocol is compared with the proposed protocol for different parameters like delay throughput.

In future this proposed protocol will be extended for the detection of various other attack of the MANET, with more detection probability and will help in secure data transmission providing mode security.

References

1. Ameza, F., Assam, N., Beghdad, R.: Defending AODV routing protocol against the black hole attack. Int. J. Comp. Sci. Inf. Secur. **8**(2), 112–117 (2010)
2. Priyanka, G., Parmar, V., Rishi, R.: Manet: vulnerabilities, challenges, attacks, application. IJCEM Int. J. Comp. Eng. Manag. **11**(2011), 32–37
3. Guo, Y., Perreau, S.: Trace flooding attack in mobile ad hoc network. In: IEEE International Conference on Intelligent Sensors, Sensor Networks and Information (2007)
4. Chris, K., Wagner, D.: Secure routing in wireless sensor networks: attacks and countermeasures. Ad hoc netw. **1**(2), 293–315 (2003)
5. Vennila, G., Arivazhagan, D., Manickasankari, N.: Prevention of co-operative black hole attack in Manet on DSR protocol using cryptographic algorithm. Int. J. Eng. Technol. (IJET) **6**(5), 2401 (2014)
6. Nan, K., Shakshuki, E.M., Sheltami, T.R.: Detecting forged acknowledgements in MANETs. In: IEEE International Conference on Advanced Information Networking and Applications (2011)
7. Irshad, U., Rehman, S.Ur.: Analysis of black hole attack on MANETs using different MANET routing protocols (2010)
8. Vennila, G., Arivazhagan, D., Manickasankari, N.: Prevention of co-operativeblack hole attack in Manet on DSR protocol using cryptographic algorithm. Int. J. Eng. Technol. (IJET) **6**(5), 2401 (2014)

Enhance the Data Security in Cloud Computing by Text Steganography

Amit Sanghi, Sunita Chaudhary and Meenu Dave

Abstract Cloud computing is new age computing model. Which gives that how to use the resources even if don't own them. It gives the facility to share the resources over the internet, without building the necessary infrastructure. This is an on demand service. This model gives faster accessing of the resources and reduces management related problems. In this paper we are going to discuss the different cloud computing models with its advantages, characteristic and various security measures in the current scenario. At last we discussed a new level of security in cloud computing for data diffusion and storage on hybrid cloud in security concern. We are presenting new linguistic text steganography approach for increasing data security. In these techniques first we encrypt our secret data by using substitution method and then embed it in some random cover media for hiding the aspect of original text. We have proposed new method of Indian script encoding method and capital alphabet shape encoding which would be hard to decipher by inarticulate person. We are applying this techniques on private cloud server when we transfer our private data to public cloud so our data will also be secure in transmission and storage on public cloud.Time overhead of this techniques is very less and also it's not taking too much space as one character of random cover is use to hide one character of encrypted secret text.

Keywords Cloud computing · Resources · Utility · Security
Steganography · CASE · ISET

A. Sanghi (✉)
Department of Computer Science & Engineering, Marudhar Engineering College,
Bikaner, Rajasthan, India
e-mail: dr.amitsanghi@gmail.com

S. Chaudhary · M. Dave
Department of Computer Science & Information Technology, Jagannath University,
Jaipur, Rajasthan, India
e-mail: er.sunita03@gmail.com

M. Dave
e-mail: meenu.s.dave@gmail.com

X.-S. Yang et al. (eds.), *Smart Trends in Systems, Security and Sustainability*, Lecture Notes in Networks and Systems 18,
https://doi.org/10.1007/978-981-10-6916-1_22

1 Introduction

Cloud computing; make the resources as a utility. For example we all are using electricity but not making it, and its infrastructure. We just only pay the bill for its usage. Same thing is applied in cloud computing in which we only use resources without taking burden of maintain in house computing infrastructure. Cloud computing is a method which gives access to shared resources over the internet. Shared resources like: printers, storage and servers can be shared via internet between the candidates who are unaware to each other. This is an on demand service. This model gives faster accessing of the resources and reduces management related problems. There are five important characteristic of cloud model are available. Cloud computing can be deployed as a private, public, hybrid and community cloud [1].

In private cloud services are provided by the main center of Business Company and other internal users use it on demand and requirement. This service will be charged or not that depends upon the policies governed by the company. This model gives central control, security and central management. In public cloud model third party service providers provide the services over the internet and charge accordingly. The charges can be according to the hour or minute basis. Users should pay for the storage, speed and frequency. Hybrid cloud model is the combination of private and public cloud computing models. It provides advantages of both the models. The services which are very critical and sensitive to security can be handled by the organization and the services which are bulky and do not need much attention in terms of safety can be hired by the public service providers. The advantage of hybrid model is that one can get the advantage of the public infrastructure and have all the central control on critical data. Our main focus is on virtualization that is related to security measures. A community Cloud can be used by many organizations working for a same theme or belonging to the same market area but they must be in collaboration with each other forming a specific community in the market. The services of this model are only for the customers those are members of the participating organizations i.e part of the same community. Others cannot access or use [2].

The main features of a cloud as defined by NIST [3] are as follows:

On demand self service. On demand self service is the highest-flying characteristic of cloud computing model. On demand means whenever the users needed and requires the service, they can access it via internet without interruption to host. This service is controlled by using an online control panel. This service enables users to demand at run time. One can demand the resources and increase its frequency time to time without interruption of the hosting device. In this scheme charges are also subject to time i.e. these are not according to the frequency and speed of using resources [2].

Broad network access. Cloud is created for an organization or a company in which available resources are accessible within the company's firewall. These resources can be all the accessible resources like: tablet, pc, laptops, Smartphone's, printers etc. Broad network access means the resources are owned by a private company and accessible by any user at any location apart from the location of the resource [4].

Resource pooling. This characteristic give the capability that service provider can give services to the multiple users at the same time and even without giving them any clue that there is another one user is also using the same resources parallel. Resource pooling can be applied to the resources like: data storage services, processing services and bandwidth provided services. [3]

Rapid elasticity. When company feels increase in the computing services it can easily scale up and at the time of low demand company can scale down the computing services [5].

Measured service. Measurement of the services in the scheme is fully transparent at both levels either customer level or the service provider level [3].

2 Delivery Models

Various models are available that demonstrate that how cloud services will be provided to the customers. Basically three standard models [3] are there and these are:

SaaS (Software as a Service). These are the service of software's required by the user. In this the service provider gives the users SaaS application software's on demand. The Main thing is that there is no need to install these software's. User can own a SaaS license and can use directly [6].

PaaS (Platform as a Service). In this model various platforms are provided to the users. A user can use a variety of programming languages, testing tools and can develop his/her own applications without being worried for the maintenance of the platforms as it becomes the only responsibility of the service providers to preserve the platforms at their own maintaining cost. The users can access these platforms through a web client easily [2].

Infrastructure as Service (IaaS). The model includes hardware and software resources both. The users are provided with various computing resources on payment basis. The payment is on the basis of the use by the customer. Computing equipments and resources can be added or removed at any time on demand [2].

Customers can use any of the models described above as required or they can use more than one model in parallel also [7] (Fig. 1).

2.1 Advantages

There are many reasons due to one or many organizations of varying size prefer a Cloud. One of the main reason is there is no need to invest in purchasing new infrastructure, training people, purchasing licensed software because all these things

Fig. 1 Delivery Model [7]

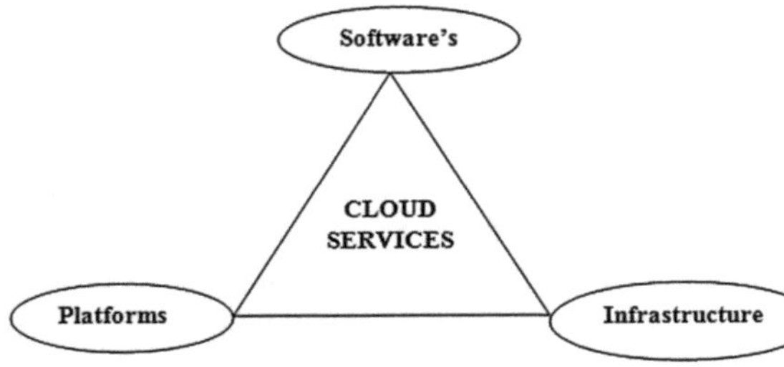

are provided by a cloud vendor. The cloud services are a quick, fast and remote operated for various IT solutions. Due to its scalability it is awesome in terms of resource utilization. There are many economical benefits as well as technical. Consumers can develop their own solution by using the vendor's infrastructure. Multi tenancy is very useful for small business and the can compete with big ones. QoS is high as responsibilities becomes of both consumer as well as the vendor. Cloud is available at anywhere anytime where there a computer network. In case of a failure or problem the backup is available at both side's customer and the service provider which is much secure.

3 Security Measures in Cloud Computing

Various security issues are concerned with a cloud. We can see two dimensions of the security in a cloud one as per the client side and other as per the service giving organization. In this security is providing in both these directions by the service provider but the main issue in this is that a service provider security design supports and integrate the clients existing security frame work. But in case of change of the security policy at either side a security issue will raise up. Most of the issues are:

Shared and virtual resources. This concern occurs when a physical infrastructure is shared by multiple customers [2].

Data privacy. If the data is centralized i.e. hosted at the data center of the service provider then there is an issue of privacy of the database of the customer. [2]

Multi tenancy. This occurs when the hosting mechanism and swap over of the data is based on contributed environment [2].

Competing share. Furthermore we can discuss many issues like in case of two parties those are competing with each other but taking services of a common vendor in shared mode and if there is a bug at the vender's implementation then the protocol of the isolation of processes will be violated and can result in stealing of data by either competing parties.

To increase security on cloud computing any of the following can be used.

Firewall. A firewall is a system that is used to protect a private network from the outside world. It helps in decreasing attacks on virtual servers. It is used to achieve isolation and it also helps in data segregation. It also prevents from the attacks like denial of service. A firewall allows different security policies for different networks.

Intruder detection and preclusion. It protects the operating system by creating a shield on the operating system which results in prevention from known and zero days' attacks or any new kind of attack being suspected on the virtual machine operating system or the applications of the virtual server can also be detected and s timely protection can be achieved. Basically the scheme is based on artificial intelligence that is why this method of protection can learn dynamically to the new vulnerabilities and can immediately react to them [8].

Integrity check. This method can monitor any kind of changes in the database as in the file system or the registry. It detects any kind of unexpected changes or malicious code that can change the whole meaning of the cloud computing. So integrity check software's must be implemented at Virtual machine level to ensure on demand or scheduled detection, extensive file property checking, directory level monitoring, flexible practical check and auditable reports [9].

Log monitoring. In log checking or monitoring, synthesis and analysis of all the logs of kernel and utilities for the security purpose is done. In this method protocols for efficient extraction of the data related to security is done. Analysis of the collected logs can be done through a standalone security system or a centralized logging server. Suspicious behavior can be detected by the log inspection software available on the cloud. This log checking mechanism must be implemented at the level of virtual machine. The phenomena are as same as the integrity check. This security scheme is highly beneficial as it protects at application level [10].

4 Steganography to Enhance Data Security in Cloud

Even though we are using firewall, log monitoring and other method there are still some limits while using the hybrid cloud computing, mainly about the security of the user's data. The users often worry about that the cloud computing provider could take use of their sensitive commercial data. So we applied the steganography techniques on our data to encrypt and hide in some cover media which is hard to figure it out using prying eyes.

We are proposing a new security layer in cloud computing for secure data storage and data transmission. In this layer we are using steganography approach for encryption of data which is fast and taking less memory space on hardware. We are using text media in steganography rather than image, audio, video as we know that cloud data are in large size so it would take long time to transfer any secret data hidden in image, audio, videos. It would also take too much space on public cloud.

We are proposing IASE (Indian Script Encoding Technique) and CASE (Capital alphabet Shape Encoding) based text steganography model for hiding our secret data in cover media [11]. In these methods to encrypt our plain text we used simple basic substitution encryption techniques by mapping of each byte of secret data with cover text by using the particular designed grouping mechanism. These grouping mechanism have two different tables. One for English language and another for Hindi language. Cover text is generated by random function which size would be Five time to our secret data. This substitution operation is very fast and this encryption can be happen in parallel as there is no dependency with previous result so this process is fast with less time overhead. After encryption techniques we will merge this enciphered text with our same random cover text according to key [11].

4.1 Algorithms for Hiding Secret Data

```
INPUT: A secrete message msg.
OUTPUT: final hidden message.
Algorithm for Message Encoding:
1. Read character one by one.
2. for each character:-
2.1 Check the read character is belonging to CASE group or
ISET group.
2.2 Find the character in the group table and convert into 8
bit binary value.
2.3 If character is compound statement (In case of Hindi
Language), means have Diacritics then we make two separate
binary stream one for character and another for diacritics or
next character.
2.3 Convert each 8 bit binary sequence into its ASCII
equivalent.
2.4 This new letter is now embedded into the randomly gener-
ated cover text according to the key.
3. Send this encrypted embedded and hidden message to public
cloud.
```

4.2 Experimental Results

This section shows the performance of proposed approaches with reference to time overhead and data storage space requirement on public cloud server.

Length Capacity of Cover Text. Length capacity is defined as how much secret data can be covered in given cover text. The capacity ratio is computed by dividing the amount of hidden bytes over the size of the cover text in bytes.

Table 1 Execution time and cover text size requires for hiding [11]

Text steganography techniques	Message text size (Bytes)	Data storage on Public cloud (Bytes)	Time overhead (ms)
CASE Text Steganography	600	1201	15-20
ISET Text Steganography	600	1210	15-20

$$\text{Capacity ratio} = (\text{amount of hidden bytes})/(\text{size of the cover text in bytes})$$

Assuming one character occupies one byte in memory, we have calculated the percentage capacity which is capacity ratio multiplied by 100.

We can calculate capacity ratio in percentage format. Let's assume we have n bytes of secret message so we require total n byte of cover text. So it requires total 2n + 1 bytes for storage and data transmission from private cloud to public cloud. We can define capacity ratio for our proposed approach (Table 1)

$$\text{Capacity ration} = n/(2n + 1)*100 = 50\% \text{ approx}$$

In our experimental result we can conclude that we require around double memory for store our encrypted data on public cloud. Time overhead of this techniques is very less and also it's not taking too much space as one character of random cover is use to hide one character of encrypted secret text.

5 Future Work

Cloud computing is new in market and it's not fully mature so there is still lot of work needs to do. After our current work we are claiming that security is the most important threat to both the users and the vendors of cloud computing. Researchers and IT security professionals are working on security issues associated with cloud computing in data transmission and storage media. Different models and tools have been proposed but still there is no proper techniques used in cloud for data security. We have proposed two steganography approaches in which we need double space memory for storage on public cloud. We are working on some other techniques in steganography in which we are trying to reduce memory space and execution for encryption and decryption [12].

6 Conclusion

Cloud computing is the latest skill in the area of computer science. There is endless list of benefits but the care about security while implementation and use is a prime factor. Virtualization cooperates highly in the cloud environment as it gives isolation in multi tenant situation but some suppositions impose a security hazard and they are not true in cloud environment [13]. To overcome from these threats ISET and CASE steganography methods gives reliable solutions.

References

1. Christodorescu, M., Sailer, R., Schales, D.L., Sgandurra, D., Zamboni, D.: Cloud Security is not (just) virtualization security. In: Proceedings of the Conference on Computer and Communication Security, ACM (2009)
2. Barde, A.S.: Cloud computing and its vision 2015!!. Int. J. Comp. Commun. Eng. **2**(4) 2013
3. Mell, P., Grance, T.: The NIST definition of cloud computing. US Department of Commerce, Jan 2011
4. Wei, J., Zhang, X., Amimons, G., Bala, V., Ning, P.: Managing security of virtual machine images in a cloud environment. In: Proceeding of the workshop on cloud computing security, ACM (2009)
5. Foster, I., Zhao, Y., Raicu, I., Lu, S.: Cloud computing and grid computing 360-degree compared. In: Grid Computing Environment Workshop (2008)
6. Armbrust, M., Fox, A., Griffith, R., Joseph, A., Katz, R., Konwinski, A., Lee, G., Patterson, D., Rabkin, A., Stoica, I., Zaharia, M.: Above the Clouds: a berkeley view of cloud computing, UC berkeley reliable adaptive distributed systems laboratory, 10 feb 2009
7. Ferriman, R.: The benefits of Cloud Computing, White Paper. Third-Brigade. White Paper: Cloud Computing Secsurity: Making Virtual Machines Cloud-Ready
8. Chadha, K., Bajpai, A.: Security aspects of cloud computing. Int. J. Comp. Appl. **40**(8), 43–47 (2012)
9. Kumar, A., Chadha, K., Asawa, K.: Framework for vulnerability reduction in real time intrusion detection and prevention systems using SOM based IDS with Net filter-Iptables IJCSIS. ISSN Int. J. Comp. Sci. Inf. **8**(4), 1947–5500 (2010)
10. Parte, S., Dehariya, N.: Cloud computing: issues regarding security, applications and mobile cloud computing. Int. J. Advanc. Res. Comp. Sci. Softw. Eng. **5**(3), 403–406 (2015)
11. Chaudhary, S., Dave, M., Sanghi, A.: An elucidation on steganography and cryptography. In: Second International Conference on Information and Communication Technology for Competitive Strategies, ACM, Udaipur, India (2016)
12. Raoa, R.V., Selvamanib, K.: Data security challenges and its solutions in cloud computing. In: Procedia of Science direct Computer Science International Conference on Computer, Communication and Convergence (ICCC), Vol. 48, pp. 204–209 (2015)
13. Khan, S.S., Tuteja, R.R.: Security in cloud computing using cryptographic algorithms. Int. J. Innov. Res. Comp. Commun. Eng. **3**(1) (2015)

A Hybrid Community Based Rough Set Feature Selection Technique in Android Malware Detection

Abhishek Bhattacharya and Radha Tamal Goswami

Abstract Feature selection is the process of grouping most significant set of features which reduces dimensionality and generates most analytical result. Choosing relevant attributes are a critical issue for competitive classifiers and for data reduction also. This work proposes a hybrid feature selection technique based on Rough Set Quick Reduct algorithm with Community Detection scheme. The proposed technique is applied in Android malware detection domain and compared with the performances of existing feature selectors. It produces highest average classification accuracy of 97.88% and average ROC values up to 0.987.

Keywords Feature selection · Android malware · Rough set
Community detection · Quick reduct

1 Introduction

Now Android smart phones are vulnerable to threats like activating malevolent services without user's knowledge, denial of services etc. Moreover, Android applications are easy targets for reverse engineering—an explicit characteristic of Java applications, which is often abused by malicious attackers. Unlike other mobile operating system, Android maintains openness and doesn't put much constraint on its users in downloading and uploading apps. Every Android app requires a set of permissions and these permissions are generally requested by applications

---

A. Bhattacharya (✉)
Institute of Engineering & Management, Kolkata, West Bengal, India
e-mail: abhishek.bhattacharya@iemcal.com

R. T. Goswami
Techno India College of Technology, Kolkata, West Bengal, India
e-mail: tamal.goswami@gmail.com

X.-S. Yang et al. (eds.), *Smart Trends in Systems, Security and Sustainability*, Lecture Notes in Networks and Systems 18,
https://doi.org/10.1007/978-981-10-6916-1_23

during installation on mobile devices. Permission vector of Android app may contain around 135 features. Some of the unnecessary permissions of an over privileged app may be leaked to malware apps. So it is quite feasible to identify malware based on the permission sets they require during installation time. But these huge data is extraordinary difficult because of dimensionality and it may slowdown learning process and learning efficiency also may be degraded. As a common technique for data mining, feature selection has been attracted much attention in recent times [1, 2]. So feature reduction techniques are highly required to reduce the dimensionally of data. Community detection stands for the process of grouping data according to certain similarity distances from weighted graph. It is one of the major tools in social network analysis, like viral marketing, sharing of information, sentiments, emotions etc. [3], but its implementation in the area of feature reduction in the context of malware detection is quite rare. In this paper, a hybrid community based rough set feature selection framework in Android malware detection is proposed where community detection and rough set Quick Reduct algorithm have been hybridized to select more prominent set of features. The reason behind hybridization is that, on many cases, a single technique may have certain limitations and always it cannot be treated as most efficient technique. Under such scenario, it is better to hybridize the technique with some other proven techniques to make the solution relatively better optimal. The rest of this paper is organized as follows. Few works related to this malware detection and feature reduction techniques are discussed in Sect. 2. Then proposed methodology on hybrid community based feature reduction technique has been discussed in Sect. 3. It is followed by results and discussion of the same in Sect. 4. We conclude in Sect. 5 discussing the future scope of this work.

2 Related Works

Different feature reduction techniques in permission based detection of Android malware have been proposed in recent times. In recent years, rough set theory (RST) has become a subject of great importance to researchers and has been applied to many data mining domains. It is possible to find a optimal subset of the attributes of a dataset with discrete attribute values using RST that are most informative and where the loss of information will be minimal. One of such works like [4], presents the application of rough set theory for feature selection in pattern recognition. Authors in [5] propose customer classification prediction system based on rough set to reduce complexity. In [6], a two phase rough set based model in multi criteria decision making has been discussed. Authors in [7] present feature selection and classification technique based on rough set approach. A new approach of feature selection based on rough set has been proposed in [8]. Authors in [9] have discussed the hybridization of rough set and statistical method (Principal Component Analysis) in feature reduction. Works in [10] discuss the hybridization of rough sets with fuzzy sets for feature reduction. Similarly, the hybridization of rough sets with meta heuristic algorithms is proposed in [11].

3 Proposed Methodology

This section deals with static framework of permission driven Android malware detection using community based feature reduction approach.

3.1 Dataset Preparation

To start with, permission sets from known malware and benign applications are extracted and permission vectors are created from a large number of app (3004 benign and 1363 malwares). Dataset 1 comprises of 504 Benign and 213 malware samples. Malwares are collected from Contagiodump Mobile Dump [12] and from different user agencies. Dataset 2 comprises of 2500 Benign and 1150 malware samples which are downloaded from Wang's repository [13] as research community is keen on using standardized dataset.

3.2 Feature Selection Technique

Huge data is extraordinary difficult because of the dimensionality. So feature reduction techniques are highly required to reduce the dimensionally of data. The basic reason behind feature reduction is that there are superfluous and unimportant attributes in datasets which may affect the performance of machine learning classifiers and may generate inaccurate results. Feature selection allows faster model construction by reducing the number of features and hence helps visualizing the trend of data. Feature reduction has been applied to permission based malware detection in order to improve its scalability, efficiency and classification accuracy [1]. One of the popular methods of feature selection is calculation of similarities between feature vectors. Various similarity measures have been used by researchers to find similarity between attributes. This work attempts to measure similarities between feature vectors through term level similarity. Correlation coefficient has been used in this work which is defined as follows.

Correlation Coefficient:
It can be used to measure the degree of similarity between feature vectors:

$$\text{Correlation Coefficient (r)} = \frac{n(\sum xy) - (\sum x).(\sum y)}{\sqrt{[n\sum x^2 - (\sum x)^2][n\sum y^2 - (\sum y)^2}}$$

The computed value lies in (0, 1). A feature is measured to be redundant if it is highly correlated with other features.

3.3 Feature Similarity Graph (FSG) Generation

The resulting similarity matrix can be considered as a weighted complete graph. This graphs are referred Feature Similarity Graph (FSG) which is an ordered pair G = (V, E) comprising a set of nodes V together with a set E of edges. Every permission vector is considered as node and between every pair of nodes a weighted edge is considered. Here weight denotes the similarity between two permission vectors. FSG contains edges between every pair of nodes, so if numbers of permission vectors be n, the number of edges is $\frac{n(n-1)}{2}$.

3.4 Community Detection

A complex network is likely to have community structure if the nodes of the network can be grouped into sets of nodes in such a way that every set of nodes are tightly connected internally. For non-overlapping community detection, this implies that the network splits naturally into sets of nodes with dense relations internally and sparser relations between sets. The idea is based on the principle that pair of nodes are more expected to be connected if they are both members of same community(s), and less likely to be connected if they do belong to different communities. Infomap [14] community detection technique is presented as follows.

3.4.1 Community Detection Through Infomap [14]

In proposed methodology, similarity of permission vectors is the basis of finding community within feature similarity graph (FSG). Infomap is a tool developed by [14]. Each community is also represented by a representative node (permission vector) and these community representatives are considered in the formation of reduct using Algorithm 1. Number of detected communities detected by Infomap for Dataset1 and Dataset 2 are 1 and 3 respectively.

Algorithm 1: Intermediate Reduct formation from Feature Selection Graph through Community Detection using Infomap

Input: *Feature Selection Graph G (V, E, w) created using term level similarity where v represents the set of vertices, E represents the set of edges and w represents the set of similarity values assigned to different edges.*
Output: *Intermediate Reduct RED*

```
Step 1: Initialize community representatives set R and set of detected communities (E)
        R ← { }                          // start with an empty set
        E ← { }
Step 2: Apply community detection through Infomap
        E ← E ∪ {e1, e2, e3,..., en }    // set of detected communities by Infomap
Step 3: Compute feature representative set
            For i=1...n                  // for each community in E
                  R ← R ∪ { ri }         // ri is the representative of community ei as
                                         // computed by Infomap
            End For
Step 4: For i = 1...n                    // for each community representatives
            For j = 1...p                // for each attribute in community representative ri
                                         // Attribute [j]represents jth attribute of community
                                         // representative ri
                If Attribute[j] =1 for some ri
                      Add that attribute k to RED
                End If
            End For
        End For
Step 5: Return RED
```

3.5 Threshold Feature Similarity Graph (FSG) Generation

As the number of communities identified by Infomap are significantly low in case of FSG generated based on Correlation coefficient, the threshold version of Feature Similarity Graph (FSG) have been considered. Threshold is computed by taking the average weight of all the edges in the graph and the edges whose weights are less than threshold weight are removed. So Threshold FSG contains less light weight edges than the original FSG. After creating threshold FSG, Algorithms 1 have been applied on that threshold FSG to generate communities. The number of communities detected by Infomap on threshold FSG generated based on mean Correlation coefficient for both Dataset 1 and Dataset 2 are 6 and 25 respectively. Community representatives are finally considered in the formation of intermediate reduct.

3.6 Hybridization Using Supervised Rough Set Quick Reduct (SRSQR) Algorithm

Feature selection methods are generally classified to three types as Filter, Wrapper and Hybrid approaches. The major difficulty of filter method is that the interface of features between themselves is not considered. The wrapper method uses a given learning algorithm to assess the feature sets. The disadvantages of previous two methods can be eradicated by hybrid approaches. Hybrid models work mainly in two stages. In first phase, any filter method is applied to remove irrelevant features. In the second phase, a wrapper method is used on filtered dataset and better classification accuracy is obtained. In hybrid models, classification result is generally same or better than that of filter/wrapper methods [15]. Rough Set Theory is used as a tool to find out the data dependencies and to decrease the number of attributes contained in a dataset [16, 17]. Like neural network and fuzzy sets, rough set theory has been also used in hybridization with conventional techniques. Some illustrious hybridization of rough sets with other techniques are: Rough sets with fuzzy sets, Rough set with neural networks and Rough set with meta heuristic algorithms [18]. In this work, Infomap based community detection technique has been hybridized with supervised rough set Quick Reduct (SRSQR) algorithm. In first phase, Infomap based community detection technique has been applied on datasets to generate intermediate reducts and in the second phase, supervised rough set Quick Reduct algorithm (SRSQR) has been executed on those intermediate reducts to produce more concrete reducts. Finally machine learning classifiers have been applied on those reducts to generate classification results. In Quick Reduct algorithm, the reduction of features is determined by comparing equivalence relations which are generated by sets of attributes. SRSQR tries to calculate a minimal reduct without exhaustively generating all possible set of subsets. As a result, it does not always guarantee to find a minimal reduct, actually it generates nearly minimal reduct. Table 1 shows the number of features present in reducts generated by proposed method and SRSQR for both Dataset 1 and Dataset 2.

Table 1 Reduct details of community detection using infomap

Datasets	Objects	Features	Reduct size (no. of features)	
			SRSQR	Proposed method
Dataset 1	717	82	38	28
Dataset 2	3650	88	37	13

4 Result and Discussions

One of the effective features of machine learning algorithms is that they improve their ability to discriminate normal behavior from anomalous behavior with experience. The performances of proposed hybrid rough set community based feature selection approach have been compared with six existing feature selection tools of Weka toolkit [19] which are used frequently in different literatures. For each existing feature selectors (Pearson coefficient, Information Gain, Gain Ratio, Chi Square, One R and Relief) top ten ranked features have been considered. The performance of Supervised Rough set Quick Reduct algorithm (SRSQR) is also compared with proposed hybrid method. Experiments carried out to evaluate the effectiveness of proposed method with eight machine learning classifiers (Bayesnet, Naïve Bayes, SMO, Decision Table, Random Tree, Random Forest, J48 and Multilayer Perceptron). Figures 1–4 depict the results of experiments carried out to evaluate the effectiveness of hybrid community based rough set feature reduct with eight machine learning classifiers. Figure 1 reveals that highest average TPR % is produced by proposed method in Dataset 1(97.13%) and in Dataset 2 (99.28%) which is greater than that of those six conventional feature rankers for both Dataset 1 and Dataset 2.

Proposed hybrid method generates highest average F1 score (0.916) in Dataset 1, whereas in Dataset 2, it produces highest average F1 score (0.985) which is presented in Fig. 2. It clearly shows that proposed hybrid feature selection method outperforms conventional feature selectors. Figure 3 depicts that in Dataset 1, highest average ROC is 0.869 through proposed hybrid method and similarly highest generated average ROC (0.987) in Dataset 2. The comparison of average Accuracy percentages with Dataset 1 and Dataset 2 is demonstrated in Fig. 4. It shows that proposed hybrid method obtains highest average Accuracy % (87.80%) with Dataset 1 and (97.88%) with Dataset 2. It clearly shows the effectiveness of proposed feature selection strategy over conventional feature rankers.

Fig. 1 Average TPR % of different reducts for Dataset 1 and Dataset 2

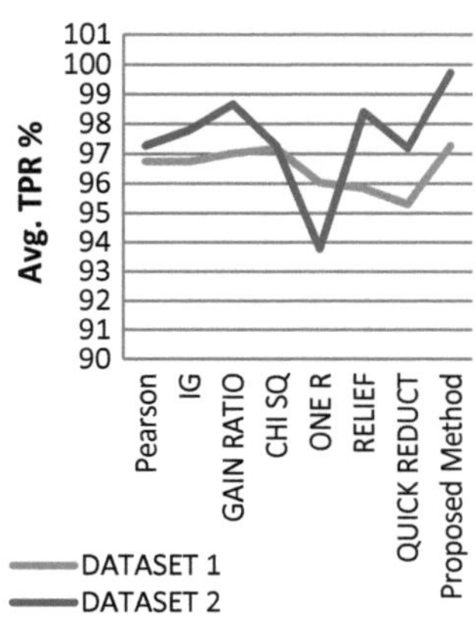

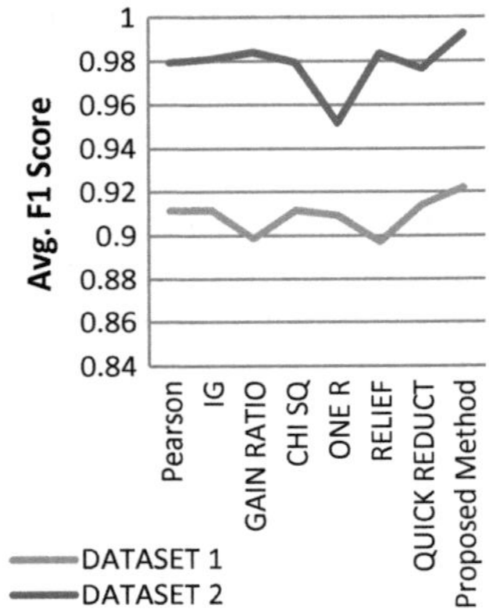

Fig. 2 Average F1 score of different reducts for Dataset 1 and Dataset 2

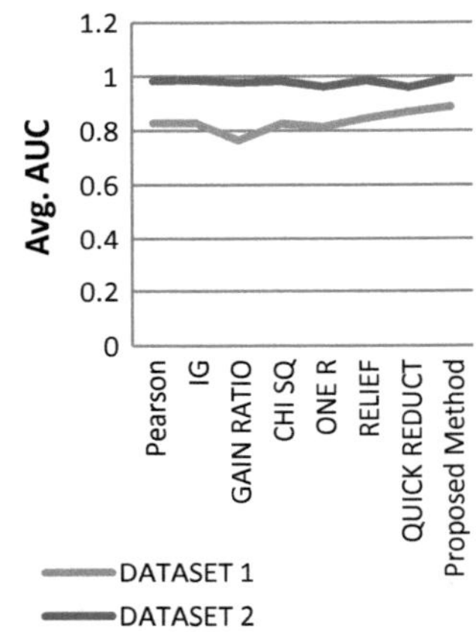

Fig. 3 Average ROC values of different reducts for Dataset 1 and Dataset 2

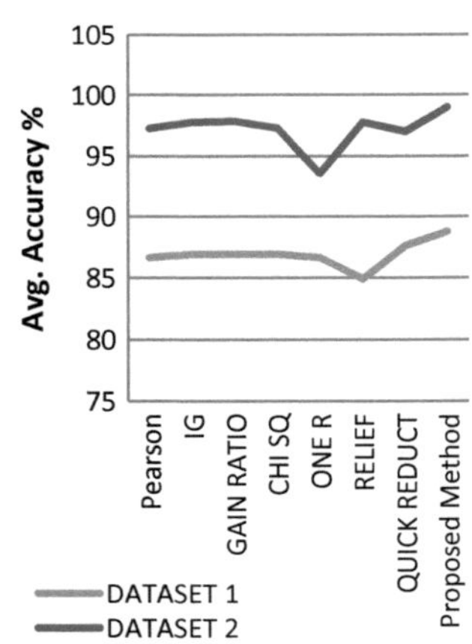

Fig. 4 Average accuracy % of different reducts for Dataset 1 and Dataset 2

5 Conclusion and Future Scope

In this work, a hybrid technique for permission based detection of Android malware through community based rough set feature selection methods has been proposed. Better classification performances have been yielded over existing feature rankers like IG, Gain Ratio, Pearson's Correlation coefficient, OneR, Chi Square and Relief for most of the machine learning classifiers. Therefore, the main contribution of this work is to exhibit that it is possible to implement hybrid community based rough set feature selection technique for filtering out malware in Android. From the experimental results, it can be concluded the proposed approach is quite comparable with the well-known feature selection algorithms. Even if a larger data set is applied, the feature selection techniques developed in this work can be used for feature selection to make the classification more efficient. In future, rough set theory hybridized with meta heuristic algorithms can be projected to design more efficient feature selection methods in permission based Android malware detection.

References

1. Hassanien, A.E., Tolba, M., Azar, A.T.: Advanced machine learning technologies and applications, Communications in Computer and Information Science, vol. 488, Springer, GmbH, Berlin/Heidelberg (2014). ISBN: 978-3-319-13460-4
2. Hu, Q.H., Yu, D.R., Xie, Z.X.: Information-preserving hybrid data reduction based on fuzzy-rough techniques. Pattern Recognit. Lett. **27**(5), 414–423 (2006)
3. Baumgarten, M.M.D., Mulvenna, M.D., Rooney, N., Reid, J.: Keyword-based sentiment mining using twitter. Int. J. Ambient Comput. Intell. (IJACI) **5**(2), 56–69 (2013)
4. Swiniarski, R.W., Skowron, A.: Rough set methods in feature selection and recognition. Pattern Recogn. Lett. **24**, 833–849 (2003)
5. Li, J., Wang, X., Xu, S.: Prediction of customer classification based on rough set theory. Procedia Eng. **7**, 366–370 (2010)
6. Chakhar, S., Saad, I.: Dominance-based rough set approach for groups in multicriteria classification problems. Decis. Support Syst. **54**, 372–380 (2012)
7. Kadzinski, M., Greco, S., Slowinski, R.: Robust ordinal regression for dominance-based rough set approach to multiple criteria sorting. Inf. Sci. **83**, 211–228 (2014)
8. Zhang, N., Yao, J.T.: A rough sets based approach to feature selection. In: Proceedings of the International Conference of the North American Fuzzy Information Processing Society, pp. 434–439 (2004)
9. Zhong, N., Dong, J.: Using rough sets with heuristics for feature selection. J. Intell. Inf. Syst. **16**, 199–214 (2001)
10. Meher, S.K.: Explicit rough-fuzzy pattern classification model. Pattern Recogn. Lett. **36**, 54–61 (2014)
11. Inbarani, H.H., Azar, T.A., Jothi, G.: Supervised hybrid feature selection based on PSO and rough sets for medical diagnosis. Comput. Methods Programs Biomed. **113**, 175–185 (2014)
12. Contagiodump mobile dump. http://contagiodump.blogspot.in/ (2016) Accessed October 31 2016
13. Wei Wang's Home Page. http://infosec.bjtu.edu.cn/wangwei/?page_id=85 (2015). Accessed 31 October 2015
14. Infomap Project. http://www.mapequation.org

15. Kishor, D.R., Venkateswarlu, N.B.: Novel hybridization of expectation-maximization and K-means algorithms for better clustering performance. Int. J. Ambient Comput. Intell. (IJACI) **7**(2), 47–74 (2016)
16. Pawlak, Z.: Rough Sets: Theoretical Aspects of Reasoning about Data. Kluwer (1991)
17. Pawlak, Z.: Rough set approach to knowledge-based decision support. Eur. J. Oper. Res. **99**, 48–57 (1997)
18. Dorigo, M. Stutzle, T.: Ant colony optimization. A Bradford Book (2004)
19. Weka Toolkit. http://www.cs.waikato.ac.nz/ml/weka/ (2016) Accessed 2016

Natural Language Interface for Ontology in Agriculture Domain

Nidhi Malik, Aditi Sharan and Jaya Shrivastav

Abstract Ontologies have recently become hot topic in research community. Ontologies have numerous applications and they are being used by researchers in almost all fields. They are primarily used for information retrieval, information extraction, knowledge representation and knowledge management purposes. This paper has two objectives. First is design and development of an ontology based on fertilizers in agriculture domain and second is to create a natural language interface for that ontology. Development of an Ontology requires both manual and expert efforts, so it requires significant amount of time as well. One of the goals behind designing and developing this ontology is to make it useful in real time scenario. Integration with other ontologies from the same domain such as soil or crop will enhance its real time usage. Information from the ontology can be fetched by making a natural language interface for it. It will accept user's questions in natural language and will generate corresponding SPARQL query.

Keywords Ontology · Knowledge · Natural language interface · SPARQL

1 Introduction

An ontology formally defines a conceptual representation of concepts and their relationships within a specific domain. Ontologies play an essential role in the semantic web by enabling knowledge sharing and exchange [1].

Efficient and accurate natural language interfaces have been the need of the hour. Though work in this direction had started very early in the 1960s but there are many issues which are still unsolved and not tackled properly [2]. Some of the key issues are:

N. Malik (✉) · A. Sharan
JawaharLal Nehru University, New Delhi, India
e-mail: nidhimalik14@gmail.com

J. Shrivastav
Indian Institute of Technology, Delhi, New Delhi, India

X.-S. Yang et al. (eds.), *Smart Trends in Systems, Security and Sustainability*, Lecture Notes in Networks and Systems 18,
https://doi.org/10.1007/978-981-10-6916-1_24

1. From linguistics point of view, natural language is highly ambiguous in nature. There have been continuous efforts to build systems which are capable of handling natural language queries and at the same time gives good precision and recall [1].
2. Knowledge representation is very challenging in case of natural language because it is not possible to generalize the knowledge representation mechanism. Different systems have different representations and it is extremely difficult to map them into one specific type [2]. Every system works on its own vocabulary/schema so it becomes difficult to find out correct mappings which can be applied universally and to a significant number of systems.
3. Systems which have their own knowledge base are developed with the aim of targeting specific needs. They are not general in nature and so their specific working cannot be adapted for general systems [3].

Most of us are using internet today. Everybody is not technically sound to write a query using some query language. Even if anybody is from technical background, it is not necessary that the person would be familiar with a language that can be used to query from an ontology. Farmers in particular are not technical persons. They will ask their questions in natural language and it has to be converted to a query automatically. So, it becomes tedious to manually translate each question. Natural language interface will facilitate this task by providing the users convenience of getting the desired information from the ontology. It will further expand the usability of ontologies and semantic web [2].

2 Related Work

Work in developing natural language interfaces has been done since 1970s. LUNAR was the first effort in this direction [1]. There have been systems developed based on two approaches: one being the full natural language interface which does not depend on any terminologies and operates in a fully natural language way. The other approach is restricted natural language interface. Systems based on this approach have to learn some specific language for querying. Each type of approach has its own advantages and disadvantages. We intend to go for the full natural language interface as a part of this work. PANTO [4] is one natural language interface for translation of natural language queries to SPARQL and is based on triplets based model for intermediate representation. There are various question answering systems available for ontologies. There are various open source solutions provided by the open source community. Quepy [5] is one such system which generate queries to be ran in the DBpedia database or the Freebase. [6] also automatically generates SPARQL queries for the given questions. It first obtains a basic interpretation of the user NL query, then interacts with the user, asking for positive and negative examples (i.e. elements which are or are not in the list of expected answers), in order to refine the initial interpretation by performing a learning algorithm. AutoSPARQL [6] follows the supervised machine learning approach for generation of SPARQL query. Positive examples will contain resources which will be present in the SPARQL query's result set and negative examples will indicate the resources which should not be in the SPARQL query's result set. [6] have also illustrated question answering process on an ontology of cinema domain. They have grouped queries into clusters where each cluster will comprise of different queries asking for the same information. Entailment based engine will then infer semantics between a new query asked by user and the clusters in the database. AquaLog [7] is a portable question-answering system which accepts queries in the form of natural language and an ontology as input, it returns answers drawn from the available semantic markup. It uses the GATE NLP platform, string metrics algorithms, WordNet and novel ontology-based similarity services for relations and classes to make sense of user queries with respect to the target knowledge base. [8] have presented a tool QuestIO for querying ontologies using unconstrained language-based queries. QuestIO requires no training for the users and it has a very simple interface. It does not require any customization and can be used with any type of knowledge base or ontology. It converts queries presented in the form of natural language to formal ontology languages. It supports queries of any length and there is no customization overhead. [9] have also discussed an approach that involves deep linguistic analysis over linked data for the purpose of question answering.

3 Proposed Architecture

The figure below shows the architecture diagram of the overall system. It will take as input question in natural language and SPARQL query will be returned as output. Template based approach is used for different classes of questions and it works well for the questions used by us. Questions were selected from various text books and three classes of questions are selected namely what/which/list/name type of questions. The overall process comprises of selecting the appropriate question template on the basis of question entered by the user, template matching and finally answer retrieval. Though the architecture is generalized, but we are not targeting each and every type of question. Attempts have been made to cover most of the general questions about fertilizers which are/can be asked normally (Fig. 1).

Lexicon: It consists of the terms that appear in the ontology such as names of the classes, subclasses, different types of properties, instances of the ontology. It is created to fasten up the overall process. It also enables the system to clearly distinguish the instances of the ontology as they are stored with their corresponding class names and properties if any. It removes the chances of any mismatches or ambiguity between the instances.

Data: We have populated the ontology with information about the fertilizers. The information has been collected from different authentic sources such as textbooks, journals of agriculture field, various journals in this field etc. Much efforts

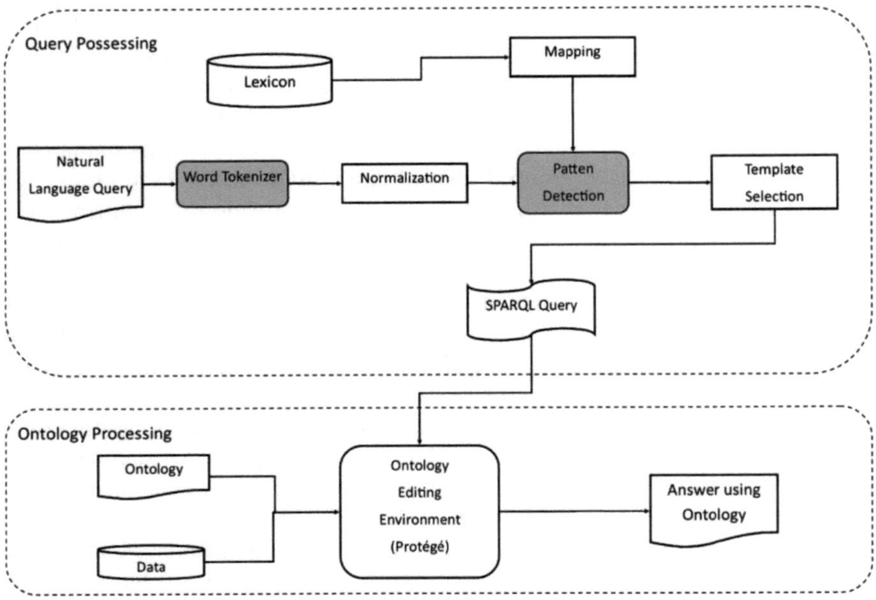

Fig. 1 Proposed architecture

have been put into correctly identify the type of information to be included in the ontology. Classes, concepts, properties, types of restrictions on different values were all decided after much discussion with domain experts.

Questions: We have gone through various sites/portals which are providing facilities to users/farmers to ask about their problems. Questions have been collected from textbooks and people from agriculture domain. Experts of the domain are also involved in validating the questions to be used for the overall process. As of now it is not possible to target every type of question. We are mainly targeting the 'wh' class of questions such as questions which have what/which/where/when/list/name keywords in them. Based on these questions, patterns have been identified. These patterns are then mapped and matched in order to generate SPARQL query. SPARQL has been used as the query language for ontology. The reason being the reasoning and inferencing capabilities providedby the language. Some of the questions are given in Table 1. More and more questions can be included in the question patterns database. It will become easy and fast to process new queries entered by users if we already have large number of patterns available with us.

Templates: After identifying the class of question and all the preprocessing activities, the raw query will be matched with the different templates available. We have created various templates for classes of questions such as what/which/when/list/name. Normalization plays an important part in identifying the correct template for conversion to query.

Table 1 Some examples of questions

	Competency question	Generalized questions
1.	What is the nitrogen content in Amide fertilizers?	What is the specific content in various specific various fertilizers?
2.	What is the application method of fertilizer?	What is the application method for specific manganese fertilizer?
3.	What should be the Nitrogen content Nitrogenous fertilizers?	What should be the specific content in specific in fertilizers?
4.	What are different types of fertilizers available for acidic soil?	What are different types of for specific soil?
5.	Which is the most concentrated fertilizer used for nutrient supply?	Which is the most concentrated fertilizer used for specific supply?
6.	Which are common Bio Fertilizers?	What are common specific fertilizers?
7.	List some of the phosphorous fertilizers alkaline soil?	List some of the specific fertilizers for specific type of soil
8.	Which chemical fertilizers can be mixed physically and used as mixed fertilizers?	Which fertilizers can be mixed specifically and used as mixed fertilizers?

1. Natural language query
   What should be the Nitrogen content in Nitrogenous fertilizers?
2. Word tokenizer
   ['What', 'should', 'be', 'the', 'Nitrogen', 'content', 'in', 'Nitrogenous', 'fertilizers', '?']
3. POS tagging
   [('What', 'WP'), ('should', 'MD'), ('be', 'VB'), ('the', 'DT'), ('Nitrogen', 'NNP'), ('content', 'NN'), ('in', 'IN'), ('Nitrogenous', 'NNP'), ('fertilizers', 'NNS')]
4. Normalization
   [('What', 'WP'), ('.', '.'), ('Nitrogen', 'NNP'), ('content', 'NN'), ('.', '.'), ('Nitrogenous', 'NNP'), ('fertilizers', 'NNS'), ('type', 'IN')]
5. Template selection
   There are different templates created for different types of questions. Once the preprocessing is done and question is normalized then it is matched with the question templates. The template matched will be then converted into the SPARQL query.
   In this case template which is for complex questions of 'what' type and identified by 'IN" will be used.
6. SPARQL query

```
SELECT ?NitrogenousFertilizers ?NContent
WHERE { NitrogenousFertilizers fer:hasPercentN ?NContent}
```

Fig. 2 Running example of question to query process

The reason for choosing competency questions is that the scope of ontology will be identified by clearly defined competency questions. The knowledge base formed in the form of ontology should be able to give answers to these questions. The ontology used is also created by us as part of this work (Fig. 2).

Experimental Setup: Programming language Python has been used to carry out all the experiments. It provides all the programming functionalities and is very simple to use. The free availability and open source access makes it highly preferred language. It provides rich support for carrying out natural language tasks.

1. Programming language Python provides very rich support for handling natural language in the form of NLTK. It provides support for many natural language processing tasks such as parsing, tokenization, stemming, classification, semantic reasoning etc. It has very simple interface which provides several libraries, lexical resources and corpora for handling natural language.
2. The input to the system will be the natural language question entered by the user.
3. The whole process will begin with the natural language query input by the user. That question will be tokenized into words by making use of the tokenizer library of NLTK. Different senses of the words that are used in the input query will be identified by the Part Of Speech tagging. After this, the query will be normalized so as to remove the unnecessary portions. Type of the question will also be identified.
4. There are different templates made for different types of questions. After identifying the type of question, template matching will be performed from the collection of templates.
5. Then the SPARQL query will be generated. The overall process is depicted in Table 1. Further, two more natural language questions and their corresponding SPARQL queries are shown in Fig. 3.

Ontology Used: Recently, many ontologies have been developed by researchers in most of the fields. Ontologies have proven to be an optimum way of expressing knowledge [10]. Though there are other also available for knowledge representation but ontologies are suitable for fulfilling the general needs of any arbitrary user as they are strong in adding semantics. There are ontologies available in the field of agriculture as well. We chose to develop an ontology for fertilizers as we did not come across one during literature survey. It was time consuming to gather

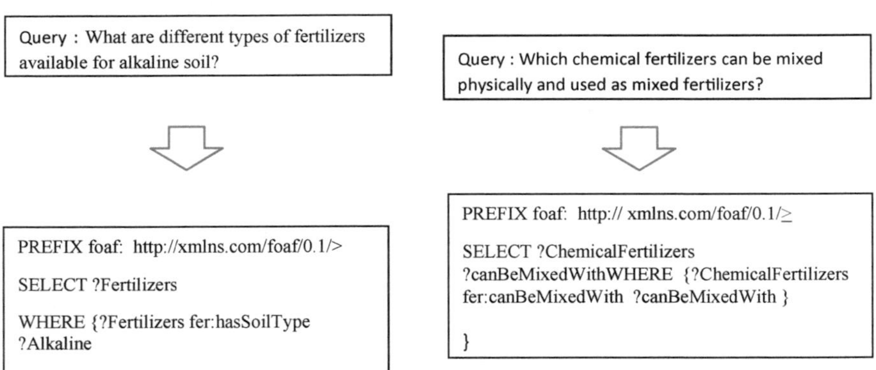

Fig. 3 Examples showing generation of SPARQLqueries for two more questions

information which is suitable to be included in the ontology. Therefore, we decided on the properties of the classes and restrictions to be applied in consultation with domain experts. Competency questions were identified after much brainstorming since these derive the overall development process. Since, we did not find any fertilizers taxonomy which can be directly mapped to our ontology, we had to structure it on our own by using a mix of knowledge available in Agriculture books, journals, magazines and expert advice. Only static information is included in our ontology as it will not be changed frequently. There are a number of methodologies available as discussed in. Every methodology has certain pros and cons. We have used the design methodology described in [11]. The reason behind using this methodology is the high degree of formalism that it provides. The objectives for development of this ontology are to represent knowledge about fertilizer domain such as type of fertilizers, nutrient contents, equivalent acidity, crops benefitted, preferred soil type, time of application, etc. and answering some general questions on fertilizers using SPARQL, rules and reasoning. As have developed the ontology from scratch, information has been incorporated carefully in the ontology in such a way that it is possible to answer basic questions such as what are the different categories of fertilizers, what should be their properties and restrictions etc. All this information has been identified and included after much discussion. It will be meaningful only if we are able to capture all such type of knowledge into the ontology. For information gathering purpose, we have referred standard books and information provided by Indian council for agriculture research. We will not go into depth of ontology development process. We will focus here on the main points only. Defining properties with more details is also an important part of the ontology building process. Protégé provides several ways to define object properties:

Inverse property: This represents bidirectional relationship as shown in Fig. 4

Symmetric property: This property defines relationships that are symmetric in nature. For example, rock phosphate can be mixed with basic slag implies that basic slag can be mixed with rock phosphate as shown in Fig. 5

In both the examples, the blue lines show the inferred relationships. Figure 4 below shows the ontology using Onto Graf in Protégé. Utmost care has been taken

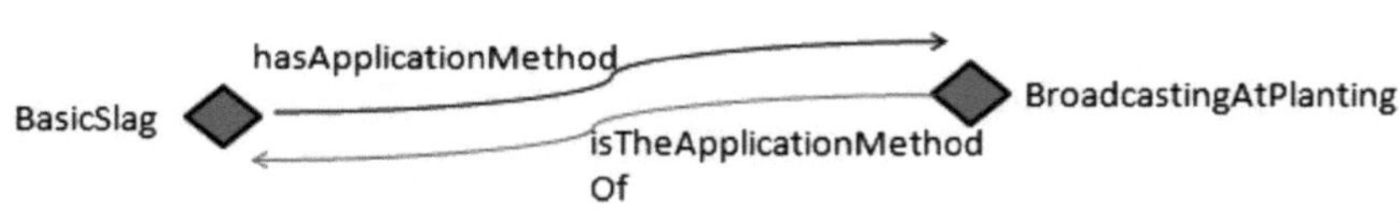

Fig. 4 Property example of Instance BasicSlag

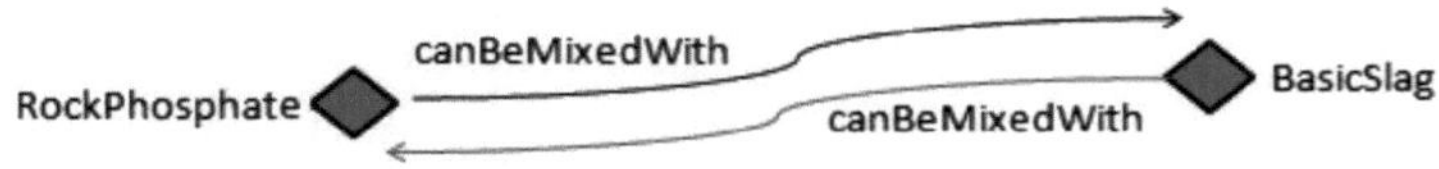

Fig. 5 Relationship between RockPhosphate and BasicSlag

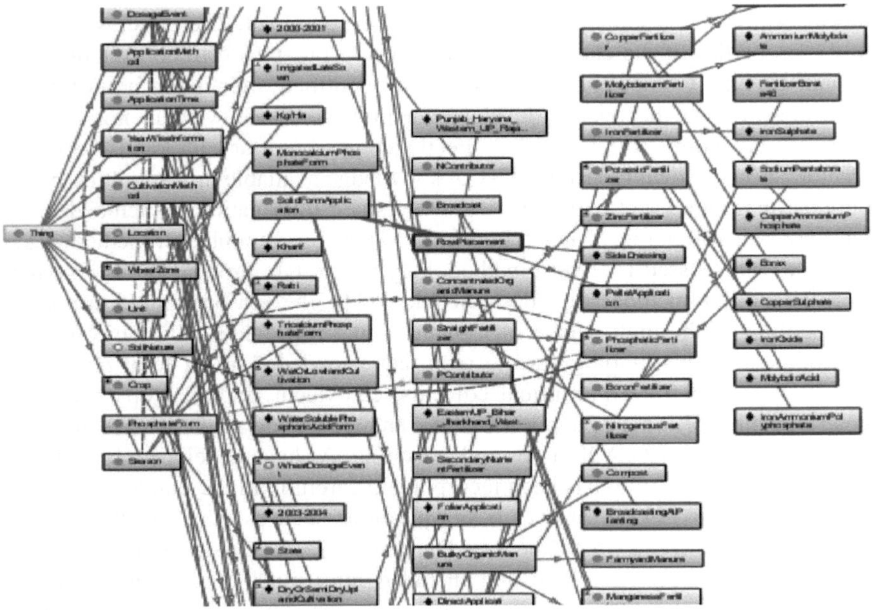

Fig. 6 Ontology view using OntoGraf

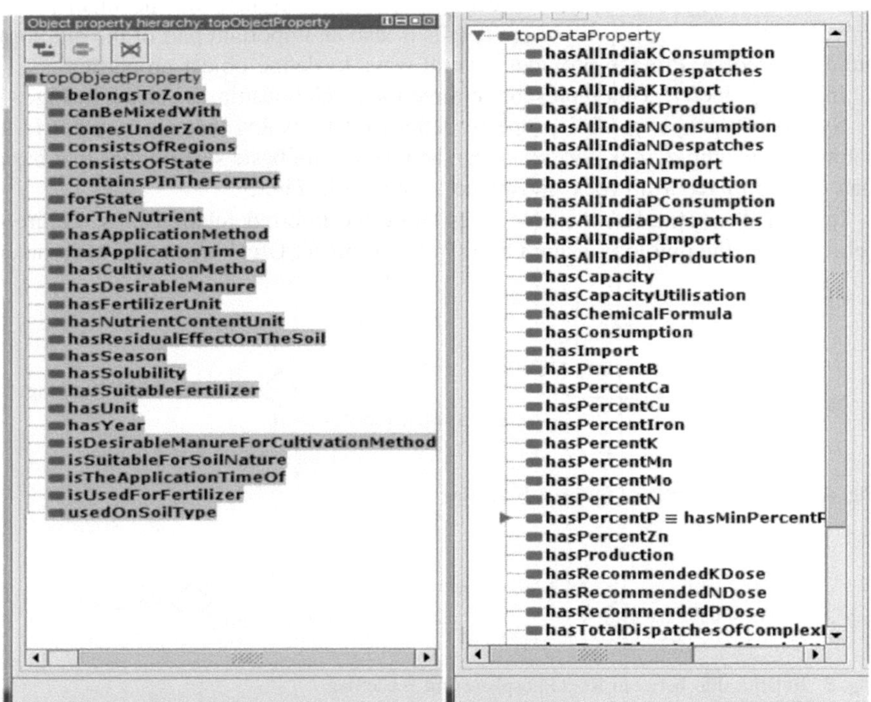

Fig. 7 Various object and data properties

in deciding the key components of the ontology. It is further possible to integrate our ontology with vocabularies such as Agrovoc or any other agriculture based taxonomy which will enhance the extensibility of our work. It can be integrated with any other ontologies of this domain soil and crop ontology which will further enable us to extract more specific information. Figure 6 gives the view of the ontology. It clearly shows different classes and their relationships. We have incorporated information on the types of fertilizers, their application methods, nutrients, year wise consumption, and state wise consumption etc. of particular fertilizers (Fig. 7).

4 Conclusion

The main contribution of this paper is the development of a prototype natural language interface for an ontology. Specifically, we wish to answer natural language questions asked by farmers using ontology as a knowledge base. For this purpose, we have developed an interface which will take question asked in natural language and will automatically return SPARQL query as output. The system makes use of an ontology that represents the domain and question templates generated to match with the questions entered by the user. The methodology followed can easily be followed for other domains also. Template based approach is able to more number of questions of different types. The fertilizers ontology is also developed as part of this work. We have evaluated the ontology internally and resolved the problems encountered in consultation with agriculture scientists. We wish to enhance our work by including more complex questions having different patterns which requires reasoning. Right now, we have focused on developing the natural language interface part. We wish to extend the coverage of the system by including more classes of questions. Our aim is to implement this system in real time scenario.

References

1. Ferrandez, O., Izquierdo, R., Ferrandez, S., Vicedo, J.L.: Addressing ontology-based question answering with collections of user queries. Int. J. Inf. Process. Manage. 175–188 (2009)
2. Athira, P.M., Sreeja, M., Reghuraj, P.C.: Architecture of ontology based domain specific question answering system. Int. J. Web Seman. Technol. **4**(4) (2013)
3. Malik, N., Sharan, A., Srivastava, J.: Conceptual framework for knowledge management in agriculture domain. In: Proceedings of the International Congress on Information and Communication Technology. Advances in Intelligent Systems and Computing, vol. 438. Springer (2016)
4. Wang, C., Xiong, M., Zhou, Q., Yu, Y.: PANTO: a portable natural language interface to ontologies. In: ESWC 2007, pp. 473–487 (2007)
5. http://quepy.machinalis.com/
6. http://aksw.org/Projects/AutoSPARQL.html

7. Lopez, V., Pasin, M., Motta, E.: AquaLog: an ontology-portable question answering system for the semantic web. In: Gómez-Pérez, A., Euzenat, J. (eds.) ESWC. Lecture Notes in Computer Science, vol. 3532, pp. 546–562. Springer (2005)
8. Danica D., Valentin V., Kalina B.: A text based query interface to OWL ontologies. In: ESWC 2010. Springer (2010)
9. Christina, U., Lorenz, B., Jens, L., Axel, C.N.N., Daniel, G., Philipp, C.: Template based QA over RDF data. In: WWW 2012. ACM (2012)
10. Brewster, C., O'Hara, K.: Knowledge representation with ontologies: present challenges—future possibilities. Int. J. Hum. Comput. Stud. **65**(7), 563–568 (2007)
11. Grüninger, M., Fox, M.S.: Methodology for the Design and Evaluation of Ontologies (1995)

3D-Based Unattended Object Detection Method for Video Surveillance

Chia-Hung Yeh, Kahlil Muchtar and Chih-Yang Lin

Abstract Inspired by 2D GrabCut for still images, this paper proposes automated abandoned object segmentation by introducing 3D GrabCut in surveillance scenario. This allows the method to produce precise object extraction without needing user intervention. Both RGB and depth input are utilized to build abandoned object detector which can resist to shadow and brightness changes. We performed the indoor experiments to show that our system obtains an accurate detection and segmentation. Both quantitative and qualitative measurements are provided to analyze the result.

Keywords GrabCut · 3D processing · Abandoned object detection

1 Introduction

One of essential low-level computer vision applications is unattended objects detection. This system is able to locate the occurrence of unusual events, especially an abandoned or removed object in real-time. Recently, the model-based approaches

C.-H. Yeh · K. Muchtar
Department of Electrical Engineering, National Sun Yat-sen University, Kaohsiung, Taiwan
e-mail: yeh@mail.ee.nsysu.edu.tw

C.-H. Yeh
Department of Electrical Engineering, National Taiwan Normal University, Taipei, Taiwan

C.-Y. Lin (✉)
Department of Communications Engineering, Yuan-Ze University, Taoyuan, Taiwan
e-mail: andrewlin@saturn.yzu.edu.tw

C.-Y. Lin
Department of Bioinformatics and Medical Engineering, Asia University, Taichung, Taiwan

X.-S. Yang et al. (eds.), *Smart Trends in Systems, Security and Sustainability*, Lecture Notes in Networks and Systems 18,
https://doi.org/10.1007/978-981-10-6916-1_25

(e.g., Gaussian Mixtures, Bayesian Inference, etc.) are greatly leveraged by building robust background (BG) model [8, 16].

A prevalent issue in intensity-based background model is that the abrupt light changes or shadow is easily incorporated. With the invention of the low-cost depth camera has enabled to conduct more notable researches in computer vision by using depth and visual (RGB) sensing. To date, the number of reliable vision applications using this sensor have been introduced such as [6, 13, 10, 12, 18], and so on. To solve the aforementioned issues, we exploit the Kinect sensor further especially for designing a robust *AO* detection system which takes RGB and depth information into account. Motivated by this challenge, we devise a new 3D GrabCut framework with its focus on *AO* event. We extend a MRF-based GrabCut [7, 14] in order to obtain final result, that is, the segmented *AO* region. In literatures, although recently there have been few attempts made to improve existing GrabCut using depth sensor [11, 15, 17], but none of them are focus on automatic and advanced surveillance problem. In recent past, there are two approaches that consider depth information in order to devise more reliable *AO* detection system. Beleznai et al. [1] proposed an approach to detect *AO* object by employing combined passive stereo depth and intensity cues. It is quite important to also note, instead using Kinect sensor, the authors used in-house developed stereo camera to obtain depth data by stereo matching process. In [5], another approach is proposed by exclusively utilizing only depth data.

This paper is structured as follows: the subsequent section provides a review of GrabCut, Sect. 3 explains the proposed framework, Sect. 4 demonstrates quantitative and qualitative results, and Sect. 5 concludes the method.

2 Related Works

2.1 Overview of GrabCut

GrabCut [14] is an iterative-based minimization method that utilizes color information (instead of gray-scale [2, 4]), in order to build the model. An image I is set to be an input frame and the $z = (z_1, \ldots, z_n, \ldots, z_N)$ of N pixels. Note that the $z_i = (R_i, G_i, B_i)$, $i \in [1, \ldots, N]$, where each channel indicates a color intensity. Suppose that the labelling is a binary problem, so we can define the array $\alpha = (\alpha_1, \ldots \alpha_N)$, $\alpha_i \in \{0, 1\}$. In other words, each pixel of input frame should be either background or foreground pixel. A trimap T is originally pre-defined by the user and separated into three regions: T_B (initial background), T_F (foreground) and T_U (unlabelled region). Finally, A GMM of K components is built for each region, that is, a background $(\alpha_i = 0)$ and foreground $(\alpha_i = 1)$ model.

3 Abandoned Object Detection

We detail our system pipeline in Fig. 1. We initially determine the set of seed for GrabCut by means of combined intensity and depth BG modelling-based *AO* detection system. Then, our proposed 3D GrabCut is triggered to segment an *AO* object, thus getting the final 3D results. The rest of the paper, the new method, which we call *AOGrabCutD* will be described.

3.1 System Overview

The main objective of our system is to greatly extend the interactive and an image-based GrabCut to an automatic 3D video-based GrabCut.

3.2 Automatic Trimap Initialization

Suppose each frame of the video as f_t, $t \in \{1, \ldots, N\}$, where N is the number of synchronized RGB and depth frames. First, we employ an *AO* detection system. Then, we determine the trimap T from the box $BBox$ returned by the *AO* detector: $T_U = \{z_i \in BBox\}$, $T_B = \{z_i \notin BBox\}$. The system contains of two independent detector; the intensity and depth-based detector which yield $T_U^{intensity}$ and T_U^{depth}, respectively.

3.3 AO Detector

The depth inputs are not affected by common intensity issues (e.g. brightness changes and shadow), that is, the detection is more robust. However, by only employing depth-based detection, it is somehow reduce the accuracy, especially in the object boundary due to the problem of invalid (missing) data in low-cost sensor such as Kinect. Hence, by combining intensity and depth-based detection, we are

Fig. 1 System pipeline of the proposed method

able to filter out the incorrect *AO* (due to noise) and lead to more accurate *AOGrabCutD* initialization.

3.3.1 Intensity-Based Detection

We build two GMM models for each pixel z, namely, model A and B (high and low learning rates), respectively. Once an object is unattended, the model A will detect it earlier than the model B. While the *AO* is steady in model A, but still unsteady in the model B, we can subtract these two models in order to fully locate the *AO*. We describe the algorithm as shown below:

Algorithm 1 Intensity-based detection algorithm

Input:
An input pixel z;
Output:
Identify the pixel either *AO* or not;
Step:
1. Obtain the first i ($i \leq k$) distributions from model A and B with decreasing order to obtain (μ_{Aj}', w_{Aj}', σ_{Aj}') and (μ_{Bj}', w_{Bj}', σ_{Bj}'), where $j = 1$ to i;
2. Normalize w_{Ai}' and w_{Bj}';
3. If $\sum_{j=1}^{i}\left[\max(w_{Aj}, w_{Bj}) \times (\mu_{Aj} - \mu_{Bj})\right] > TH_1$, and $\sigma_{Aj}' < TH_2$, and $\sigma_{Bj}' < TH_2$, set the pixel as an *AO* candidate;
4. Furthermore, if the frequency of occurrence (TH_p) is satisfied, then the pixel is set to be an *AO* pixel. Finally, set pixel of an *AO* as $T_U^{intensity}$;

where μ, w, and σ are the mean, weight estimation, and standard deviation of each model, respectively. Note that, The *AO* is maintained according to the scheme explained in our previous method (the scheme is clearly discussed in [9]).

3.3.2 Depth-Based Detection

Motivated by [1], we build the depth *BG* model using the accumulate running average. First, the depth data is reprojected into 2D image space. Next, the corresponding *BG* model is designed as follows:

$$BD_t = (1 - \alpha)BD_{t-1} + \alpha f_t \tag{1}$$

where BD_t, α, and f_t are *BG* model in frame t, learning rate and incoming frame t, respectively. If spatial changes occurs, the aggregator will increment the corresponding entries. Furthermore, if the accumulator entry of certain areas is above threshold N, it will be regarded as abandoned item. We set the corresponding areas as T_U^{depth}. Finally, once the $T_U^{intensity}$ and T_U^{depth} acquired, the following combination scheme is used to obtain the final bounding box $BBox$:

$$T_U^{final} = \frac{\#pixel(T_U^{intensity} \cap T_U^{depth})}{\#pixel(T_U^{intensity} \cup T_U^{depth})} \tag{2}$$

If T_U^{final} is greater or equal to threshold TH_{BBox} then an AO is detected, and therefore, the automatic *AOGrabCutD* is initiated. We extend a hybrid GrabCut proposed by Sallem and Devy [15] in order to segment a RGB-D point clouds that contain a detected AO in automatic manner. The coordinate of T_U^{final} is auto-initialized on RGB image (for simplicity), then the *AOGrabCutD* will be performed combining color and geometric information.

4 Experimental Results

Our input is a recorded stream of depth and color images. We use a Kinect for Windows camera, which streams VGA resolution color and depth images at 30 fps. To the best of our knowledge there is no public dataset/test scenario for Kinect-based AO detection. The dataset comprises of one small and one medium boxes. As shown in Fig. 2, our method can combine the detectors and auto-stimulate the GrabCut to produce more precise results. Furthermore, we provide the quantitative measurements that depict the accuracy of proposed method.

4.1 Qualitative Results

4.1.1 Dataset: 1 Small and 1 Medium Boxes (Frame no.: 632 and 1128)

4.2 Quantitative Results

We compare our proposed method with intensity-based detection (connected component is used to specify the ROI). All the scores fall within the range of 0–1. Hence, the higher values are, the greater accuracy will be. The detailed comparisons are presented in Table 1.

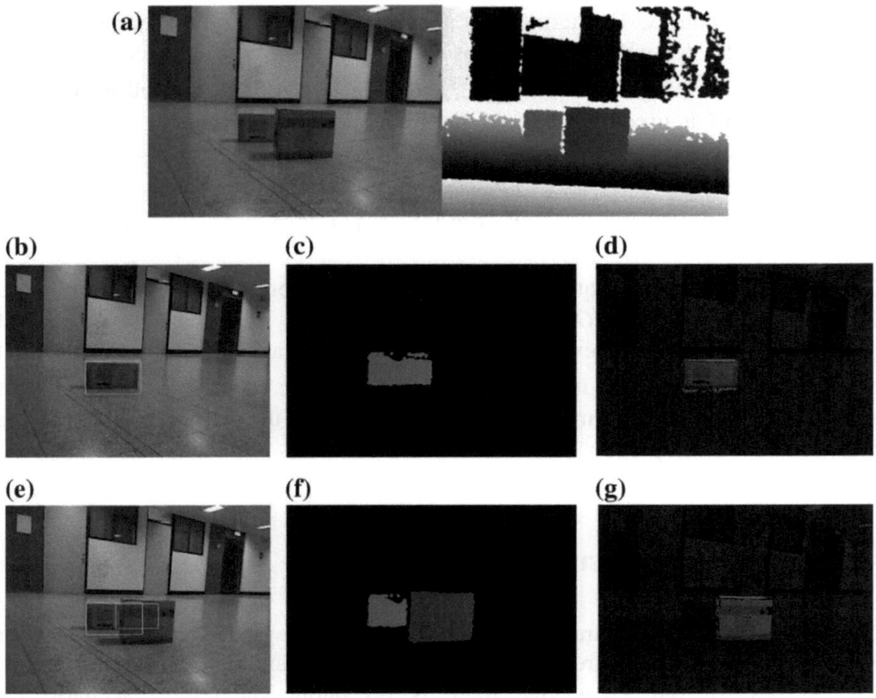

Fig. 2 **a** RGBD input, **b** frame 632: result of intensity-based detection, **c** frame 632: result of depth-based detection, **d** result of *AOGrabCutD* in frame 632, **e** frame 1128: result of intensity-based detection, **f** frame 1128: result of depth-based detection, and **g** is the result of *AOGrabCutD* in frame 1128

Table 1 Quantitative measurement of indoor sequences

Dataset				
Frame: 632	*Recall*	*Precision*	*Similarity*	*f-Measure*
Intensity-based [8]	1	0.825144	0.825144	0.904196
Proposed method	0.999441	**0.857109**	**0.856698**	**0.922819**
Frame:1128	*Recall*	*Precision*	*Similarity*	*f-Measure*
Intensity-based [8]	0.390043	0.924219	0.377955	0.548574
Proposed method	**0.97758**	**0.960635**	**0.939927**	**0.969033**

5 Conclusion

We have introduced a robust method for detecting and segmenting unattended object in surveillance scenario. In addition, we include qualitative and quantitative analysis to show the competitiveness and accuracy.

Acknowledgements This research was supported by National Science Council, Taiwan, under Grants MOST 105-2218-E-468-001 and MOST 105-2221-E-468-008.

References

1. Beleznai, C., Gemeiner, P., Zinner, C.: Reliable left luggage detection using stereo depth and intensity cues. In: Computer Vision Workshops (ICCVW), 2013 IEEE International Conference on. Sydney, Australia (2013)
2. Boykov, Y., Funka-Lea, G.: Graph cuts and efficient ND image segmentation. Int. J. Comput. Vision **70**(2), 109–131 (2006)
3. Boykov, Y., Kolmogorov, V.: An experimental comparison of min-cut/max-flow algorithms for energy minimization in vision. Patt. Anal. Mach. Intelligen. IEEE Trans. **26**(9), 1124–1137 (2004)
4. Boykov, Y.Y., Jolly, M.-P.: Interactive graph cuts for optimal boundary & region segmentation of objects in N-D images. In: Computer Vision, 2001. ICCV 2001. Proceedings. Eighth IEEE International Conference on, vol. 1. Vancouver, BC, 2001, pp. 105–112
5. Collazos, A., Fernández-López, D., Montemayor, A.S., Pantrigo, J.J., Delgado, M.L.: Abandoned object detection on controlled scenes using kinect. In: Natural and Artificial Computation in Engineering and Medical Applications. Springer, Berlin Heidelberg. pp. 169–178 (2013)
6. Gallo, L., Placitelli, A.P., Ciampi, M.: Controller-free exploration of medical image data: experiencing the Kinect. In: Computer-Based Medical Systems (CBMS), 2011 24th International Symposium on. IEEE, Bristol, United Kingdom (2011)
7. Geman, S., Geman, D.: Stochastic relaxation, Gibbs distributions, and the Bayesian restoration of images. Patt. Anal. Mach. Intelligen. IEEE Trans. **6**(6), 721–741, November (1984)
8. Lin, C.-Y., Chan, C.-S., Kang, L.-W., Muchtar, K.: Left-object detection through background modeling. Int. J. Innovat. Comput. Informat. Control **9**(4), 1373–1388, April (2013)
9. Lin, C.-Y., Muchtar, K., Yeh, C.-H.: Robust techniques for abandoned and removed object detection based on Markov random field. J. Visual Commun. Image Represent. **39**, 181–195, (2016)
10. Machida, E., Cao, M., Murao, T., Hashimoto, H.: Human motion tracking of mobile robot with Kinect 3D sensor, SICE Annual Conference (SICE), 2012 Proceedings of. Japan, Akita (2012)
11. Meyer, G.P., Do, M.N.: 3D GrabCut: interactive foreground extraction for reconstructed 3D scenes. In: Proceedings of the 2015 Eurographics Workshop on 3D Object Retrieval, pp. 1–6 (2015)
12. Munaro, M., Basso, F., Menegatti, E.: Tracking people within groups with RGB-D data. In: Intelligent Robots and Systems (IROS), 2012 IEEE/RSJ International Conference on. Vilamoura-Algarve, Portugal: IEEE (2012)
13. Ren, Z., Yuan, J., Meng, J., Zhang, Z.: Robust Part-based hand gesture recognition using kinect sensor. IEEE Trans. Multimed. **15**(5), 1110–1120 (2013)
14. Rother, C., Kolmogorov, V., Blake, A.: GrabCut: interactive foreground extraction using iterated graph cuts. ACM Trans. Graph. (TOG), **23**(3), pp. 309–314, August (2004)
15. Sallem, N.K., Devy, M.: Extended GrabCut for 3D and RGB-D Point Clouds", in Advanced Concepts for Intelligent Vision Systems, pp. 354–365. Springer International Publishing, Poznań, Poland (2013)
16. Tian, Y., Feris, R.S., Liu, H., Hampapur, A., Sun, M.-T.: Robust detection of abandoned and removed objects in complex surveillance videos. Syst. Man Cybernet. Part C Appl. Rev. IEEE Trans. **41**(5), 565–576, September (2011)

17. Vaiapury, K., Aksay, A., Izquierdo, E.: GrabcutD: improved grabcut using depth information. In: Proceedings of the 2010 ACM workshop on Surreal media and virtual cloning, Firenze, Italy, pp. 57–62 (2010)
18. Whelan, T., Leutenegger, S., Salas-Moreno, R.F., Glocker, B., Davison, A.: ElasticFusion: dense SLAM without a pose graph. Proc. Robotics Sci. Syst. Rome, Italy, (2015)

Electrical Energy Output Prediction Using Cuckoo Search Based Artificial Neural Network

Sankhadeep Chatterjee, Nilanjan Dey, Amira S. Ashour and Cornelia Victoria Anghel Drugarin

Abstract The environmental ever demanding improvement along with the increasing demand of electricity attracted researchers in designing efficient, accurate and robust models. Such models are used mainly to predict the energy output of combined steam and gas turbine mechanisms. The applicability of these systems depends on their sustainability. It is inevitable to predict the combined mechanisms output energy in order to produce more trustworthy mechanisms. Since the acceptability of the aforesaid turbine systems is judged in terms of their profitability, the output energy prediction plays a vital role. In machine learning, the neural network (NN) based models has been proven to be a trustworthy in critical prediction tasks. However, the traditional learning algorithms in the NNs suffer from premature convergence to local optima while finding the optimum weight vectors. Consequently, the present work proposed a Cuckoo Search (CS) supported NN (NN-CS) and a Particle Swarm Optimization (PSO) supported NN (NN-PSO) to efficiently predict the electrical energy output of the combined cycle gas turbines. In the current study, five features are extracted, namely the ambient temperature, relative humidity and ambient pressure in gas turbines and exhaust vacuum from a steam turbine. The results established the improved performance of the CS based

S. Chatterjee (✉)
Department of Computer Science and Engineering, University of Calcutta, Kolkata, India
e-mail: chatterjeesankhadeep.cu@gmail.com

N. Dey
Department of Information Technology, Techno India College of Technology, Kolkata, India
e-mail: neelanjandey@gmail.com

A. S. Ashour
Faculty of Engineering, Department of Electronics & Electrical Communications Engineering, EGYPT College of CIT, Tanta University, Taif University, Taif, Kingdom of Saudi Arabia
e-mail: amirasashour@yahoo.com

C. V. A. Drugarin
Department of Electrical and Informatics Engineering, "Eftimie Murgu" University of Resita, Reşiţa, Romania
e-mail: c.anghel@uem.ro

X.-S. Yang et al. (eds.), *Smart Trends in Systems, Security and Sustainability*, Lecture Notes in Networks and Systems 18,
https://doi.org/10.1007/978-981-10-6916-1_26

NN compared to the multilayer perceptron feed-forward neural network (MLP-FFN) and the NN-PSO (particle swarm optimization) in terms of root mean squared error. Proposed NN-CS achieved an average of 2.58% the mean square error (RMSE).

Keywords Artificial neural network · Particle swarm optimization
Cuckoo search · MLP-FFN · Combined cycle · Electrical energy
Gas turbines

1 Introduction

Computational intelligence and optimization have significant role in various applications, where resources, time and energy are limited. In contemporary design and implementation applications, a paradigm shift in systems' design to achieve energy-saving and to find optimal solutions become compulsory. Nevertheless, finding optimal solutions is non-trivial and complex. One of the critical, demanding problems all over the world, especially in the developing countries is the electricity. In order to tackle this problem, it is essential to engage most advanced and efficient energy producing mechanisms. Recent trends pointed out Gas Turbines (GT) as a potential solution. It is currently used in several power plants around the globe. Although, it is important to ensure that such systems are reliable and sustainable to cope with the current challenges of the power demand effectively. Reliability and sustainability of the GT depend on the output energy accurate prediction while subjected to profit. Thus, it becomes imperative to predict the GT power generation based systems accurately and effectively. Traditional attempts to propose a simulation method for performance analysis of different GTs have been addressed [1]. The stage-stacking method was utilized for compressor, while the stage-by-stage model was adopted for the turbine.

Simulation based method was quite accurate in real life scenario. The effects of different ambient conditions such as ambient temperature and ambient pressure are well studied [2–7]. The artificial NN (ANN) has been successfully employed in determining the industrial GTs performance as well [8–11]. Muhammad et al. [12] proposed a recurrent neural network based approach to predict the GTs performance in a very short time. Though, application of machine learning, especially the NN in output energy prediction is still at its primitive level. Few research works considered ANN based method for this task [13]. It has utilized a primitive MLP-FFN trained with back-propagation strategy and achieved reasonable accuracy. Chen et al. [14] revealed that the ANN trained with traditional learning algorithms can prematurely converge into local optima during optimization. The problem can be overcome by employing meta-heuristics algorithms in the training phase of ANN. Recently, there are several nature inspired meta-heuristic algorithms [20] including the Artificial bee colony (ABC), Ant colony optimization (ACO), Bacterial foraging optimization algorithm (BFOA), Cuckoo search (CS), Particle swarm optimization (PSO) and Firefly algorithms (FA). Meta-heuristic trained NNs performed well in predicting real life problems to a greater extent [15, 16].

From the preceding literatures, the present work realized an accurate and trustworthy model for electrical energy output prediction of the GTs. Since the CS is a very efficient meta-heuristic optimization technique that recently developed [17]. A model based on NN trained with Cuckoo Search (CS) optimization algorithm [18] is proposed in the current work. Three different models for the prediction of electrical energy output of GTs are introduced. These models are (i) an MLP-FFN trained with scaled conjugate gradient descent algorithm [19], (ii) NN trained with Particle Swarm Optimization (NN-PSO) and (iii) the NN-CS.

The remaining sections are arranged as follows. Section 2 introduces the CS algorithm along with the NN for training. Section 3 describes the experimental method followed by the experimental results' discussion. Finally, the conclusion is included in Sect. 4.

2 Proposed Cuckoo Search Based Neural Network

2.1 Methodology

One of the effective, nature-inspired meta-heuristic algorithms is the Cuckoo search (CS), which employed the Lévy flights scheme. It is inspired from the aggressive reproduction strategy of the cuckoo birds. Typically, the CS algorithm achieves improved convergence toward global optimization compared to other global search algorithm. The standard rules for the CS algorithm are: (i) at a time, each cuckoo lays one egg (solution) in a randomly selected nest; (ii) the best nests, including high-quality eggs are carried to the next generation; (iii) there exist a fixed number of the host nests and (iv) a probability $\mathbf{P_a} \geq (0, 1)$ is used during the discovery of the egg laid by a cuckoo. In order to solve the optimization problem, a fitness function of each nest should be determined.

In the present work, the CS algorithm is employed to determine the ANN optimal weight for the GTs electrical energy output prediction. The block diagram of the CS procedure is illustrated in Fig. 1.

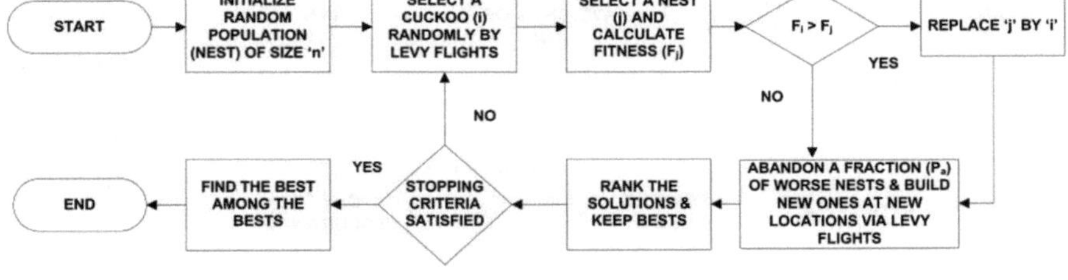

Fig. 1 Flowchart of Cuckoo search algorithm

The proposed CS based NN system assumed each Cuckoo nest is represented by the weight vector *w*, where the number of populations in the CS algorithm is set to 20, whereas other algorithm parameters are $\boldsymbol{\alpha} = 0.25$, $\mathbf{P_a} = 0.01$ and $\boldsymbol{\lambda} = 1.3$. The parameters are set by running the algorithm 50 times with different configurations and the best configuring is taken. The range of the weight *w* values is used within the interval [−1.0, 1.0]. The fitness function is the Root Mean Squared Error (RMSE) of the NN, which is determined for each nest. After every iteration, all cuckoos move toward the nest using the Lévy flight scheme. The fitness of the solution should be linked to the objective function and start iterating to generate a new nest by using the Levy flight while keeping the current best. A new nest is produced along with the best nest of the previous iteration. Once the CS algorithm reaches its maximum iteration, it stops and gives the global optimum solution for the weight vector *w*, which are the optimized of values using the CS algorithm. In the current work, the maximum iteration is set to 300. These optimal weights *w* are selected to predict the electrical energy output of the GTs. Prediction of the electrical energy output of GTs using CCPP dataset is performed. The dataset used in the current study is composed of 9568 data points collected while the combined cycle plant is set to work with full load over 674 different days. The dataset spans a variety of ambient conditions over 6 years of operation. The dataset is freely available in [20].

2.2 Proposed System

The CS algorithm based optimization of the NN weights is used in the proposed system for prediction of the electrical energy output of GTs. In addition, the NN training phase using the PSO is employed for comparative analysis. The general proposed training phase using different meta-heuristic algorithms are illustrated in Fig. 2.

Figure 2 illustrated that the CS algorithm is employed to optimize the input weight vectors of the NN by minimizing the RMSE of the NN. Finally, the test phase is performed, followed by performance metrics calculations. The fitness

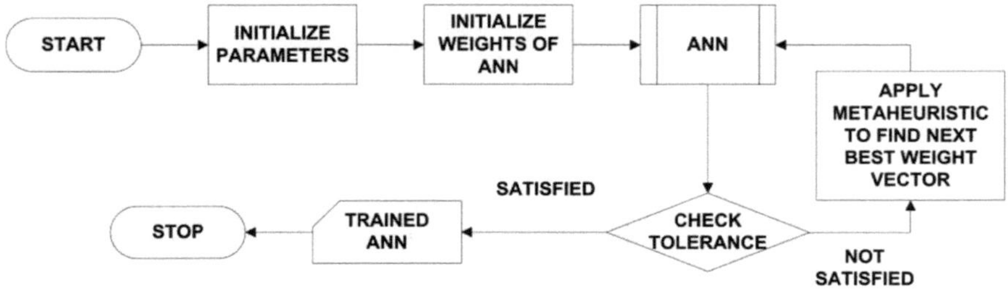

Fig. 2 Flow chart of training phase of ANN using different meta-heuristic algorithms

function of the CS and PSO is calculated using the difference between the predicted values by the NN and the actual discovered values. The RMSE of a prediction model with respect to the computed variable v_{c_k} is determined using the following expression:

$$RMSE = \sqrt{\frac{\sum_{k=1}^{n} \left(v_{d_k} - v_{c_k}\right)^2}{n}} \tag{1}$$

where, v_{d_k} denotes the originally observed value of the kth data instance and v_{c_k} denotes the predicted value of the prediction model. The experimental setup for the NN-PSO based model has been reported in [15].

3 Results and Discussion

The current article proposes a NN-CS model for predicting the electrical energy output of a GT from a combined cycle power plant. The testing of the proposed models is based on a 5 × 2 cross validation scheme [21]. Thus, the dataset is randomly shuffled to produce five instances, each of which is employed in a 2-fold cross validation scheme. The experimental results for each of the tests are tabulated in Table 1 for the RMSE of each model.

Table 1 establishes that the average RMSE of MLP-FFN is 3.92%, while the NN-PSO improved the result with average RMSE of 2.94%, even though, the proposed NN-CS based model outperforms these models with average RMSE of 2.58%. These results are illustrated in Fig. 3 showing the superiority of NN-CS.

Figure 3 establishes that the NN-CS outperforms the other NN-CS as well as the MLP models in terms of the RMSE. The NN-PSO performed well in all the ten

Table 1 Comparison of experimental results for MLP-FFN, NN-PSO and NN-CS

	MLP-FFN (%)	NN-PSO (%)	NN-CS (%)
Experiment 1	3.58	2.88	2.25
Experiment 2	4.15	2.95	2.84
Experiment 3	3.95	3.22	2.59
Experiment 4	3.65	2.84	2.68
Experiment 5	4.25	3.15	2.14
Experiment 6	4.36	3.06	2.89
Experiment 7	3.68	2.88	2.65
Experiment 8	3.89	2.78	2.49
Experiment 9	3.82	2.68	2.61
Experiment 10	3.91	2.96	2.68
Average	3.92	2.94	2.58

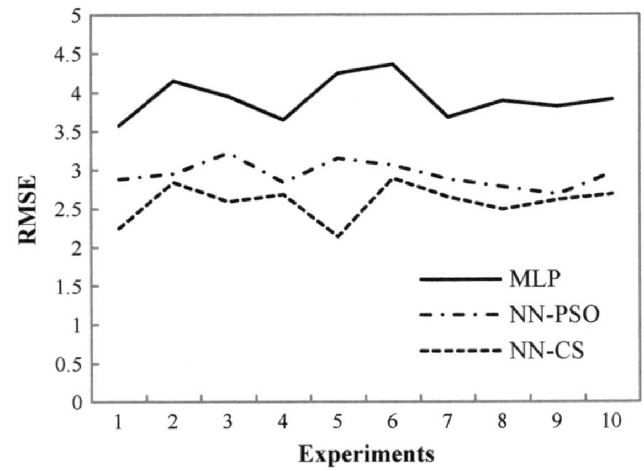

Fig. 3 RMSE of different experiments for different models

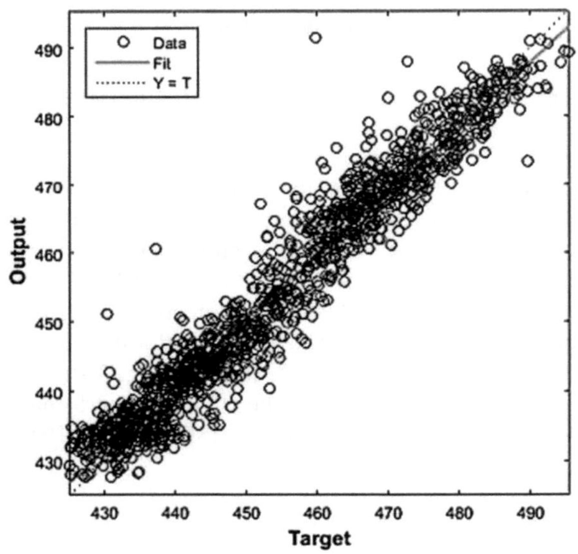

Fig. 4 Prediction electrical energy output of the NN-PSO based model

experiments compared to the MLP-FFN model in terms of the RMSE. Although, the NN-CS based model reduces the RMSE further in almost all the experiments and establishes its ingenuity. Furthermore, Figs. 4 and 5 illustrate the prediction of output electrical energy by the NN-PSO and the NN-CS models; respectively. The comparison of both figures reveals that the predicted output energy of the NN-CS model is more dense and closer to the perfect fit line (dashed), while several predicted values have a significant deviation from the expected output in the case of NN-PSO. This establishes the superiority of NN-CS model over NN-PSO in predicting the output electrical energy of GTs.

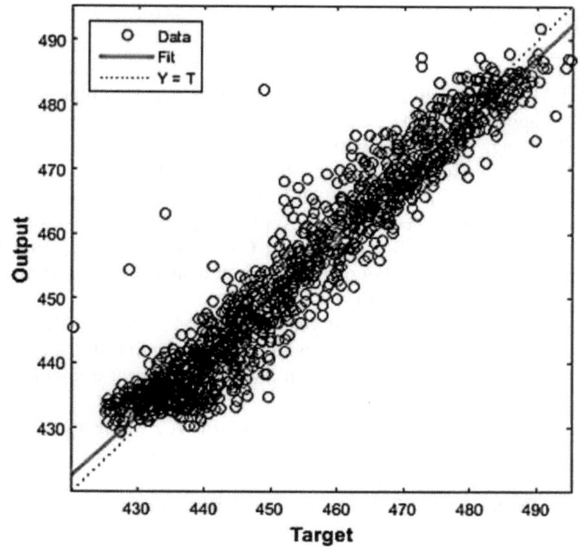

Fig. 5 Prediction electrical energy output of the NN-CS based model

The preceding results establish that the predicted output by NN-CS based model is highly closed to the expected output in the testing phase. This further establishes the ingenuity claim of the NN-CS based model. Consequently, the proposed method can be used and compared with other optimization algorithms that have been used in several applications as in [22–27].

4 Conclusion

Accurate prediction of the electrical output energy is inevitable for the success of GTs. Subsequently, in the present work, a CS based NN model has been proposed along with a well-known NN-PSO based model to predict the electrical energy output of a GT depending on different ambient features. Three different models have been tested on a dataset collected from a combined cycle power plant at different times of six years of span. The features considered in the current study were the ambient temperature, relative humidity and ambient pressure in gas turbines and exhaust vacuum from a steam turbine.

The proposed model has been compared with the NN-PSO and the MLP models in terms of RMSE. Experimental results have suggested that the MLP-FFN achieved poor performance for predicting the output energy; while the NN-PSO reduced the RMSE to some extent. Average RMSE achieved by the NN-PSO model was 2.94%, while the proposed NN-CS based prediction model achieved an average of 2.58% RMSE. The study has revealed that the NN-CS based model is superior to the traditional machine learning algorithms. Nevertheless, other meta-heuristics based NN can be tested for more accurate results.

References

1. Lee, J.J., Kim, T.S.: Development of a gas turbine performance analysis program and its application. Energy **36**(8), 5274–5285 (2011)
2. Yang, W., Su, M.: Influence of moist combustion gas on performance of a sub-critical turbine. Energy Convers. Manag. **46**, 821–832 (2005)
3. Arrieta, F.R.P., Lora, E.E.S.: Influence of ambient temperature on combined-cycle power-plant performance. Appl. Energy **80**, 261–272 (2005)
4. Erdem, H.H., Sevilgen, S.H.: Case study: effect of ambient temperature on the electricity production and fuel consumption of a simple cycle gas turbine in Turkey. Appl. Therm. Eng. **26**, 320–326 (2006)
5. Lugo-Leyte, R., Zamora-Mata, J.M., Toledo-Velázquez, M., Salazar-Pereyra, M., Torres-Aldaco, A.: Methodology to determine the appropriate amount of excess air for the operation of a gas turbine in a wet environment original research article. Energy **35**, 550–555 (2010)
6. Basrawi, F., Yamada, T., Nakanishi, K., Naing, S.: Effect of ambient temperature on the performance of micro gas turbine with cogeneration system in cold region. Appl. Therm. Eng. **31**, 1058–1067 (2011)
7. De Sa, A., Al Zubaidy, S.: Gas turbine performance at varying ambient temperature. Appl. Therm. Eng. **31**, 2735–2739 (2011)
8. Maren, A.J., Harston, C.T., Pap, R.M.: Handbook of Neural Computing Applications. Academic Press, San Diego (2014)
9. Baughman, D.R., Liu, Y.A.: Neural networks in Bioprocessing and Chemical Engineering. Academic press, San Diego (2014)
10. Rojas, R.: Neural Networks: A Systematic Introduction. Springer Science & Business Media, Berlin (2013)
11. Dash, R.N., Subudhi, B., Das, S.: A comparison between MLP NN and RBF NN techniques for the detection of stator inter-turn fault of an induction motor. In: 2010 International Conference on Industrial Electronics, Control and Robotics (IECR), 251–256 (2010)
12. Muhammad, M., Mohammadreza, T.B., Karim, Z.A.: Methodology for short-term performance prognostic of gas turbine using recurrent neural network. In: 2015 IEEE International Conference in Industrial Engineering and Engineering Management (IEEM), 787–791 (2015)
13. Fast, M., Assadi, M., De, S.: Development and multi-utility of an ANN model for an industrial gas turbine. Appl. Energy **86**(1), 9–17 (2009)
14. Chen, J., Do, Q.H., Hsieh, H.: Training artificial neural networks by a hybrid PSO-CS algorithm. Algorithms **8**, 292–308 (2015)
15. Chatterjee, S., Sarkar, S., Hore, S., Dey, N., Ashour, A.S., and Balas, V.E.: Particle swarm optimization trained neural network for structural failure prediction of multistoried RC buildings. Neural Comput. Appl. 1–12 (2016)
16. Chatterjee, S., Ghosh, S., Dawn, S., Hore, S., Dey, N.: Forest Type Classification: a hybrid NN-GA model based approach. Inf. Syst. Design Intel. Appl. 227–236 (2016)
17. Walton, S., Hassan, O., Morgan, K., Brown, M.: Modified cuckoo search: a new gradient free optimisation algorithm. J Chaos Solitons Fractals **44**(9), 710–718 (2011)
18. Yang, X.S., Deb, S.: Cuckoo search via Lévy flights. In: Proceedings of World Congress on Nature and Biologically Inspired Computing, 210–214 (2009)
19. Nanda, S.J., Panda, G.: A survey on nature inspired metaheuristic algorithms for partitional clustering. Swarm Evolut. Comput. **16**, 1–18 (2014)
20. Tüfekci, P.: Prediction of full load electrical power output of a base load operated combined cycle power plant using machine learning methods. Int. J. Electr. Power Energy Syst. **60**, 126–140 (2014)
21. Dietterich, T.G.: Approximate statistical tests for comparing supervised classification learning algorithms. Neural Comput. **10**, 1895–1923 (1998)

22. Dey, N., Samanta, S., Chakraborty, S., Das, A., Chaudhuri, S.S., Suri, J.S.: Firefly algorithm for optimization of scaling factors during embedding of manifold medical information: an application in ophthalmology imaging. J. Med. Im. Health Inform. **4**(3), 384–394 (2014)
23. Dey, N., Samanta, S., Yang, X.S., Das, A., Chaudhuri, S.S.: Optimisation of scaling factors in electrocardiogram signal watermarking using cuckoo search. Int. J. Bio-Inspir. Comput. **5**(5), 315–326 (2013)
24. Kaliannan, J., Baskaran, A., Dey, N.: Automatic generation control of thermal-thermal-hydro power systems with PID controller using ant colony optimization. Int. J. Serv. Sci. Manag. Eng. Technol. (IJSSMET) **6**(2), 18–34 (2015)
25. Samanta, S., Acharjee, S., Mukherjee, A., Das, D., Dey, N.: Ant weight lifting algorithm for image segmentation. In: 2013 IEEE International Conference on Computational Intelligence and Computing Research (ICCIC), pp. 1–5 (2013)
26. Jagatheesan, K., Anand, B., Samanta, S., Dey, N., Santhi, V., Ashour, A.S., Balas, V.E.: Application of flower pollination algorithm in load frequency control of multi-area interconnected power system with nonlinearity. Neural Comput. Appl. 1–14 (2016)
27. Jagatheesan, K., Anand, B., Dey, N., Ashour, A.S.: Artificial intelligence in performance analysis of load frequency control in thermal-wind-hydro power systems. Artif. Intel. **6**(7) (2015)

ASMAN Framework: A Framework for Comparison and Selection of SaaS Services

Mamta Dadhich and Vijay Singh Rathore

Abstract Cloud computing is an amazing technology in the present era, in all services provided by the service provider. Users only need a web browser and an Internet connection. As part of its definition of a theory of no server needs to be run in the hosted hardware and transfer data to your computer or device is only increased their own connection to the Internet. The user is required for each of the services used. It includes three different types of models have different uses, naming the parts of assigned amount [Service Platform] [IAAS] infrastructure as a service, software-as-a-service [service] software. The aim of this research is to support the saas consumers in choosing the right saas provider. In recent times, the list of service providers that offer various services to consumers. Consumer demand to select suitable saas providers to fulfill their needs. Asman the framework provides the advantages of the user in choosing the best SAAS SERVICES when several provider SAAS-based products. It includes the Compare the few parameters; Cost, speed, ease of use, reliability and availability. Asman algorithms depend on the selected suitable saas services based on the priority of the selected parameters for a comparison of the desired and the parameters provided.

Keywords Cloud computing · Software-as-a-service
Cloud computing service providers · User parameters

M. Dadhich (✉)
Department of Computer Science, IIS University, Jaipur, India
e-mail: mamtadadhich76@gmail.com

V. S. Rathore
CSE, JECRC, Jaipur Engineering College & Research Center, Jaipur, India
e-mail: vijaydiamond@gmail.com

X.-S. Yang et al. (eds.), *Smart Trends in Systems, Security and Sustainability*, Lecture Notes in Networks and Systems 18,
https://doi.org/10.1007/978-981-10-6916-1_27

1 Introduction

In this year's cloud computing has become a major trend in the IT industry. In this calculation of service model, all the servers, networks, applications and other elements that are associated with the data center enables the IT staff and end users may be accessed through the Internet, to enable them to acquire only the types and quantities of computing services, their actual needs. The main services provided by cloud computing service mode can be divided into three main categories of clearly articulated infrastructure as a service (IAAS), platform as a service (PAAS), and software as a service (SAAS). Software as a service (SAAS) is the most widely used the known and is mainly used for forms of cloud computing. In the Service Provider allows consumers to use only their applications, with without the headaches of maintaining the applications available for backup, recovery, software failures, [2] in the saas software distributed model promotes a of the needs of the users of the software and provides easy access to the application software and the database is not complete, and the congregation of those resources. [3] Consequently, SAAS customers evaluate saas applications in accordance with their needs and SAAS provider can also facilitate the comparison of different saas services so that they should have the means to monitor the use of resources and assessment of different QOS targets can compare different basic parameters to their customers and meet SLA requirements. [4] the growth of service users and service providers and a rising proportion. But still in the saas services are cumbersome and select Software Service Cloud computing, software services are fit for their business needs. Therefore, to select a suitable saas models become a real problem for users of cloud computing. The basic idea of this research papers 5 Assessment of the basic parameters and place them in accordance with the needs of the user.

1.1 The Relevant Work

This work has been done by many researchers to select a suitable saas services. If we focus on the work of previous studies, we observe that there are different methods to select a comparison of SAAS-based Quality of Service parameters such as cost, availability, ease of use, security, etc. as well as Godse Yooal [5], describes the various parameters "File method, and select the appropriate SAAS offerings cloud" must consider SAAS products [analyze layered process] technology. This work is recommended to use a hierarchy of the quantitative techniques to solve this problem [6].

Elarbi Badidi submitted the architectural saas selection and configuration. The framework relies on cloud computing services broker. This study algorithm SAAS provider, that can meet consumer demands for its services. The algorithm to take into account only the non-functional (mainly the question of QOS selection process). The proposed SAAS Provider selection algorithm uses a linear aggregation utility functions [4].

Mehul Mahrishi proposed an algorithm to select the appropriate form of cloud service cloud service provider [CSP], the study will include an algorithm that made possible the cloud middleware to determine the features of the CSP using rough set analysis base level of satisfaction with the services [7].

Nacera Boussoualim Yooal [7], recommended levels of vague measurement to quantify the software parameters and extract the best software, different software. In fact, their methods use a hierarchical impartial analysis and extraction of different parameters, and then create a hierarchy between attributes in the report submitted by the Service Agent to solve this problem by providing a proxy to connect to the server, they can modify the service [8].

In the Evaluate cloud service providers in the comparison of the author, Jagadeesh, Babu with Saikiran analysis of certain service providers at different levels provider; salesforce, IBM, HP, rack space, and the at&t synaptic. They observed that what were the different needs of different consumers and, on that basis, they also made a comparison. [9]

Professor Deepak Kapgate [10] a predictive broker algorithm is based on the weighted (WMAFM) the Moving Average forecast model. They have developed a new approach to balance the load of the data center and also minimizes response time. Amrutha et al., "the international journal of advanced research in computer science and software engineering recommendations 2014 Proxy algorithm", said an algorithm to assess appropriate cloud Agent Select the best cloud service provider based on its performance. The Broker service provider based on certain limits (cost and performance). They developed a row of cloud computing infrastructure in all registered brokers cloud contact consumers and service providers are more appropriate services, are many [11, 12].

The conclusions of the literature review, we note that a small number of the focus of the work of the research is a common and universal frame to select the cloud services rather than the specific service IAAS, PAAS, SAAS, selection, and proposed a general comparison extensive technical and non-technical parameters. Some of the research work is based on the use of the Framework is a middleware is cloud agent can provide the best of both Cloud Service Provider selection from a large amount of commerce service provider. High Quality of Service parameters provides better the CSP selected are very popular.

We recommend that the development of the system framework named "Asman" [the appropriate selection of the SAAS model necessary] is to select the appropriate software as a service allows organizations to meet their needs, to provide its services. This framework provides the best selection of software as a service a large number of the SSP. High Quality of Service parameters provides better choose SSP is a very popular. Research relies on the following parameters:

1. Cost
2. Speed
3. Ease of Use
4. Reliability
5. Availability.

2 Research Methods

The Software as a service (SaaS) Model is a technology offers the same software saas different consumer needs internet connection and a browser. Saas applications are provided to consumers through the network. This phenomenon is not only in the business as well as the environment but also in almost every field. All of these services are hosted saas provider's data center and not on a customer's individual computers. This SAAS model embedded smart, analytical and innovative distributed technology saas provider is responsible for the maintenance of the software. The client sends a request to to access the services required, including a separate license for each user that will use the services provided by the saas providers [13]. SAAS software reduces the cost of deploying and maintaining the burden of the service provider, and freeing up resources for other projects. Saas is an easy-to-access of cloud computing and the great ways to make a lot of Saas Applications [14]. This approach allows consumers to access the needs of Saas entire completed application from the saas vendor applications there is simply no cloud environment and can access from any browser and suppliers must be aware of all responsibility is associated with the application and SAAS services. Thus, the end user must pay per use of services the customer's saas providers of end users [5].

The aim of the research is to review a specific speed, availability, cost, availability and reliability of the parameters. The proposed framework. The objective of the study is to provide the benefits of the user selects the best saas services to meet their needs, because when several providers of SAAS-based products, select the product has become an important analysis of the issues involved in the Select the parameter and the services that it provides product saas providers. Such a study would become a framework for the work, will assist in the decision to select the appropriate SAAS offerings.

A. **Asman framework**: [in accordance with the need to select the low ebb SAAS model for user intervention is required.]

Asman of the purpose of the framework is to find and compare the best saas services in Enterprise/Person entity. Added to the common functions and features, in accordance with the provisions of the mandatory parameters such as speed, reliability, ease of use and availability of Saas services for a real-time, can help the user to find the best SAAS Services Data as shown in the figure of the user is captured using the algorithm, a work based on the program, the formula to calculate the average time and more information about the required computing performance parameters; speed, ease of use, reliability, availability and cost-effectiveness. These are called the mandatory parameters Asman framework. In addition to these parameters, administrators can add additional parameters to let users know more about their saas service needs.

B. **Recommendations of the saas comparison algorithm [SCA]**:

The description of the algorithm for assessing the performance parameters comparison saas services. These parameters and their values are estimated stored in the database and the final outcome of the parameters for the analysis. For each saas services. This helps the user to easily select the desired saas services on the basis of the results of the comparison after asman framework.

Start

1. Ssp offers all the services, users and their cost.

2. The computation speed of individual SSP i. Average Response Time = small TI/n

In TI's time the user request, and provide services that n is user requirements 2, Maximum response time the response time of the SSP commitment to the customer.

iii. response time fault = (xn)*/1000

Where x is the number of time the response time is greater than the maximum value of the response time.

3. Compute Availability parameters for individual SSP

I. Effectiveness = no task has been successfully completed the task/General under the * 100.

2. Efficiency = Work Time = End Time − Start Time.

4. The calculation of the reliability parameters for individual SSP. Reliability MTBF =/1 + Mean Time Between Failure

Where MTBF Mean Time Between Failure.

5. Compute Availability parameters for individual SSP

Availability = Mean Time Between Failures MTBF/MTTR+

Where MTBF MTBF and MTTR mean time to repair.

6. Select the appropriate SSP in cost, speed, ease of use, availability, and reliability.

7. In accordance with the parameters of the SSP for comparison and the result is fetched.

8. Repeat steps 2–7. End

C. **Featured performance parameters**:

Performance parameters are required for each saas services. The calculator in accordance with the algorithm output the data required for the user's screen. In comparison, it will also collect responses from the algorithm and the data for these services compared with each other and the data displayed to the user. The following is added data every saas services", "Acknowledged" of the saas service provider, we have updated the details of a back-end, use of the following screen:

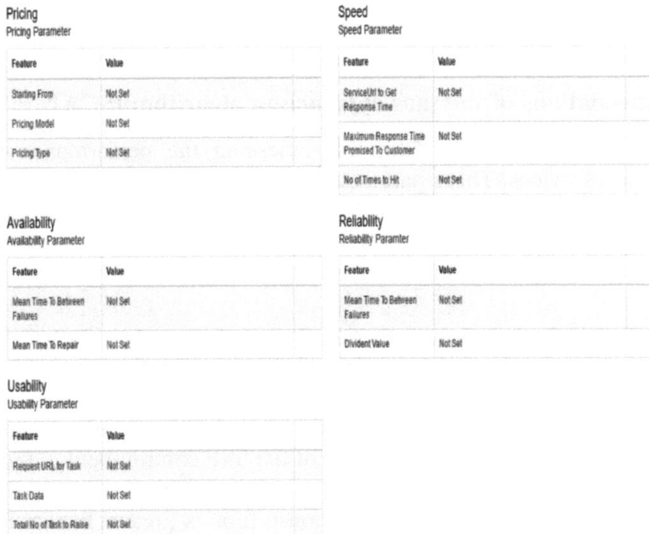

We can see in the image above, for mandatory parameters, we will add this data as part of the HTTP response against the company SAAS Services server to get the data is accurate.

D. **The user interface of the saas services**:

Asman framework as follows NET MVC and API mode. All user requests are routed through the NET engines for the controller. Controller model to such a request and begin collecting detailed information related to the sending of data model in the light of the observations adopted by the response returned to the user. In response is received, view and model will be sent to the user. Now, at the end of the Company Calculated—The above said parameter data, the administrator's initial setting is this service search. After we receive the response, we create the charts and tables to allow the user to view real-time data for saas service requests.

E. **User request processing**:

When users search for saas services, he entered the required parameters. Press the nature of his request, asman framework in the list generated relevant search list saas services. The user can choose to make a comparison of the three largest saas services. The user can also provide ratings and reviews any saas services of his intentions. In this connection, he has to log into asman framework.

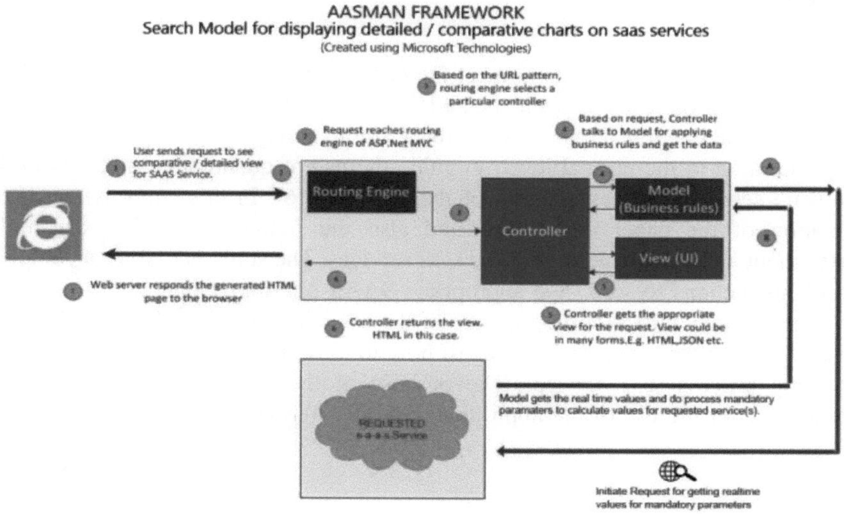

Asman architecture design in the top ASP.NET MVC 4.5 architecture. The incoming Web request through the ASP.NET rpicnickers who controls the engine, the processing of all web requests and select the appropriate controller classes. The controller is a special class controls the interaction between the model and the view and send the response back to the user via the routing engine. It performs on the interactive data model object. It receives input and the implementation of operational changes the operating status of the data model.

On receipt of a request through the Routing Engine, ASP.NET MVC Schema Select the appropriate controller. The Controller and the model to access and prepare the result of user requests. There is something which a dynamic view of the user. This view will be displayed to the user provide accurate information, as well as the parameters required for running view, AJAX request through the model and the view updates and the value of the mandatory parameters, such as availability, speed, reliability, etc. in this request, the user can see the progress made and the experience with the needs of can time and experience, he can in order to use this service.

The model of the design in the framework of the Cayman and logical, are creating a separate class file. The class files using the logic of creating a database identity db context use code the first method. This means that all data tables you will need to run this application automatically created the first time you run or when using the data migration using Package Manager Console. Only such a need is to update the database connection in a parameter file configuration files.

3 Conclusions

Choose the best saas products to meet the needs of nearly any customer is difficult to analyze and there is no such methods are available for saas product selection used in business applications to sales force automation.

In addition to the normal level, to guide the use of quantitative methods in the software selection and assessment of the available [6]. In this study is the determination of the paper and SAAS and also the framework within which the assessment and classification of these saas services. The study is to form a complete set of instructions to select the appropriate SAAS offerings cloud.

It is also concluded the importance of the selection and the adoption of appropriate SAAS offerings and also suggest how to these instructions will improve the selection process saas services. The purpose of the study was to implement a framework for the use of the appropriate SAAS offerings cloud.

References

1. Hoi Chak, J.: Everything you need to know about cloud computing security and compliance, gartner, research, ID. https://www.ukessays.com/essays/information-technology/the-outline-of-cloud-computing-information-technology-essay.phpG00168345 (2009)
2. http://www.lexology.com/
3. Badidi, E.: Framework, software-as-a-service to select and configure. Int. J. Comput. Netw. Commun (IJCNC). Section 5, **5**(3) (2013)
4. xxx
5. Pijanowski, K.: Understand public cloud: IAAS, PAAS, SAAS. Keith Pijanowski blog (2011)
6. Godse, M., Mulik, S.: The method of selecting S Software-as-a-Service (SAAS) products. In: The IEEE International Conference on the Cloud Computing, pp. 155–158 (2009)
7. Mahrishi, M.: A novel approach, the selection of an appropriate software as a service on a cloud computing. In: Magazine and Industrial Engineering Management of JIEM, vol 70–26, May 2013
8. Boussoualim, N., Aklouf, Y.: The evaluation and selection of SAAS offerings according to user preferences. In: Third International Technical Progress Made With Regard to Electrical, Electronic and Computer Engineering 2015, Beirut, Lebanon (2015)
9. Babu, J., Saikiran, P.: Evaluate cloud service providers to provide the comparison. Int. Mag. Adv. Res. Comput. Sci. Softw. Eng. **3**(5). ISSN: 1282277 X (2013)
10. Kapgate, D.: Weighted moving average forecast model based on service broker algorithm cloud computing. Int. J. Comput. Sci. Mob. Comput. Article 3 of the 2 Feb (2014)
11. Amrutha, K., Madhu, K.: B R 3519 technology, "an effective way to find the best cloud providers use proxy". In: Magazine Advanced Research in Computer Science and Software Engineering Research Papers, vol. 4, Issue 7 (2014)
12. Shanley, M.N.: What is SAAS. (2011). http://www.wisegeek.com
13. Biddick, M.: Why you need saas strategy. Information Week (2010)
14. Tiwari, P.K.: Cloud computing: a new computing model. The Journal **9**, 8–14 (2009) (compliance)
15. Tarzey, B.: The distinction between SAAS, PAAS and IAAS. (2010). http://www.computerweekly.com

16. Bear in K and Perros, H.: The performance of the services and analysis of cloud computing. In: Service—My World Conference, pp. 693–700 (2009)
17. Chang, S.: Simplified Chinese, 10; 10, Mr Fok Kin-wah shipped, cloud computing research and development trend. In: The Network of the Future, the Second International Conference, pp. 93–97
18. Dhaka, V.S., Vyas, S.: Analysis server performance with a different technology virtual database. J. New Trends Comput. Inf. Sci. 794–797 (2014)
19. Vyas, S.: Analyze performance of virtual and non-virtual database. J. Global Res. Comput. Sci. Technol. Forum Fellow (JGRCST) III, chapter VIII (2015)

Author Biographies

Mamta Dadhich completed her Postgraduate degree in information technology in 2010. She has 5 years of experience in teaching. She is currently a Research Fellow (computer science) and her area of interest including networking and cloud computing. She had published a paper in the international journal.

Dr. Vijay Singh, Rathore received his doctorate degree in Computer Science (in 2009). He has published more than 18 papers in national and international Conference 50. He is a member of the Committee on the UGC-funded institutions, IGNOU, RTU, UORS, UOK, VMOU, JNU, IISU, SGVU, JNVU, UOJ, JRNRVU, and so on. He has completed his Bachelor in 1998 and a master degree in 2001. He is currently working as a professor of the CSE Jaipur engineering college research center, Accident & Emergency Departments, Jaipur, Chairman, Association of Christian (Computer Jaipur, India chapter. He was 16 years of teaching experience and his interests including network security, operating system, and system software. He also brandished his pen 10 books were subsequently published.

Estimation of Heterogeneity to Improve Robust Anisotropic Diffusion

Rohit Kamal Chatterjee and Avijit Kar

Abstract *Anisotropic diffusion (AD)* become a prominent image enhancement and de-noising method after its introduction in 1987. However, anisotropic diffusion requires selection of a diffusion function whose definition remained ad hoc. Additionally, AD requires determining a scale parameter on which the final result is intimately related. Typically this parameter is chosen in ad hoc basis which makes difficult to compare the final output. In literature, Median absolute deviation (MAD) is proposed as a heterogeneity scale for using with *robust anisotropic diffusion* (RAD). Despite its strong statistical foundation, diffusing an image by RAD results in undesirable *staircasing effect* and artefacts are created in presence of noise. In this paper, we propose a *robust L-estimator scale* that correctly demarcates the homogenous and heterogeneous regions of an image and smoothens the image with minimal staircasing effect. Experimental results show that this scale remains effective in presence of heavy noise.

Keywords Anisotropic diffusion · Adaptive enhancement · Adaptive transforms · De-noising · Staircasing effect · Robust L-estimator scale

1 Introduction

Since Perona and Malik (P-M) [1] introduced celebrated *anisotropic diffusion* (AD) model, it has become a widely used technique for adaptive image enhancement with a well understood computational theory (see Weickert [2]). Besides simple scalar diffusion (edge-stopping) function introduced by Perona-Malik,

R. K. Chatterjee (✉)
Department of Computer Science and Engineering, Birla Institute of Technology, Mesra, Ranchi, India
e-mail: rkchatterjee@bitmesra.ac.in

A. Kar
Department of Computer Science and Engineering, Jadavpur University, Kolkata, India

X.-S. Yang et al. (eds.), *Smart Trends in Systems, Security and Sustainability*, Lecture Notes in Networks and Systems 18,
https://doi.org/10.1007/978-981-10-6916-1_28

many diffusion functions are proposed by several authors [3–6] and modified for application in numerous fields.

Despite these impressive results, many authors [7, 8] have shown that the Perona–Malik model is ill-posed i.e. it is unstable with respect to initial perturbations of the image. Interestingly, numerical implementation of P-M model shows that *staircasing* effect is the only instability found in practical experiments. Fundamentally, *anisotropy* in the P-M equation is introduced by using a non-constant diffusion ("edge stopping") function that controls the degree of local smoothing of grey values. Additionally, all these functions depend on one or more parameters (e.g. scale parameters or heterogeneity scale) that highly influences the final result, but most cases these parameters are tuned in some ad hoc basis. A static choice of heterogeneity scale always blurs the image if the number of iteration is large enough. None the less automatic detection of the scale is still an open problem. Black et al. [9] proposed a *robust anisotropic diffusion (RAD)* and suggested to use Median Absolute Deviation (MAD) of the gradient image as a heterogeneity scale. Despite its strong statistical foundation, experiments show that problem of *staircasing effects* cannot be avoided using RAD. Further, in presence of heavy noise, MAD fails to determine the correct scale, thus, noises are not smoothed satisfactorily with an undesirable introduction of artefacts.

This paper is an extension of Black et al. work [9]. In this paper, we propose a robust *L-estimator* scale that correctly determines the noise level or heterogeneity level. The proposed scale does not depend on any *location* measure. In presence of highly skewed distribution of noise the scale remains robust with minimum staircasing effect. Interestingly, this scale converges to a constant value when diffusion process continues, so it determines the *stopping condition* of the process.

2 Anisotropic Diffusion

The model proposed by Perona and Malik [1] is a generalisation of the heat equation by introducing a diffusion function that moderates diffusivity at the locations where likelihood of edges are high. This celebrated model is given by the following nonlinear parabolic PDE

$$\begin{aligned} \frac{\partial I(s,t)}{\partial t} &= div(g(||\nabla I||)\nabla I) \\ I(s,0) &= I_0 \end{aligned} \tag{1}$$

where $||\nabla I||$ is the gradient magnitude of the image *I*, with initial image I_0. *g(x)* is the adaptive diffusion (edge-stopping) function, that satisfy $g(x) \rightarrow 0$ when $x \rightarrow \infty$. This means that diffusion is suppressed across edges so that edge like structures are preserved. Two possible choices of diffusivity introduced by Perona and Malik are

Fig. 1 Perona-Malik diffusion with static scale k = 70 **a** original image, **b** after 60 iterations, **c** after 120 iterations. **d** Canny edge of (**c**)

$$\left.\begin{aligned} & g(x) = \left(1 + \tfrac{x^2}{k^2}\right) \\ \text{or} \quad & g(x) = e^{-\left(\frac{x^2}{k^2}\right)} \end{aligned}\right\} \qquad \text{(2a and 2b)}$$

where k is taken as a positive constant. From Eq. (2a and 2b) suggests that k plays an important role in determining the heterogeneity that demarcates the low and high contrast areas. For large values of k, all discontinuities disappear, and the result will be same as smoothing by a linear diffusion equation. On the other hand, if the value of k is small, no smoothing is performed and all the edge like discontinuities along with noise are preserved. Traditionally this parameter is set by hand and remains static for all iterations. Hence comparison of the results of different anisotropic diffusion process becomes difficult for the same image with different k values.

Figure 1 shows the result of using Eq. (1) on an image. Observe that on 120 iteration edges and other structures become almost blurred. Canny edge finds only the coarse edges.

3 Robust Interpretation of Anisotropic Diffusion

When an image is considered as a piecewise continuous regions separated by edges or some other structures and if the noise present in the image is *additive Gaussian white noise*. With this simplified assumption, Black et al. [10] observed that distinguishing homogeneous region and heterogeneous structure of an image I can be interpreted as an estimation problem of finding outliers. Intensity differences $(\mathbf{I_p} - \mathbf{I_s})$ in an image between pixel 's' and its neighbours 'p' is small if the region is homogeneous. On the other hand, if edge or texture is present, the local mean of intensity differences will be highly *influenced* by high/low intensity and can be considered as coming from different distributions. Depending on above principle Black et al. devised a robust statistical formulation of P-M model as an M-estimator problem given by

$$\min_{I} \sum_{s \in I} \sum_{p \in \eta} \rho(I_p - I_s, \sigma) \tag{3}$$

where $\boldsymbol{\rho}(.)$ is a criterion function of an *M-estimator* with $\boldsymbol{\sigma}$ as scale parameter and η is the neighbourhood of currently considered pixel s. So the appropriate choice of a good edge-stopping function is nothing but the choice of a *robust ρ-function* (criterion function) in an *M-estimator.* Discrete formulation of Eq. (3) given by

$$I_s^{t+1} = I_s^t + \frac{\lambda}{|\bar{\eta}|} \sum_{p \in \eta} \psi(I_p - I_s^t, \sigma) \tag{4}$$

where $\psi(s, \sigma) = \frac{\partial \rho(s, \sigma)}{\partial s}$ is known as the *influence function* in robust statistics literature and t denote time (iteration). Defining diffusion coefficient $g(x)$ in Eq. (2a) in P-M model as $g(x) = \rho'(x)/x$, this can be shown that $\rho(x)$ rapidly decreases at tails, but never touches zero, so small amount of smoothing still results in image structure for higher upscale. For restoration of the edges, a more robust choice of ρ is required that descend rapidly at the boundary [9]. The most useful choice will be the redescending M-estimators available in the robust statistics literature. Amongst them, Tukey's biweight function is chosen by Black et al. due to its algebraic nature. Biweight function is given by

$$g(x, \sigma) = \begin{cases} 1/2\left(1 - (x/\sigma)^2\right)^2, & |x| \le \sigma \\ \mathbf{0}, & \textbf{otherwise.} \end{cases} \tag{5}$$

it rapidly descends to zero at tails. The diffusion function in Eq. (1) is replaced by Eq. (5) in robust anisotropic diffusion. But both the functions (2 and 5) depends on the correct estimation of scale (k or σ).

4 Heterogeneity Scale Estimation

As mentioned above estimation of heterogeneity scale in P-M model as well as in robust AD becomes a challenging problem. In this section, we discuss our proposed heterogeneity scale and the modified version of Black et al. robust diffusion model.

4.1 *MAD as a Heterogeneity Scale*

For automatic estimation of heterogeneity scale k, Black and Shapiro et al. [9] propose to use a well-known robust scale $k^t = \text{cMAD}(\nabla I^t)$. Where MAD (x) = median(||x-median(||x||)) and c = 1.4826 and k^t is calculated dynamically in each iteration 't'. In spite of its robustness, MAD assumes a symmetric view of the underlying distribution, because at first a central location (the median) is computed and about this centre, it attaches equal importance to positive and negative deviations, which is not very natural approach in case of an asymmetric distribution. As Rousseeuw et al. [11] observed, "*Of course, there is nothing to stop us from using the MAD at highly skewed distributions, but it may be rather inefficient and artificial to do so*". So, in the next section, we present an alternative *location independent L-estimator scale* that expeditiously segregates the *homogenous* and the *heterogeneous* regions of an image with minimum *staircasing* effect. Not only that, this scale converges to a constant value when diffusion process continues, so it determines the stopping condition of the process.

4.2 Proposed Heterogeneity Scale and Diffusion Function

L-estimators is a linear combination, computed using the *order statistics* $x_{(n:1)} \leq x_{(n:2)} \leq \ldots \leq x_{(n:n)}$ corresponding to the observations $x_1, x_2, \ldots, x_n$ [12, 13]. The general *L-estimator* is given by the following equation

$$T_n = \sum_{i=1}^{n} c_{ni} h(x_{n:i}) + \sum_{j=1}^{k} a_j h^* \left(x_{n:\,[np_j]+1} \right) \tag{6}$$

where c_{ni} and $a_j \geq 0$ are given coefficients, $[np_j]$ denotes integer part of the argument and $0 < p_1 < p_2 < \ldots < p_k < 1$. The functions $h(.)$ and $h^*(.)$ are determined accordingly to make the estimate robust. An important point to be noted in the Eq. (6) is that first summation on the right takes into account all observations, but the second summation chooses only a finite number of sample quantiles. In the proposed scale, the first summation on the right Eq. (6) is taken as zero, i.e. $c_{ni} = 0$ for $0 < i < n$. In this case, Eq. (6) is known as *Type-II* L-estimator (*Type-I* is when $a_j = 0, \forall j$). The remaining function $h^*(.)$ is calculated as follows

$$h^* \left(x_{[np_j]} \right) \begin{cases} Med\{x_i < [np_j]\}, when\ 1 \leq i \leq [np_j] - 1 \\ Med\{x_i \geq [np_j]\}, when [np_j] \leq i \leq n \end{cases} \tag{7}$$

where $j = 1, 2$; value of $a_j = 1/2, \forall j$ and $\frac{1}{n} \leq p_j \leq \frac{n-1}{n}$. Thus, $1 \leq [np_j] \leq n-1$ and the simplified version of T_n will look like

$$T_n([np_j]) = \frac{1}{2}\left(Med\{x_i < [np_j]\} + Med\{x_i \geq [np_j]\}\right) = \frac{1}{2}\left(\mathbf{L}_{[np_j]} + \mathbf{H}_{[np_j]}\right) \tag{8}$$

where $\mathbf{L}_{[np_j]} = Med\{x_i < [np_j]\}$ and $\mathbf{H}_{[np_j]} = Med\{x_i \geq [np_j]\}$. Hence when p_j varies T_n gives average location of the median for all $x_i < [np_j]$ and $x_i \geq [np_j]$. Now to get the heterogeneity scale parameter, an objective function is created for optimisation as

$$\widehat{\tau} = \min_{\tau} \sum_{[np_j]} |T_n([np_j]) - \tau| \tag{9}$$

The advantage of this estimator is that it is *location independent*, because with increasing value of the threshold $[np_j]$, on each iteration, T_n is calculated as an average of the medians above and below the threshold. This means that, T_n does not depend on any central location. In the proposed algorithm $\widehat{\tau}(t)$ depends on the iteration 't'. On each iteration $\widehat{\tau}(t)$ will be calculated, until it converges, i.e. $|\hat{\tau}(t) - \hat{\tau}(t-1)| < \in$, where $\epsilon > 0$ is small enough real number.

Most of the anisotropic diffusion methods use local gradient ∇I_s in the diffusion function as criteria to measure local edge. But we find local variation (C_s^t) gives a better alternative as proposed by Aja-Fernandez et al. [14]. Chao et al. [10] observe that for additive noise local variance gives a good estimate of heterogeneity. Accordingly, modified Tukey's biweight function is given by

$$g(C_s^t) = \begin{cases} \frac{1}{2}\left(1 - \left(\frac{C_s^\tau}{\tau_c^t}\right)^2\right), & when\, |C_s^\tau| \le \tau_c^t \\ 0 & otherwise \end{cases} \tag{12}$$

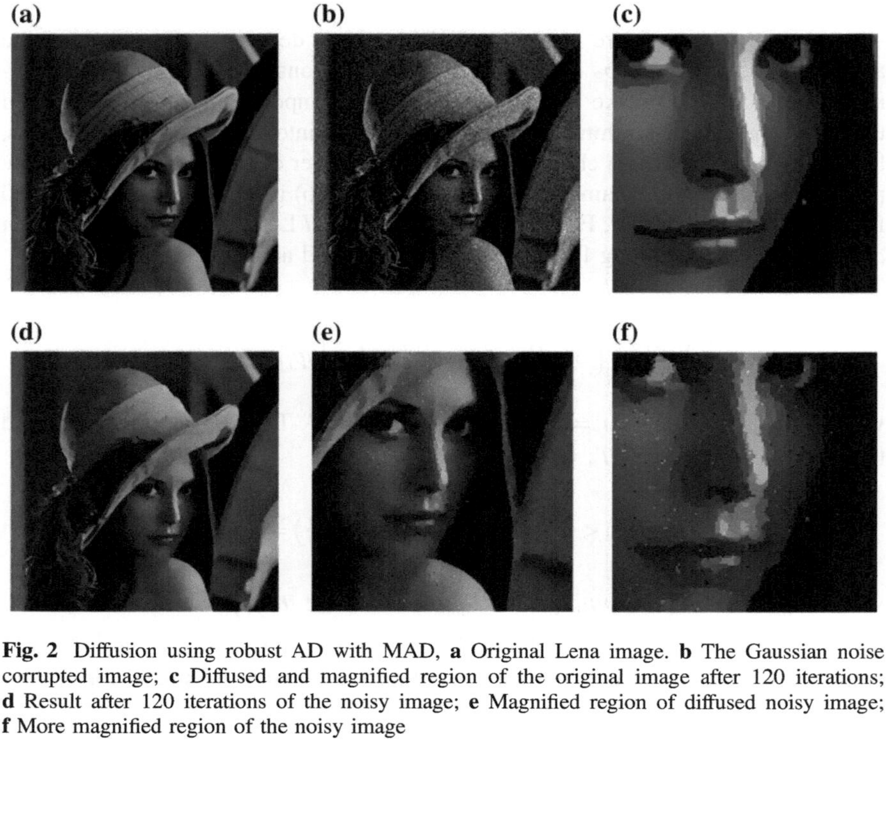

Fig. 2 Diffusion using robust AD with MAD, **a** Original Lena image. **b** The Gaussian noise corrupted image; **c** Diffused and magnified region of the original image after 120 iterations; **d** Result after 120 iterations of the noisy image; **e** Magnified region of diffused noisy image; **f** More magnified region of the noisy image

where C_s^t is local variance in the neighbourhood of s and τ_c^t is proposed robust scale. We replaced x in Eq. (5) by C_s and σ by τ_c.

5 Discussion on Experimental Results

In this section test results of the proposed algorithm is presented and compared with other diffusion algorithms. Figure 2a shows the original Lena image as our test image. 2% Gaussian noise is added to this image as shown in Fig. 2b. Next, we use robust diffusion algorithm of Black et al. with MAD as a scale on the original image and noisy image and the results are shown in Fig. 2c–f. Staircasing effect is clearly visible from the magnified images in Fig. 2c, f.

Figure 3 shows the results of robust anisotropic diffusion with our proposed scale. Excellent smoothing with minimal staircasing effect is visible from the magnified images in Fig. 3b, c, e and f. Further, in presence of noise, the proposed scale is much more effective than MAD. This can be observed if a comparison is done between the second rows of Figs. 2 and 3.

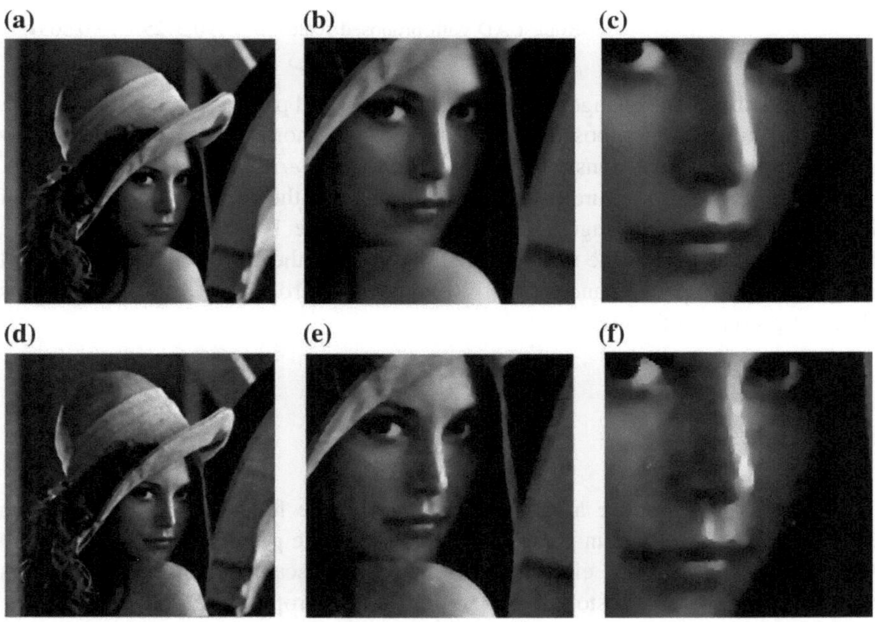

Fig. 3 Diffusion using robust AD with proposed scale, **a** Diffusion of original Lena image after 120 iterations **b** Magnified region of (**a**); **c** More magnified region of (**a**); **d** Result after 120 iterations of the noisy image; **e** Magnified region of diffused noisy image; **f** More magnified region of the noisy image

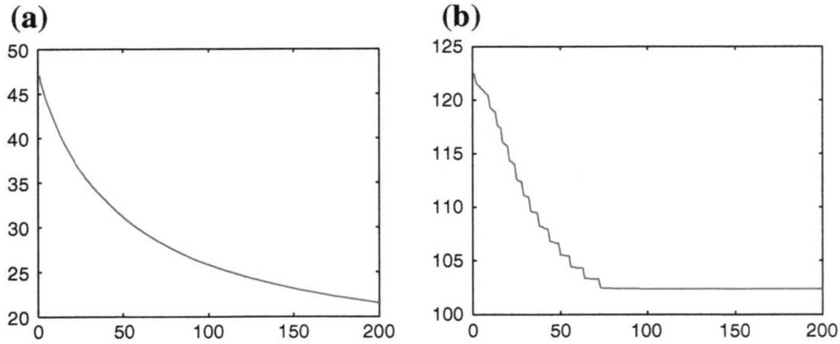

Fig. 4 Graph showing the change in value of the heterogeneity scales during diffusion process applied on the Lena image: **a** for MAD **b** for proposed scale

Table 1 FOM values of the diffusion algorithms

Diffusion algorithm	FOM	
Number of iteration ➔	60	120
Perona-Malik (P-M) (scale k = 75)	0.2436	0.2240
P-M with MAD	0.3520	0.3825
P-M with proposed scale	0.4732	0.5321
Robust AD with MAD	0.4572	0.4932
Robust AD with proposed scale	0.5125	0.5634

Figure 4 shows the change in the value of MAD and proposed scale on iteration. Figure 4b shows that proposed scale converges to a nonzero value after few iteration. This value can be considered as the *stopping criteria* in anisotropic diffusion.

We also use Pratt's figure of merit (FOM) [15] as the performance measure of these algorithms. FOM ranges from 0 to 1, with value 1 when all ideal edges are detected. Table 1 shows the results of FOM. We used the proposed scale with P-M model to see the performance. Clearly proposed heterogeneity scale outperforms MAD in presence of noise.

6 Conclusion

In this paper, we propose a robust L-estimator scale for using with anisotropic diffusion. It is shown that in presence of noise the scale performs better than MAD with minimal staircasing effect. Additionally, this scale stabilizes on iteration processes and provides a stopping condition in anisotropic diffusion.

References

1. Perona, P., Malik, J.:Scale-space and edge detection using anisotropic diffusion. IEEE Trans. Pattern Anal. Mach. Intell. **12**(7) (1990)
2. Weickert, J.: Anisotropic diffusion in image processing. B.G. Teubner Stuttgart (1998)
3. Aboulaicha, R., Meskinea, D., Souissi, A.: New diffusion models in image processing. J. Comput. Math. Appl. **56**, 874–882 (2008)
4. Acton, S.: Multigrid anisotropic diffusion. IEEE Trans. Image Proc. **7**, 280–291 (1998)
5. Weickert, J., Scharr, H.: A scheme for coherence-enhancing diffusion filtering with optimized rotation invariance. J. Visual Comm. and Image Repr., Spec. Issue On PDEs Img. Proc., Comp. Vis. and Comp. Grap. **13**(1/2), 103–118 (2002)
6. Chen, Q., Montesinos, P., Sun, Q.S., Xia, D.S.: Ramp preserving perona-malik model. Signal Process. **90**, 1963–1975 (2010)
7. You, Y., Xu, W., Tannenbaum, A., Kaveh, M.: Behavioral analysis of anisotropic in image processing. IEEE Trans. Image Proc. **5**(11), 1539–1553 (1996)
8. Catt´e, F., Lions, P.L., Morel, J.M., Coll, T.: Image selective smoothing and edge detection by nonlinear diffusion. SIAM J. Numer. Anal. **29**, 182–193 (1992)
9. Black, M.J., Sapiro, G., Marimont D.H., Heeger, D.: Robust Anisotropic Diffus. IEEE Trans. Image Proc. **7**(3) (1998)
10. Chao, S., Tsai, D.: An improved anisotropic diffusion model for detail- and edge-preserving smoothing, Patt. Recognit. Lett. **31**, 2012–2023 (2010)
11. Rousseeuw, P.J., Croux, C.: Alternatives to the Median Absolute Deviation. J. Am. Stat. Assoc. **88**(424) (1993)
12. Huber, P., Ronchetti, E.: Robust Statistics, 2nd edn. Wiley
13. Jureckova, J., Picek, J.: Robust Statistical Methods with R. Taylor & Francis Group, LLC (2006)
14. Aja-Fernández, S., Ferrero, G., Martín-Fernández, M., Alberola-López, C.: Automatic noise estimation in images using local statistics. Additive and multiplicative cases. Image Vis. Comput. **27**, 756–770 (2009)
15. Pratt, W.K.: Digital Image Processing, 4th edn. Wiley

Efficient Economic Profit Maximization: Genetic Algorithm Based Approach

Sankhadeep Chatterjee, Rhitaban Nag, Nilanjan Dey and Amira S. Ashour

Abstract Economic profit is the main governing power of any industrial and socio-economic growth. It is imperative to maximize profit related with economic systems to retain economic stability. Traditional methods involving Lewis model has been found to be unsuitable in terms of computational complexity. Motivated by the recent developments and successful application of meta-heuristic algorithms in achieving potent solutions, the present work proposed efficient meta-heuristic algorithm to support the profit maximization formalism. The genetic algorithm (GA) is employed to maximize the profit in terms of the total revenue (TR) and total cost (TC). The real parameter objective function depicting profit has been gradually optimized by GA. Experimental results suggest that the GA based profit maximization method is extremely fast, accurate and robust.

Keywords Profit maximization · Lewis model · Meta-heuristic Genetic algorithm

---

S. Chatterjee (✉)
Department of Computer Science and Engineering, University of Calcutta, Kolkata, India
e-mail: chatterjeesankhadeep.cu@gmail.com

R. Nag
Department of Economics, Raja Peary Mohan College, Uttarpara, Hooghly, India
e-mail: nag.rhitaban@gmail.com

N. Dey
Department of Information Technology, Techno India College of Technology, Kolkata, India
e-mail: neelanjandey@gmail.com

A. S. Ashour
Faculty of Engineering, Department of Electronics and Electrical Communications Engineering, Tanta University, Tanta, Egypt
e-mail: amirasashour@yahoo.com

X.-S. Yang et al. (eds.), *Smart Trends in Systems, Security and Sustainability*, Lecture Notes in Networks and Systems 18,
https://doi.org/10.1007/978-981-10-6916-1_29

1 Introduction

Economic theory is based on the reasonable people's perception under different constraints in order to support business holders to achieve their businesses successfully. In several businesses, profit maximization is a feasible goal, given the aspiration of the industrial stockholders to drive their challengers out of business. In economic, the term profit refers to the variance between a firm's total income and its total cost opportunity. Profitability and profit increase without loss are the main drive of most businesses. Extra profit is necessary for business, especially in today's competitive market. It helps in expanding the business further, attracts investors, and increases the confidence of labors. Generally, there are two methods can be applied in order to increase profit (revenue), namely: increase the customers' number, raise the prices, or expand the sold products' number. Mathematically, profit is calculated by subtracting cost from revenue. The objective of every firm is to maximize profit in order to expand the business. Higher profit ensures a perpetuation of growth in the future. It has been expected that the future revenue would exceed the present cost as the shareholders expect more from dividend and the banks or the financial institutions expect more from the loan they gave out. More profit tests the business efficiency.

In order to prevent uncertain destabilization due to customer choice and the governmental policies, extra profit could play a vital role. Consequently, the profit maximization is an imperative task to support sustainable economic development. Typically, the economic problems can be framed as an optimization problem, such as cost minimization, and revenue maximization. Thus, efficient optimization methods are required to deliver accurate results. Consequently, several optimization algorithms can be involved to solve the required maximum profit level, such methods include the genetic algorithm (GA), particle swarm optimization (PSO), firefly algorithm, and cuckoo search algorithm (CS). Several studies have established the ingenuity and accuracy of GA [1–5]. Generally, the GA has numerous advantages including its flexibility to model the problem's constraints, and its easy convergence to the optimal solution inspired by Darwinian principle [6].

Nicoară [7] revealed about the GA relevance compared to the traditional methods for manufacturing structure optimization. Geisendorf [8] employed the GA to solve Resource Economic problem using two different assumptions to calculate the optimal extraction rate in order to achieve optimal benefits. Arifovic [9] solved decision rules of future production and sales by employing the GA in a competitive cobweb model in a market of single product. The simulation results indicated that the GA is capable of capturing different features from the experimental nature of the subjects under consideration. Riechmann [10] established that the GA can be connected with the evolutionary game theory. Hommes et al. [11] reveled that GA can converge to a series of near Nash equilibrium solutions, where the heterogeneous agent behavior has been modeled using GA.

Consequently, in the current study the profit maximization is framed as an optimization problem. The objective function in the current study is designed in terms of the total revenue and total cost. Different genetic operators of GA are set by running the experiments 100 times with different setups and the best setup has been selected. The objective function is tested with different search ranges to establish the GA stability in profit maximization. Finally, the economic significance of the experimental results is discussed.

The remaining sections are arranged as follows. Section 2 explains the profit maximization prominence followed by the mathematical deductions and mathematical analysis of the objective functions employed in profit maximization approach. Section 3 briefly introduces the GA. In Sect. 4, the experimental results are reported and the economic significance is discussed. Finally, the conclusion is depicted in Sect. 5.

2 Economic Profitability

According to Lewis model there coexist two sectors, namely: the rural sector and the modern sector having the following assumptions:

(a) In the rural sector, there exist surplus labors that migrate from rural to the modern sector.
(b) In the rural sector, the workers are paid according to their average product; while in the modern sector, the workers are paid according to their marginal product.
(c) The wage in the modern sector (W_m) is higher than the wage in the rural sector (W_r), thus $W_r < W_m$.
(d) There is diminishing return of the output in the modern sector.
(e) The main objective of the modern sector is to maximize the profit.
(f) The profit gained from the modern sector is further reinvested.

Figure 1 depicted the relationship between output and labor in both the modern and rural sector [12].

The plot at the top right corner of Fig. 1 illustrated the relationship between the total output and labor in the rural sector. It is observed that in the short run, any increase in the labor will increase the output till the intersection of the perpendiculars drawn from TP_A and L_A. After the intersection point, the labors (surplus labors) are not important for employment since they are not contributing for the production of the output till the point mentioned earlier. At the point mentioned earlier, MP_L is equal to 0. Another graph of average product of labor (marginal product of labor) versus labor is plotted and the perpendicular passing through L_A is extended below, which indicates the total employment of labor in the plot at right bottom corner. In addition, the plot at top left corner of Fig. 1 demonstrated the relationship between the total product and labor in the modern sector. Initially, the modern sector is producing TP_{M1} amount of product and hiring L_1 amount of labor. A curve ($TP_M(K_{M1})$) is plotted that illustrated the total product keeping the labor

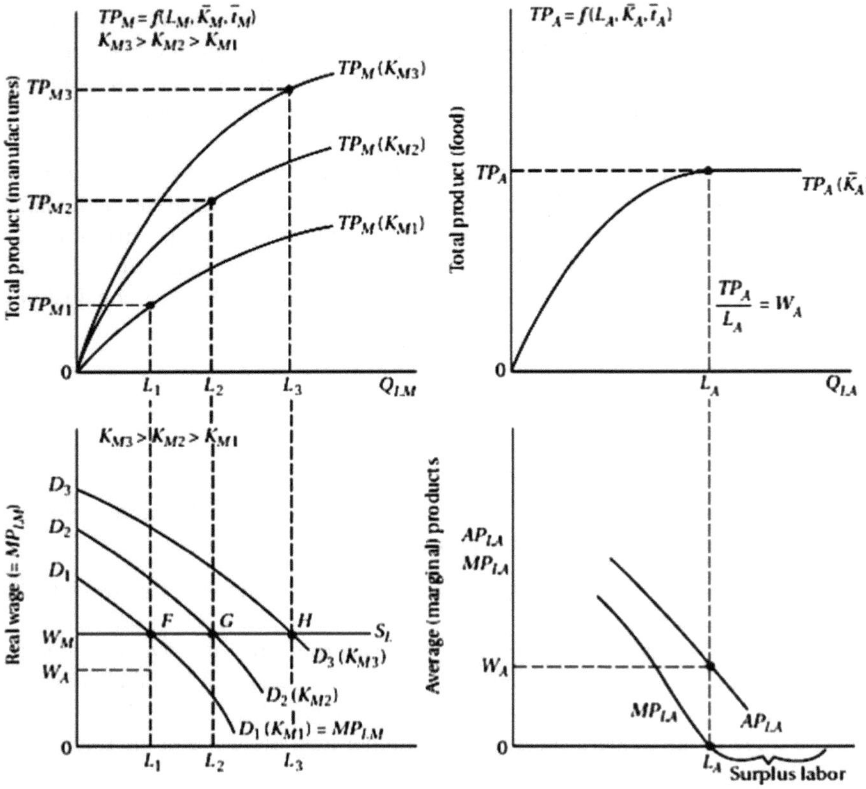

Fig. 1 Plot of labor versus the output based on Lewis model

equals to L_1. Below this graph, a plot of wage versus labor, which demonstrated the perpendicular passing through L_1 is extended below, which cuts the perpendicular parallel to the x-axis at 'F'. The same is done for L_2 and L_3. In left bottom corner plot, the *MPs* of total products are drawn individually for each of the total products. The *MPs* shown the demand curve and the perpendicular drawn parallel with the x-axis from W_M is called Supply Curve.

The area covered by the rectangle OW_MFL_1 is the total wage paid to the labors. In addition, the area covered by OD_1FL_1 depicted the total production. Thus, the profit is equal to the difference between the areas mentioned earlier i.e. $OD_1FL_1 - OW_MFL_1 =$ area covered by W_MD_1F. Furthermore, the plot revealed that the re-investment of the profit results increased the profit in the future. Consequently, the governing parameters for maximizing the profit have a vital role in economy. Apparently, deciding the exact values for such parameter is important. Thus, in the current work, the GA based formalism is applied to achieve the same. The mathematical formulation of the profit maximization model is as follows:

$$\pi = TR - TC \tag{1}$$

$$\pi = p * q - C(q) \tag{2}$$

where, π is the profit, R is the revenue, p denotes the price and q denotes the quantity. Differentiating (1) with respect to q:

$$\frac{d\pi}{dq} = \frac{d(TR)}{dq} - C'(q) \tag{3}$$

At the maxima:

$$\frac{d(TR)}{dq} - C'(q) = 0 \tag{4}$$

$$\frac{d(TR)}{dq} = C'(q) \tag{5}$$

where, $\frac{d(TR)}{dq}$ indicates marginal revenue, and $C'(q)$ denotes marginal cost. Again, by differentiating Eq. (3) with respect to q, the following expression is obtained:

$$\frac{d^2(TR)}{dq^2} - C''(q) < 0 \tag{6}$$

$$\frac{d^2(TR)}{dq^2} < C''(q) \tag{7}$$

where, $\frac{d^2(TR)}{dq^2}$ denotes the marginal revenue slope, and $C''(q)$ denotes the marginal cost slope. The main condition to maximize the profit is to achieve equal marginal revenue and marginal cost. In order to verify the maximization criteria, the slope of the marginal revenue will be less than the slope of the marginal cost.

3 Proposed Genetic Algorithm Based Profit Maximization

3.1 Genetic Algorithm

The GA is a parallel optimization procedure that relies on evolution for optimizing a group of solutions at once [6]. The model is highly inspired from Darwinian principle of Natural Evolution, and involves a population which participates in finding the solution of a particular problem under consideration. In the proposed work, the GA is applied to determine the optimal profit value, where the GA proves its effectiveness for superior convergence toward global optimization compared to other global search algorithm. In GA, the individuals' population illustrated by some chromosomes is iteratively updated by considering significant operators of selection, mutation and crossover in order to solve the problem under concern. Each individual is evaluated using a fitness function for optimization, which is performed using natural swap of the genetic material between parents, where the offspring's are formed from the parent genes. In a natural selection process, the fitness values are used to select the potential solutions to be considered for the next generation. Mating of parents is characterized by crossover and mutation operations. The overall GA algorithm is as follows [13]:

Algorithm: Genetic Algorithm
Start
Generate random chromosomes' population that represents solutions
Evaluate the population-fitness
Create new population using the following steps:
   Select two parent chromosomes form the population based on their fitness
   Crossover the parents to generate new offspring
   Mutate new offspring at each locus
   Place new offspring in the new population
Use new generated population
Test: If the end condition is satisfied, stop, and return the best solution in present population
Stop

As reported in the preceding algorithm, it is clear that each member of the population represents a potential solution of the problem under concern. Further, each of them is associated with a fitness value indicating the superiority of that particular solution. Solutions at any stage participate in Darwinian reproduction, survival of the fittest and other genetic operations to produce successors. The GA tries to find out the best solution at every generation by breeding the best solutions from the previous generation. The candidate solutions under consideration are actually artificial chromosomes which are inspired from the DNA structure. In practice, these chromosomes have fixed length strings (binary/real coded), where each location (gene) holds information about one of the variables associated in the problem. For a multidimensional search involving optimization of certain objective function, these locations can be the value of a particular component. A multidimensional point can be represented by putting each of these values together. This is one of various available techniques to map search space solutions to the chromosome representation. It manipulates the population (initially random) by applying genetic operators already defined to generate new solutions from them. The genetic operators such as crossover, mutation and several other versions of these two are popular [14, 15]. The offspring's fitness values are generally determined by the concerned objective function to be optimized. After calculating the fitness, some of the superior solutions kept for participating in the next level of evolution. Thus, after certain number of generations, the superior solution will obtained. In addition, after achieving satisfactory fitness of the best solution in the population, this process can be terminated. The selection process of the candidate solutions can also differ [12, 16].

3.2 Genetic Algorithm Based Profit Maximization

Based on the preceding genetic algorithm procedure, Fig. 2 demonstrated the proposed approach for profit maximization using the GA.

4 Results and Discussion

The used parameters of the GA are described in Table 1, which are obtained by running the GA 100 times with different parameters.

Consequently, the best parameters' setup is extracted and is used further to test the model, which consists of 300 population size, 0.15 crossover probability with a Gaussian type mutation, single point crossover and Roulette selection scheme. The experiments are carried out using Intel Core i3 4 GB Machine using randomly selected price and quantity range of market products to test the GA stability. Figure 3 demonstrates the objective function for $50 \leq p \leq 70, 100 \leq q \leq 170$.

The range for p, q can be set to other values depending on the context of profit maximization problem. The problem is converted to a minimization problem, and

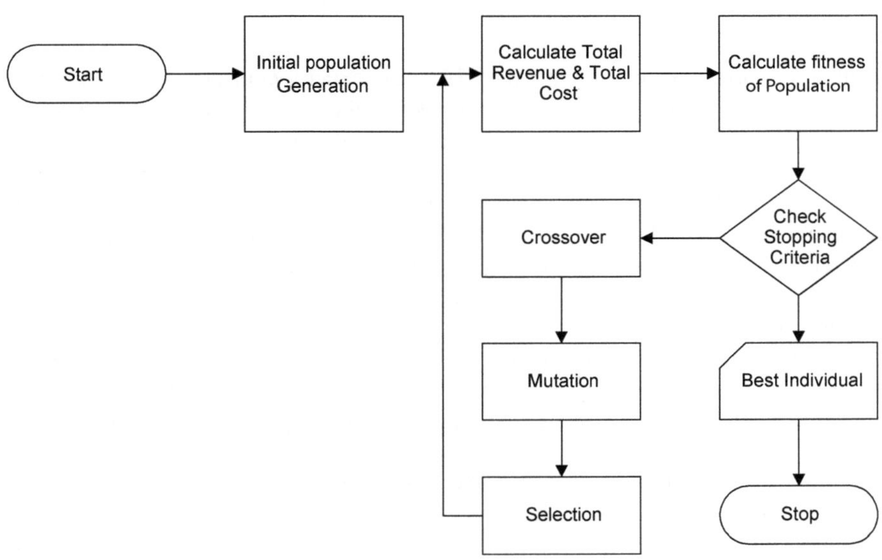

Fig. 2 The profit maximization process using Genetic Algorithm

Table 1 Genetic algorithm setup for input weight vector optimization

Maximum number of generation	800
Population size	300
Crossover probability	0.15
Mutation	Gaussian
Crossover	Single point crossover
Selection	Roulette
Stall time limit	45 s

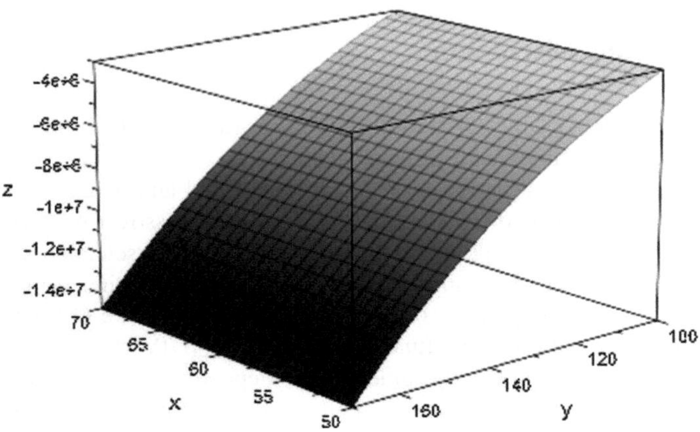

Fig. 3 Objective function with $50 \leq p \leq 70, 100 \leq q \leq 170$

then it is solved by employing the GA. The experiments are conducted using different ranges of p, q. Figures 4, 5 and 6 depict the objective function plots versus the generation during the optimization process for different setups of ranges of p, q.

Figure 4 reports that the GA converges after about 77 iterations when using $50 \leq p \leq 70, 100 \leq q \leq 170$, while Fig. 5 reports that the convergence occur after about 80 iterations when $50 \leq p \leq 100, 50 \leq q \leq 100$. In addition, Fig. 6 depicts the convergence with $50 \leq p \leq 70, 50 \leq q \leq 100$. Table 2 reports the details of the experiments and error during the optimization process. For every setup of p, q the GA has been run for 100 times. The error is calculated in terms of mean error $\pm$. andard Deviation form in calculating the maximum profit. Four different setups were utilized to test the proposed model.

Table 2 establishes that decreased range for both p, q minimizes the mean error and the standard deviation as well. Table 2 indicates the economic significance of the proposed model. It reveals that the optimal results obtained by the GA are almost 100% accurate with a negligible amount of error in achieving the global optima. This establishes the stability of GA based profit maximization and further shows that the results are economically significant.

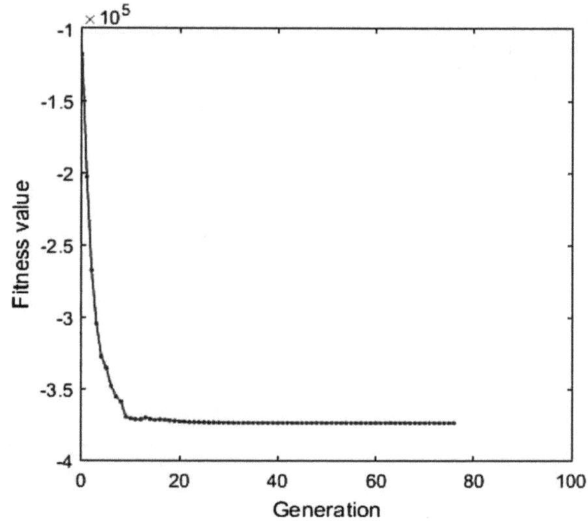

Fig. 4 Convergence of GA during optimization $50 \leq p \leq 70, 100 \leq q \leq 170$

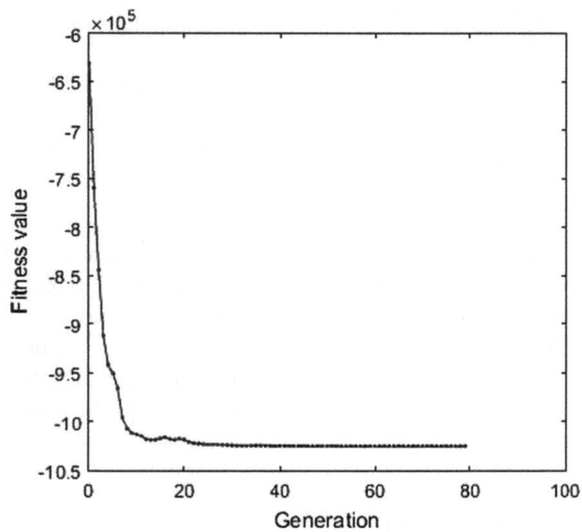

Fig. 5 Convergence of GA during optimization $50 \leq p \leq 100, 50 \leq q \leq 100$

Generally, in the current work, the input is the objective function and input variables range is given in Table 2. These are the only inputs required, as no additional explicit data is required in the proposed optimization like the inputs required in machine learning such as in the neural networks. From the preceding results, the proposed method proves its efficiency to achieve maximum profit, as the mean error for all the objective function is almost negligible along with equally negligible standard deviation, using the GA over the standard method [17] that has been used to solve profit maximization problem.

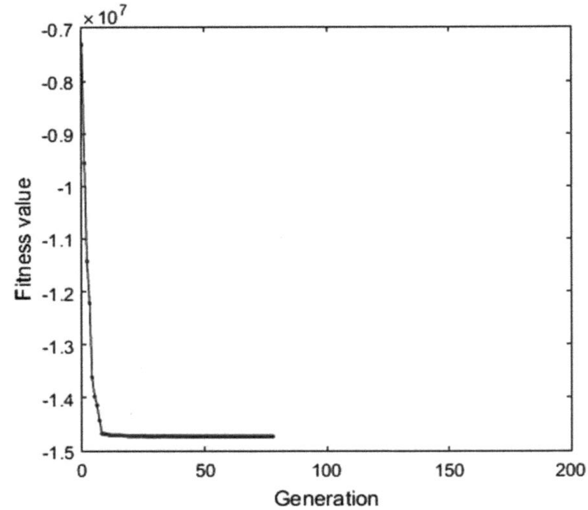

Fig. 6 Convergence of GA during optimization $50 \le p \le 70, 50 \le q \le 100$

Table 2 Performance analysis of GA

Range of p	Range of q	Mean error ± standard deviation
[50 70]	[100 170]	0.257 ± 0.005%
[50 100]	[50 100]	0.192 ± 0.043%
[30 250]	[400 600]	0.385 ± 0.006%
[50 70]	[50 100]	0.095 ± 0.002%

Such traditional methods required complex and time consuming methods. Consequently, the traditional method is not suitable for large and complex problem instances. The current study involved a global search based meta-heuristic GA to solve the same. It is recommended as a future scope is to employ other meta-heuristics [18–26] to improve the accuracy further and to compare the GA with such optimization algorithms based profit maximization.

5 Conclusion

The current work has employed Darwinian Evolution inspired meta-heuristic GA in order to maximize the economic profit by utilizing the total revenue and total cost based objective function. The traditional methods followed in the economical design are inefficient in terms of the computational complexity and accuracy as suggested. However, the proposed GA based maximum profit using Gaussian type mutation coupled with single point crossover and roulette selection scheme. The optimal parameter setup is obtained by using an iterated method. The simulation results indicated that the GA based profit maximization is extremely accurate and efficient in comparison with traditional methods, where 0.095± 0.002% of the mean error and standard deviation were obtained; respectively with [50 70], and [50 100] ranges for p and q; respectively.

References

1. Reeves, Colin R.: A genetic algorithm for flowshop sequencing. Comput. Oper. Res. **22**(1), 5–13 (1995)
2. Beasley, John E., Chu, Paul C.: A genetic algorithm for the set covering problem. Eur. J. Oper. Res. **94**(2), 392–404 (1996)
3. Orero, S.O., Irving, M.R.: Economic dispatch of generators with prohibited operating zones: a genetic algorithm approach. IEEE Proceedings-Gener. Transm. Distrib. **143**(6), 529–534 (1996)
4. Wang, Ran-Zan, Lin, Chi-Fang, Lin, Ja-Chen: Image hiding by optimal LSB substitution and genetic algorithm. Pattern Recogn. **34**(3), 671–683 (2001)
5. Shin, Kyung-Shik, Lee, Yong-Joo: A genetic algorithm application in bankruptcy prediction modeling. Expert Syst. Appl. **23**(3), 321–328 (2002)
6. Goldberg, David E., Holland, John H.: Genetic algorithms and machine learning. Mach. Learn. **3**(2), 95–99 (1988)
7. Nicoară, E.S.: Applying genetic algorithms to optimization problems in economics. In: Petroleum-Gas University of Ploiesti Bulletin, Technical Series 67, no. 3 (2015)
8. Geisendorf, S., Genetic algorithms in resource economic models. Santa Fe Institute, NM, USA, Working Papers, pp. 99–08 (1999)
9. Arifovic, Jasmina: Genetic algorithm learning and the cobweb model. J. Econ. Dyn. Control **18**(1), 3–28 (1994)
10. Riechmann, T., Genetic algorithms and economic evolution. Fachbereich Wirtschaftswiss, University (1998)
11. Hommes, C., Makarewicz, T., Massaro, D., Smits, T.: Genetic Algorithm Learning in a New Keynesian Macroeconomic Setup (2015)
12. Hore, S., Chatterjee, S., Santhi, V., Dey, N., Ashour, AS., Balas, V.E., Shi, F.: Indian sign language recognition using optimized neural networks. In: Information Technology and Intelligent Transportation Systems, pp. 553–563. Springer International Publishing (2017)
13. Malhotra, R., Singh, N., Singh, Y.: Genetic algorithms: concepts, design for optimization of process controllers. Comput. Inf. Sci. **4** (2011). https://doi.org/10.5539/cis.v4n2p39
14. Walters, David C., Sheble, Gerald B.: Genetic algorithm solution of economic dispatch with valve point loading. IEEE Trans. Power Syst. **8**(3), 1325–1332 (1993)
15. Azadeh, A., Ghaderi, S.F., Nokhandan, B.P., Sheikhalishahi, M.: A new genetic algorithm approach for optimizing bidding strategy viewpoint of profit maximization of a generation company. Expert Syst. Appl. **39**(1), 1565–1574 (2012)
16. Chatterjee, S., Ghosh, S., Dawn, S., Hore, S., Dey, N.: Forest Type Classification: A Hybrid NN-GA Model Based Approach. In: Information Systems Design and Intelligent Applications, pp. 227–236. Springer, India (2016)
17. Archibald, G.C., Lipsey, R.G.: An Introduction to a Mathematical Treatment of Economics, 3d edn. (London: Weidenfield & Nicolson, 1977), chap. 17; AC Chiang, Fundamental Methods of Mathematical Economics (1967)
18. Kausar, N., Palaniappan, S., AlGhamdi, B.S., Samir, B.B., Dey, N., Abdullah, A.: Systematic analysis of applied data mining based optimization algorithms in clinical attribute extraction and classification for diagnosis of cardiac patients. In: Applications of Intelligent Optimization in Biology and Medicine: Current Trends and Open Problems, pp. 217–231 (2016)

19. Karaa, W.B.A., Ashour, A.S., Sassi, D.B., Roy, P., Kausar, N., Dey, N.: MEDLINE text mining: an enhancement genetic algorithm based approach for document clustering. In: Applications of Intelligent Optimization in Biology and Medicine: Current Trends and Open Problems, pp. 267–287 (2016)
20. Cheriguene, S., Azizi, N., Zemmal, N., Dey, N., Djellali, H., Farah, N.: Optimized tumor breast cancer classification using combining random subspace and static classifiers selection paradigms. In: Applications of Intelligent Optimization in Biology and Medicine: Current Trends and Open Problems, pp. 289–307 (2016)
21. Dey, N., Ashour, A.S., Beagum, S., Pistola, D.S., Gospodinov, M., Gospodinova, E.P., Tavares, J.M.R.S.: Parameter optimization for local polynomial approximation based intersection confidence interval filter using genetic algorithm: an application for brain MRI image de-noising. J. Imaging **1**, 60–84 (2015)
22. Chakraborty, S., Dey, N., Samanta, S., Ashour, A.S., Balas, V.E.: Firefly algorithm for optimized non-rigid demons registration. In: Yang, X.S., Papa, J.P. (eds.) Bio-Inspired Computation and Applications in Image Processing, September 2016
23. Day, N., Samanta, S., Chakraborty, S., Das, A., Chaudhuri, S.S., Suri, J.S.: Firefly algorithm for optimization of scaling factors during embedding of manifold medical information: an application in ophthalmology imaging. J. Med. Imaging Health Inform **4**(3), 384–394 (2014)
24. Kumar, R., Rajan, A., Talukdar, F.A., Dey, N., Santhi, V., Balas, V.E.: Optimization of 5.5 GHz CMOS LNA parameters using firefly algorithm. Neural Comput. Appl. (NCAA) 1–15 (2016)
25. Ashour, A.S., Samanta, S., Dey, N., Kausar, N., Abdessalem karaa, W.B., Hassanien, A.E.: Computed tomography image enhancement using cuckoo search: a log transform based approach. J. Signal Inf. Process. **6**(4) (2015)
26. Dey, N., Samanta, S., Yang, X.-S., Chaudhri, S.S., Das, A.: Optimisation of scaling factors in electrocardiogram signal watermarking using cuckoo search. Int. J. Bio-Inspir. Comput. (IJBIC) **5**(5), 315–326 (2013)

DLNEx: A Tool to Automatically Extract Desired Learning Nuggets from Various Learning Materials

Jyoti Pareek and Maitri Jhaveri

Abstract The current educational ecosystem of untethered, anytime anywhere learning pedagogy has accelerated individualized and personalized learning. With the ubiquitous 24/7 access to educational resources each learner wants the learning contents to be filtered and manifested as per his/her desire. These short capsules of knowledge, referred to as learning nuggets, improve time efficiency of learners by distilling and targeting specific utilizable information. DLNEx tool automatically extracts learning nuggets such as examples, problem solving sums, formulae, lines of code, theorems, diagrams, charts, graphs and readability score from three types of files: Pdf, doc and HTML. For a given topic/keyword this tool also facilitates student/instructor to view their desired nugget from various learning materials at one place instead of pondering through all different materials. Experiments show that the average F-measure over three types of files in learning nugget extraction is 90%.

Keywords Learning nuggets • Preferred reading • Learning material
Knowledge extraction • Pattern matching

1 Introduction

Technology enhanced E-learning has emerged as a preferred alternative to traditional classroom learning. It can allow students to learn and study at their convenient times. Key features of this centralized, simple and blended online learning are convenience, quick results, improved retention, better access to learning, affordability of materials and introduction to global market space. Today's learner prefers

J. Pareek (✉)
Department of Computer Science, Gujarat University, Ahmedabad, India
e-mail: drjyotipareek@yahoo.com

M. Jhaveri
Banasthali University, Vanasthali, Rajasthan, India
e-mail: maitrijhaveri@yahoo.com

X.-S. Yang et al. (eds.), *Smart Trends in Systems, Security and Sustainability*, Lecture Notes in Networks and Systems 18,
https://doi.org/10.1007/978-981-10-6916-1_30

Table 1 Importance of various learning nuggets for students and evaluators

Learning nuggets	Importance
Tables and images	Student:—Easy reading, larger amount of information gain in a shorter time span
	Evaluator:—Can frame question with caption
Figures such as graph, maps and charts	Student:—Represent sizeable numerical or statistical data in a time- and space-effective manner
	Evaluator:—Can frame question with caption
Lines of code	Student:—Steps for solving a problem
	Evaluator:—Can ask question such as "completion of code"
Theorem	Student:—Proof of a hypothesis
	Evaluator:—Can ask to prove the theorem
Formula	Student:—Exact solution to a given problem.
	Evaluator:—Can ask to give formula by supplying the LHS
Numerical	Student:—Better understanding of mathematical concept/formula/theorem Evaluator:—Can ask to solve the numerical
Example	Student:—For clarity of concepts and to visualize different perspective of corresponding theory.
	Evaluator:—Can ask to give example of a concept. For "for example" template the concept can be found in the previous sentences
Case/scenario	Student:—To understand different alternatives to a situation/concept
Long sentences/paragraphs/words	Student:—short sentences/paragraphs/words indicate that the text is easier to read
Active words and assertive sentences	Student:—Indicates the LM is straightforward and communicates the concepts efficiently
Audio/video	Student:—Best for auditory/visual learner who learns best by hearing/visualizing

reading selective learning nuggets instead of longer reading materials. Learning nuggets refer to the concept of delivering content in small and segregated portions that are convenient to consume and whose meaning can be grasped quickly. A learning material[LM] in computer science domain is composed of multiple learning nuggets such as diagrams, tables, lines of code, examples, problem solving exercises, formulae, expressions, charts, maps, graphs, media files, interactive dialogues, simulation, games etc. Table 1 shows the possible benefits gained from these learning nuggets. According to bloom's taxonomy [1] synthesis and analysis level questions can be generated out of these automatically extracted nuggets.

In order to gain deep learning and understanding of concepts, reading must not be confined to a single learning material. Different learning materials capture different insights given by different authors into each topic. Some definitions and examples may appear in one LM and may not in other. Theorems and problem solutions may appear differently showing multiple ways to solve them. Different learning materials may provide different exercises from different domains. Thus knowledge acquisition from multiple learning materials enhances the development of thinking skills such as conceptual awareness, problem solving, reasoning and critical evaluation. A teacher can generate multiple assessment questions if s/he has a complete bank of learning resources from different materials.

Thus DLNEx can be utilized for

1. Reading selected nugget from one/more learning materials.
2. Generation of a new learning material having combination of preferred nuggets.
3. Recommending the topic having preferred nuggets.
4. Recommending the learning material having preferred nuggets.
5. Generation of questions as mentioned in Table 1.
6. Generation of answer content.

In following sections, Sect. 2 describes the study of various learning management systems, learning repositories and methods of extracting knowledge from unstructured text. Section 3 discusses our approach for automatic extraction of learning nuggets from learning materials. Section 4 discusses the experiments performed and the results obtained. Section 5 gives the evaluation of the experiments performed. Conclusion is drawn in Sect. 6 followed by the references.

2 Literature Review

According to a EDUCAUSE Core Data Service (CDS) survey on the Current Ecosystem of Learning Management Systems[LMS] in Higher Education done by EDUCAUSE CENTER FOR ANALYSIS AND RESEARCH, LMS are in extensive use since eight years and is expected to approach nearly $8 billion learners in 2018 [2]. Major tasks executed by these LMS are

- Online materials such as PDFs, PowerPoint presentations, and courses are delivered.
- Quick and efficient creation of online training material.
- Maintenance of attendance in classrooms and other learning events.
- Measuring the understanding of student in subject content using pre and post assessment.
- Reporting on learning history and training activity.
- Interest list can be known by flagging the corresponding course
- Maintaining training history, certification history, and curriculum history.
- Inclusion of course details such as name, code, cost, course description, target audience, prerequisites and other information.
- Resources can be searched by Keyword, Type (Link or Document and Media) or Category.

Higher education "customers" are seeking efficient, effective, and strategically priced systems that meet their institutional standards (for highly agile, interoperable

Table 2 Metadata displayed by various learning repositories

	Repository	Metadata displayed for Computer Science Learning material[LM]
1	Khan academy [12]	Learning material description
2	Merlot [13]	Category, language, type, format
3	ISO repository of teaching materials [14]	Type, description, author, publisher, keywords, language, target audience
4	Connexions [15]	Name, id, language, summary, subjects, keywords, author, publisher
5	MIT open Course ware [16]	Level, type of material, description
6	Brightspace [17]	Ppts and videos of learning material
		No metadata
7	Consortium for Educational Communication [18]	All videos of learning material
		No metadata
8	ITknowledgeBase PennState [19]	Containing any of the specified words
		Containing specific phrase
		Containing none of the specified words

systems) and user expectations (to have mobile-friendly, personalized, customizable, intuitive, and integrated systems designed to enhance student learning) [2]. We have analyzed few famous learning repositories with respect to the metadata of resources displayed. Table 2 gives the metadata displayed by 8 such repositories. Almost similar kind of metadata is displayed by other not listed repositories. The type of resources considered from these repositories are textbook, article, case study, presentation, teaching cases, web based material.

Content management is very important for any learning management system and learning object repositories [LOR]. We have shown various features of these LMS and LOR. Whenever a Learning Object is displayed to the student it is always followed by its metadata such as Name, id, language, summary, subjects, keywords, author, publisher Category, format, type and description. Nowhere individual nuggets are displayed. These nuggets can be individual images, tables, lines of code, examples, numerical etc. moreover the ease of English language is also not displayed which may be required by some students from regional background. It is very necessary for a student to be aware of presence of his/her preferable learning nuggets in the learning material [LM] s/he wants to read. It helps to choose which LM to study out of the list of LMs present in the repositories for the subject of choice. More recent research in traditional classrooms has also shown the benefits of letting students learn at their own pace and focus on topics that interest them [3]. Our work extracts learning nuggets from the LMs and displays them to the students when required. Several techniques exist to extract knowledge from unstructured text. They can be categorized as NLP, machine learning, clustering, heuristic based and statistical methods. K-Extractor [4] uses plain text, HTML, XML, PDF, MS Office and e-mail as input; extracts concepts, named entities, instances, concept hierarchy, generic relationships, user-defined relationships, events, modality, tense, entity linking, event linking, sentiment using NLP, machine learning, and heuristic rules. OntoSyphon [5] uses HTML, PDF and DOC as input; extracts concepts, relations and instances using NLP, statistical methods. Text2Onto [6] takes plain text, HTML and PDF as input; extracts concepts, concept hierarchy, non-taxonomic relations, instances, and axioms using NLP, statistical methods, machine learning and rule-based methods. Yi-Fang et al. [7] describes a Key phrase Identification Program (KIP), which extracts document key phrases by using prior positive samples of human identified phrases to assign weights to the candidate key phrases. Towards an efficient and scalable nugget-based evaluation, Matthew et al. [8] have studied the applicability of semi-automatic nugget extraction in the context of the ongoing NTCIR One Click Access (1CLICK) task. In their paper [9] Eleni Stroulia et al. describe EduNuggets as an intelligent framework that enables the semantic organization of multimedia material so that students can access coherent, authoritative information on a subject area. Raymond Y. K. Lau et al. [10] gives illustration of a novel concept map generation mechanism which is underpinned by a fuzzy domain ontology discovery algorithm. The proposed mechanism can automatically construct a concept map based on the messages posted to an online discussion board. Wen-Bin Yu et al. [11] presents a study that is able to assist the evaluators of the E-learning system to obtain the summarized key terms of major "concerns" without going through potentially huge amount of survey data. The knowledge extracted by above methods is concept, concept hierarchy, named entities, relationships, events, instances and axioms. But when unstructured text is cited as a learning material, above extracted knowledge does not suffice. If the learning style of the computer science student is in focus then the important knowledge contents needed to be extracted are tables, images, examples, numerical, formulae, case studies, theorems, lines of programming language code, audios, videos, simulation, animation, interaction etc. These existing methods are still not exploited for extracting such entities from the unstructured LMs. Moreover the concept of display of only desired nugget from multiple available learning materials is yet not touched upon. We propose to use a hybrid approach which incorporates the NLP, pattern based extraction, rule based extraction and grouping for choice based extraction and display.

3 Our Approach

Figure 1 shows the working of DLNEx. The learning material can be supplied and uploaded by the learner/instructor or s/he can choose to read from the existing repository. The learning material can be of type doc, pdf or HTML. Three parsers are designed for parsing Doc, pdf and HTML files. These parsers provide raw text, tables and embedded images from the files. The method for extraction of the remaining knowledge from the raw text is based on pattern matching of related keywords, programming symbols, conditional symbols, latex symbols, syntax, programming variables, prefix variables, data types, postfix variables, and media

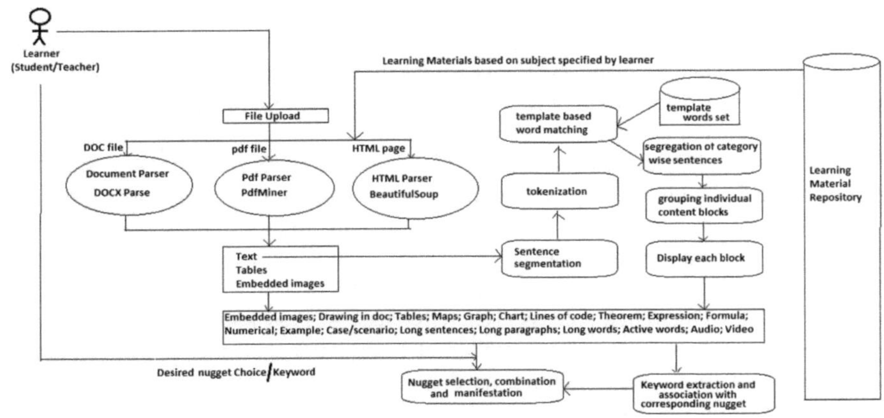

Fig. 1 Architectural diagram of DLNEx

file. This pattern set is consolidated as a result of comprehensive study of various corresponding subject materials. Student/instructor provides the topic and type of the nuggets to view. The nuggets are then displayed accordingly.

The entire architecture can be explained in following 5 steps.

Step 1. DOC, Pdf and HTML files are parsed.
Step 2. Automatic nuggets are extracted from the text obtained from step 1.
Step 3. Keywords are extracted from the text of the LM specifically from images, theorems, LOC, example sentences; sentences appearing before/after lines of code, images, tables, numerical and theorem; table/figure captions, paragraph heading, formula heading, LM title and sentences in which following patterns appear {'fig#', 'figure#', 'table#', '*#.#'}.
Step 4. Keywords are associated with corresponding nuggets.
Step 5. Displaying the learning nuggets as per desire.

The methods used in individual nugget extraction are as follows.

Nugget 1: **Lines of Code**:

The identification is based on the presence of programming variables, prefix variables, data types, access modifiers and conditional variables from languages such as c, python and java in text extracted by the parsers. Change in font style along with above rule adds confirmation.

1. Fetch each line of code
2. Group these lines into corresponding code snippet based on their group occurrence. The code is assumed to terminate as soon as non programming sentences start to appear.
3. Display these individual code snippets.

Nugget 2: **Theorem**

1. Identify beginning of the theorem using keyword such as 'theorem'.
2. Identify end of the theorem with keywords such as 'hence', 'proves' or notation such as solid circle or Solid Square.
3. Fetch lines in between.
4. Display these lines as part of theorem.

Nugget 3: **Expression/Formulae**

A formula can be an expression or formula. Presence of latex symbols and regular expression is used to identify equations/formula. We have applied the logic that if more variables (alphabets) appear when compared to constants (numbers) then the equation is a formula; otherwise it is a simple expression.

Nugget 4: **Numerical**

A numerical can appear as

1. A single chunk of numbers and expressions
2. Values in tables as a result of application of some formula
3. Numbers appearing on figures (in form of images) like finding minimum spanning tree based on weighted edges of a graph.

For identification of numerical we need to

1. Identify chunk of numbers appearing together
2. Group them into different single whole sum numerical solution/procedure.

Nugget 5: **Example**

Categories of examples

1. One sentence only
2. Multiple sentences in one paragraph
3. Multiple sentences across paragraphs
4. One complete paragraph
5. Figure
6. Table
7. A combination of text, figures and tables.

We identify the presence of example by presence of any of the example words. Apply co reference resolution to identify till when the example is extended. Fetch these lines and display. Through keyword/nugget-image/table association any figure/table referred in the example also gets displayed along with the example text.

Nugget 6: **Case/scenario**

This can be identified through the occurrence of keywords such as case or scenario followed by a number.

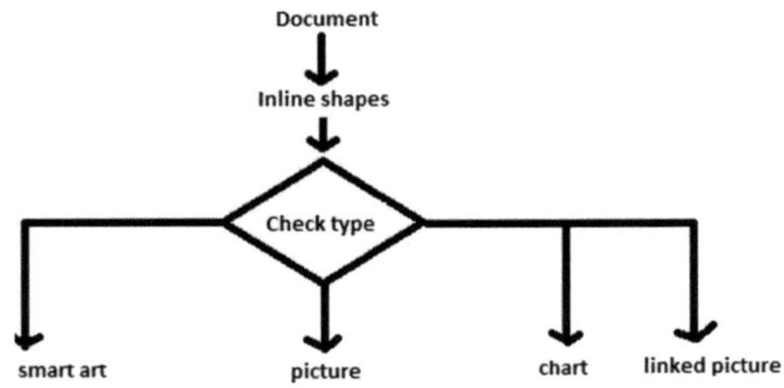

Fig. 2 Shape extraction

Nugget 7: Shapes

Inline shapes can be extracted and can be categorized as smart art, picture, chart and linked picture as shown in Fig. 2. Their coordinates can then be processed to generate and display the required shape.

4 Experiments

Our repository contains learning materials prepared by the experts in corresponding streams. Html documents do not contain unnecessary images or media files which are not related to the subject. These materials contain combination of the learning contents in different proportions. The total capacity of tested materials is approximately 60 MB. Python Natural Language Tool Kit is used for DLNEx tool development. DOCX parser is used to extract information from word documents. PDFMiner is use for extracting information from PDF documents. Beautiful Soup parser is used for scanning the learning materials of HTML type. The process of optical character recognition is used for extracting keywords from the text contained in the images. From word file chart is detected as an inline shape. Then based on the x and y coordinates java script is used to display the chart in HTML file. The desired nuggets are composed and presented in an HTML file. Learning materials from different topics such as Software Requirement Specification, Software Testing, Travelling salesman problem, Data structures and COCOMO model are used for testing. These materials are written by subject experts and are assumed to follow good technical writing guidelines. Readability score is obtained by studying the relative presence of active words, assertive sentences, long words, long sentences and long paragraphs in the learning materials.

DLNEx can answer queries like

- List all figures from LM18 and LM20 related to keywords such as address space, address translation, physical memory, page table, paging, segments and page directory.
- From the LM1 "software engineering. pdf" List all figures, examples, code, numerical, formulae, tables, theorems, expressions and case/scenario.

- List numerical of cocomo model from LM1, cocomo1, cocomo2 and cocomo3.
- List code of whitebox testing from LM1, WB1, WB2 and WB3.
- List figures for DFD from LM1, SRS1 and SRS2.
- List formulae from LM1, cocomo1, cocomo2 and cocomo3.
- List formulae for software estimation.
- List theorems, examples and figures from TS1, TS2, TS3 and TS4.
- List tables/charts related to scheduling.
- List all information related to 'cocomo' keyword from all available learning material in the repository.

Following are few of the screenshots of the queries executed with the output getting displayed in Google chrome. Figure 3 displays two images one for displaying list of formulae for a given learning material and the second displays a numerical with corresponding figure. Figure 4 displays numerical for 'stack' keyword and example for 'cocomo' keyword.

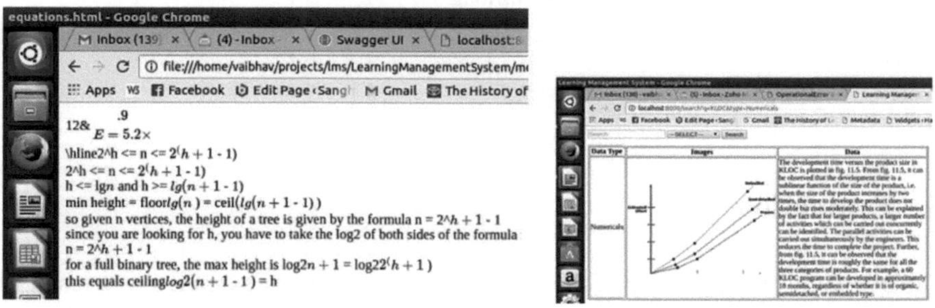

Fig. 3 Screen shot displaying formulae for a given learning material and a numerical with corresponding figure quoted in the text of numerical

Fig. 4 Numerical for 'stack' keyword and example for 'cocomo' keyword

Table 3 Accuracy in nugget extraction

Nugget	Average F1 measure in content extraction		
	DOC	Pdf	HTML
LOC	1	1	1
Theorem	1	1	1
Expression	1	1	1
Formula	1	1	1
Numerical	0.68	0.68	0.68
Example	0.82	0.82	0.82
Case/scenario	1	1	1
Charts	1	Extracted as image	Extracted as image
Embedded images	1	1	1
Tables	1	1	1
Active words/assertive sentences/long words/long sentences/long paragraphs	1	1	1
Audio	1	1	1
Video	1	1	1

5 Evaluation

Table 3 shows the accuracy of the completeness of nuggets extracted as a result of experimentation on various DOC, pdf and HTML files. Since all three files are converted into raw text using corresponding parsers, the results are same irrespective of the file type. We have used F1 score for measuring the test's accuracy. The F_1 score is a weighted average of the precision and recall. F_1 score reaches its best value at 1 and worst at 0.

F1 = 2*precision*recall/(precision + recall)

Precision = number of correct positive results/number of all positive results

Recall = number of correct positive results/number of positive results that should have been returned

6 Conclusion

In the next generation of digital learning environment selective display of learning nuggets is a perfect solution to the learners who want to read only a selected/preferred portion without having to sift through reams of information. To achieve this we have tried to extract different learning nuggets from a given learning material. This material can be of type doc, html and pdf. First the count of the each learning nugget in the material is presented. If the student wishes to see them individually then individual nugget with a combination of referred nuggets are displayed separately. A student/teacher can also view the selected nuggets such as tables or figures or formulae from all available LM in repository belonging to a selected topic/keyword. This can help instructor in development of question bank for a given topic. This nugget can also be revised manually to form an answer. As of now all the available learning materials in the repository are considered. No filtering is applied on selection of LM. We can extend this model to choose only those materials for a student which matches his/her learning style. We need to work more on the accuracy of identification of nuggets such as example and numerical. And lastly we still need more accurate parsing to improve the F-measures for the same.

References

1. http://www.bloomstaxonomy.org
2. Dahlstrom, E., Brooks, D.C., Bichsel, J.: The current ecosystem of learning management systems in higher education: student, faculty, and IT perspectives. Research Report. Louisville, CO: ECAR, September 2014
3. Rose, D., Meyer, A.: Teaching Every Student in the Digital Age: Universal Design for Learning. Association for Supervision and Curriculum Development (2002)
4. K-Extractor, Balakrishna, M., Moldovan, D.: automatic building of semantically rich domain models from unstructured data. In: Proceedings of the Twenty-Sixth International Florida Artificial Intelligence Research Society Conference (FLAIRS), pp. 22–27 (2013)
5. McDowell, L.K., Cafarella, M.: Ontology-driven information extraction with ontosyphon. In: Proceedings of the 5th international conference on The Semantic Web, Volume 4273 of the series Lecture Notes in Computer Science pp. 428–444 (2006)
6. Cimiano, P., Völker, J.: Text2Onto-A Framework for Ontology Learning and Data-driven Change Discovery. In: Proceedings of the 10th International Conference on Applications of Natural Language to Information Systems, vol. 5, pp. 15–17
7. Brook Wu, Y.-F., Li, Q., Bot, R.S., Chen, X.: Finding nuggets in documents: a machine learning approach. J. Am. Soc. Inf. Sci. Technol. **57**(6), 740–752 (2006)
8. Ekstrand-Abueg, M., Pavlu, V., Kato, M.P., Sakai, T., Yamamoto, T., Iwata, M.: Semi-automatic nugget extraction for Japanese one click access evaluation. In: Proceedings of the 36th International ACM SIGIR Conference on Research and Development in Information Retrieval, pp. 749–752 (2013)
9. Stroulia, E., Jari, K.: EduNuggets: an intelligent environment for managing and delivering multimedia education content. In: Proceedings of the 8th International Conference on Intelligent user Interfaces, pp. 303–306 (2003)
10. Lau, R.Y.K., Chung, A.Y. K., Song, D., Huang, Q.: Towards fuzzy domain ontology based concept map generation for E-Learning. In: Proceedings of the 6th International Conference on Advances in Web Based Learning, pp. 90–101 (2007)
11. Yu, W.-B., Luna, R.: Exploring user feedback of an e-learning system: a text mining approach. In: Proceedings of the 15th International Conference on Human Interface and the Management of Information: Information and Interaction for Learning, Culture, Collaboration and Business, vol. Part III, pp. 182–191 (2013)
12. Khan Academy. https://www.khanacademy.org/search?referer=%2Fsearch&page_search_query=binary+search

13. MERLOT: Multimedia Educational Resources for Learning and Online Teaching. https://www.merlot.org/merlot/advSearchMaterials.htm
14. ISO repository of teaching materials. http://www.iso.org/iso/home/standards/standards-ineducation/education_materials-higher-edu.htm
15. Connexions. https://cnx.org/browse
16. MIT open Course ware. https://ocw.mit.edu/courses/find-by-topic/#cat=mathematics&subcat=mathematicalanalysis
17. Brightspace. https://www.d2l.com/en-apac/
18. Consortium for Educational Communication. http://ceclms.nic.in/
19. ITknowledgeBase. http://kb.its.psu.edu/

A Novel Approach to Optimize Subqueries for Open Source Databases

Bhumika Shah, Jyoti Pareek and Darshit Kanziya

Abstract Query Optimization is an important process in Relational databases. Here, we present an overview of the Query Optimization process, with a focus on providing insight of *MySQL sub-queries to optimize Join Operation.* Efficient join processing for sub queries is one of the most fundamental and well-studied tasks in database research. For a given query, there are many plans that can be considered, though the output will be the same, but amount of time required is a key consideration. In this work, we examined existing algorithms for the way the sub queries are joined over many relations and described a novel sorting based algorithm to process these queries optimally. The algorithm is implemented and the results are compared with current strategies of MySQL. The proposed sorting based algorithm outperforms current algorithms without applying DBMS' advanced techniques like parallel processing, hashing etc.

Keywords Sorting · Hashing · Query optimization · Sub-queries
Joins

1 Introduction

Query optimization is a growing research area for more than a decade now, but query optimizers are still among the largest and most complex modules of database systems. The situation becomes challenging when large datasets are used in form of Online Transactions and in the commercial Database systems. These challenges in

B. Shah (✉) · J. Pareek
Department of Computer Science, Gujarat University, Ahmedabad, India
e-mail: bhumikags@yahoo.co.in

J. Pareek
e-mail: drjyotipareek@yahoo.com

D. Kanziya
Amdocs Ltd, Pune, Maharashtra, India
e-mail: darshit.kanziya@gmail.com

X.-S. Yang et al. (eds.), *Smart Trends in Systems, Security and Sustainability*, Lecture Notes in Networks and Systems 18,
https://doi.org/10.1007/978-981-10-6916-1_31

turn demand new database technologies, such as new query languages and new and effective query processing techniques, which are quite different from those in traditional transaction processing applications. A database management system has the options of several plans for a given query. All the plans produced will have same output but their cost would differ (duration to run the query). This type of query optimization is absolutely necessary for any database system. The variation in cost between two different plans can be enormous. Our primary focus, however, is on *Query optimization* for subqueries.

2 Query Processing

Query Processing is the scientific art of getting the expected information from a database system in a reliable manner. This is the method of converting a high level query into such a plan that will execute and retrieve the data from the database (Fig. 1). It consists of four phases namely: Query decomposition, Query optimization, Code generation and Run time query execution [1, 2].

a. **Query decomposition**

In this phase, the syntactic correctness of a query is checked, and if it is incorrect, the error message is generated.

b. **Query optimization**

Query Optimization is all about selecting the best query execution plan from set of plans which takes minimum amount of time to run.

Fig. 1 The stages of query processing [1]

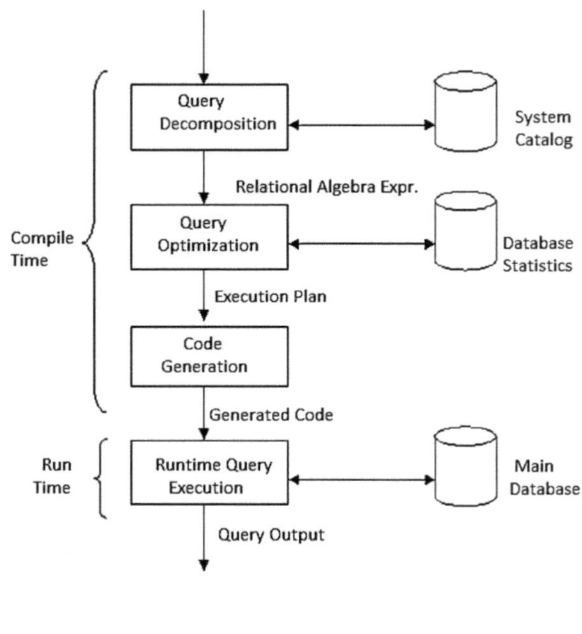

c. **Code generation**

Once the optimizer selects the optimum plan for execution, the code generator will create the equivalent code for the plan which will be sent to internal architecture level of the database for execution.

d. **Query execution**

Here, the code generator will interact with the database and will retrieve the data for the process which initiated the query. Hence, the query processing activity will act as an interface between the querying individual/process and the database. The efficiency of the query optimizer will determine the number of resources to be used and the role played by DBMS for important real time applications.

3 Query Optimization

The method of choosing the most efficient and reliable execution strategy for query execution is called Query optimization [1] and it is one of the most important tasks of any RDBMS. Ramakrishnan and Gehrke [3, 4] observed that SQL which is a de facto standard for data definition and data manipulation in RDBMSs has a variety of ways in which a user can express a query, which is evaluated by the system. Hence query optimizer is responsible for finding the best execution strategy which results in fewer resources for data retrieval.

3.1 Query Optimizer Architecture

The above figure displays an abstraction of the query optimization process in any Database management system. For any provided database and a query on it, various execution plans exist which are used to evaluate the query. All the alternatives should be taken into consideration which will help to decide the best estimation plan for efficient performance [5–8] (Fig. 2).

3.2 Query Optimization in Relational Systems

The query languages in RDBMSs gives a very high-level "declarative" interface which is used to store and retrieve data [9]. Now, SQL has become a standard for relational query languages [9–11]. The two most important components of query evaluation in a database system are the Query Optimizer and the Query execution engine.

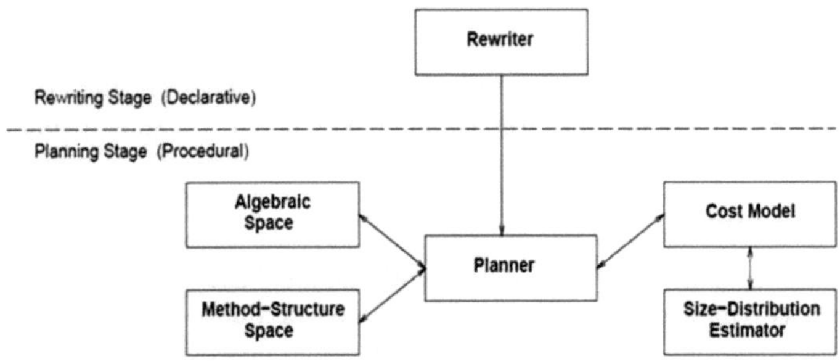

Fig. 2 The architecture of query optimizer [1]

Fig. 3 Abstract representation of operator tree [12]

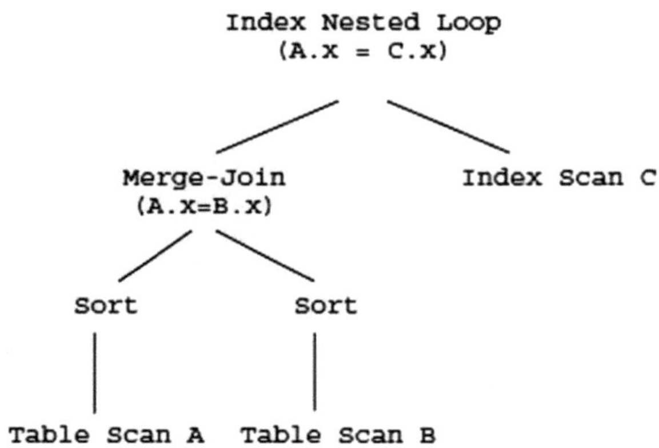

The query execution engine uses physical operators for implementation. The physical operators are those tiny pieces of code which make the query execution possible.

The execution of physical operator tree is shown in Fig. 3. The edges in the operator tree show the flow of data within the physical operators. The execution engine takes care of the execution of the plan and generates the results by giving answers to the query. Hence the efficiency of query execution engine will decide the structure of the operator trees which are feasible.

The query optimizer takes the responsibility for input generation for execution engine. A parsed SQL query is given as input and it generates an efficient execution plan for that given query.

The work of any optimizer is very important as for any given SQL query, there will be many operator trees available:

- The algebraic representation of any query may be transformed into many other logically similar representations.
  e.g., Join (Join (X, Y), Z) = Join (Join (X, Y), Z)

- There will be various operator trees which will implement the algebraic expression for any given algebraic representation.

Moreover, the throughput of the execution of such plans also would widely differ. Hence we can say that Query optimization is a difficult search problem and to solve the said problem, we will have to consider the following:

- Search space, also known as Space of plans.
- The estimation of resources which will be required for the execution of plan.
- An enumeration algorithm, which will be able to scan through the execution space.

A useful optimizer is the one which considers:

1. The space of plan has low cost
2. The technique used is accurate
3. The algorithm used is efficient

All the listed processes above are very important with reference to any database system and hence building a good optimizer is very important [12–14].

4 Subqueries and Join Optimization

Subquery is many a times referred to as inner or nested query. It is a query within a query. A subquery can be used anywhere an expression is allowed. It is a query expression, enclosed in parentheses.

Subqueries are used with Select and all the DML (Insert, Update, Delete) statements and.

e.g. SELECT * FROM emp WHERE emp.id in (SELECT id FROM dept);

4.1 Correlated Subqueries

A correlated subquery uses values from the outer query in its WHERE clause. The subquery is evaluated every time a row is processed by the outer query [15].

```
SELECT * FROM t1
WHERE column1 = ANY (SELECT column1 FROM t2
WHERE t2.column2 = t1.column2);
```

5 MySQL Subquery Processing

MySQL converts the sub-query to a semi-join and makes a cost based choice from the following strategies:

(1) The subquery is converted to a join query and then the query is run as inner join between outer tables and subquery tables.
(2) Remove Duplicates: Run Semi-join for the query as if it was join and remove duplicates using temporary table.
(3) FirstMatch: While scanning entire table, if multiple values are found, choose one rather than returning all of them.
(4) LooseScan: Use Indexes, which help to choose a single value from each group.
(5) *The subquery is materialized into a temporary table using an index and temporary table to perform join.*

After MySQL 5.6.5, optimizer uses Subquery Materialization for efficient subquery processing [16–18].

If materialization is not used, the optimizer sometimes rewrites a noncorrelated subquery as a correlated subquery.

For example, the following `IN` subquery is *noncorrelated* (`where_condition` involves only columns from dept and not emp):

SELECT * FROM emp
WHERE emp.id IN (SELECT dept.id FROM dept WHERE *where_condition*);

The optimizer might rewrite this as an EXISTS correlated subquery: (The subquery here is informed that common rows are only those where inner_expr is equal to outer_expr)

SELECT * FROM emp
WHERE EXISTS (SELECT dept.id FROM dept WHERE *where_condition* AND emp.id = dept.id);

For any such query, MySQL will always scan outermost table first and then will execute the subquery on the Innermost table for each row. If outer table has many rows and inner table has lesser rows than the query will not **be as fast as** it should be.

This analysis shows that MySQL still needs to optimize its Subqueries. MySQL is using rewriting approach for subquery which proves less efficient with multiple tables. As Subquery uses more than one table to fetch the results, there is an immense need to optimize the way tables are getting **joined**. To address this, it is worth exploring work related to join algorithms.

6 Optimizing Join Queries

6.1 *Analysis of Existing Algorithms*

Here we describe various algorithms for calculating the join of relations and then analyse their costs and finally propose our sorting based algorithm and compare it with the existing algorithms.

6.2 *Nested Loop Join Algorithm*

Nested loop join is considered the simplest algorithm to computer the theta join, $r\ \theta\ s$, of two relations *r* and *s*. This algorithm is known as the **nested-loop join** algorithm, as it forms pair of nested **for** loops. Relation r is termed as the outer relation and relation s is termed as the inner relation of the join, since the loop for *r* encloses the loop for *s*. The algorithm uses the notation $tr \cdot ts$, here *tr* and *ts* are tuples; $tr \cdot ts$ denotes the tuple built by integrating the attribute values of tuples *tr* and *ts* [19].

6.3 *Block Nested Loop*

This algorithm shows that each block of innermost relation is joined with each block of the outermost relation. And internally, each tuple in a block is joined with every other block to get the generated pairs. All those pairs which fulfil the join condition are appended to the result block [19].

6.4 *Hash Join Algorithm*

The hash join algorithms are basically used to implement natural and equi-joins. A hash function h is used here to partition the tuples of both the relations. The goal of this function is to partition the tuples of every relation into various other relations that have similar hash value on the joined attributes (Fig. 4).

The above figure depicts the working of the **hash join** algorithm, which is used to derive the natural join of relations *r* and *s*. The merge join algorithm shows tr ∞ ts as the integration of the attributes for tuples tr and ts, which is followed by projecting out repeating attributes. Hence, post partition, a separate indexed nested loop join is performed on every partitioned pairs [19] (Table 1).

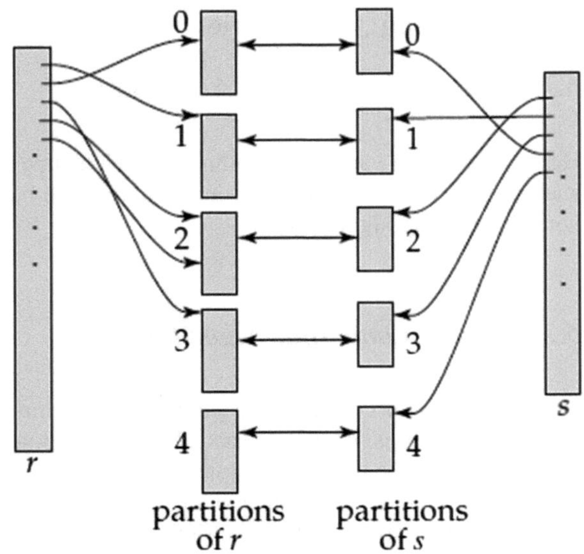

Fig. 4 Hash partitioning of relations

Table 1 Comparison of Join algorithms

Nested loop	Hash join
Here, the outer loop consists of fewer entries and then for each entry, inner loop is processed	The optimizer has to use smaller of 2 tables to build hash table in memory and compares larger tables with hash values to find joined rows
Used for smaller tables	Used for larger tables
Efficient use of Buffer space	Uses Buffer space for temporary tables

6.5 MariaDB and Semijoin Materialization

MariaDB is an enhanced, drop-in replacement for MySQL [9]. Semi-join Materialization is a special kind of subquery materialization used for Semi-join subqueries in MariaDB. Consider a query that finds countries in Europe which have big cities:

```
select * from Country
where Country.code IN (select City.Country
        from City
        where City.Population > 7*1000*1000)
  and Country.continent='Europe'
```

The subquery is uncorrelated, that is, we can run it independently of the upper query. The idea of semi-join materialization is to do just that, and fill a temporary table with possible values of the City.country field of big cities, and then do a join with countries in Europe.

Comparison of Subquery in MySql and MariaDB (Practical implementation)

EXPLAIN SELECT Name FROM Country WHERE Country.Continent = 'Europe' AND Country.Code IN (SELECT City.CountryCode FROM City WHERE City.ID! = Country.Capital AND Population > 100000);

From the above analysis, we can conclude that Mariadb processes very less number of rows compared to MySql. We have attempted to further enhance subquery execution performance by devising algorithm which uses sorting and merging techniques to achieve the goal (Figs. 5, 6).

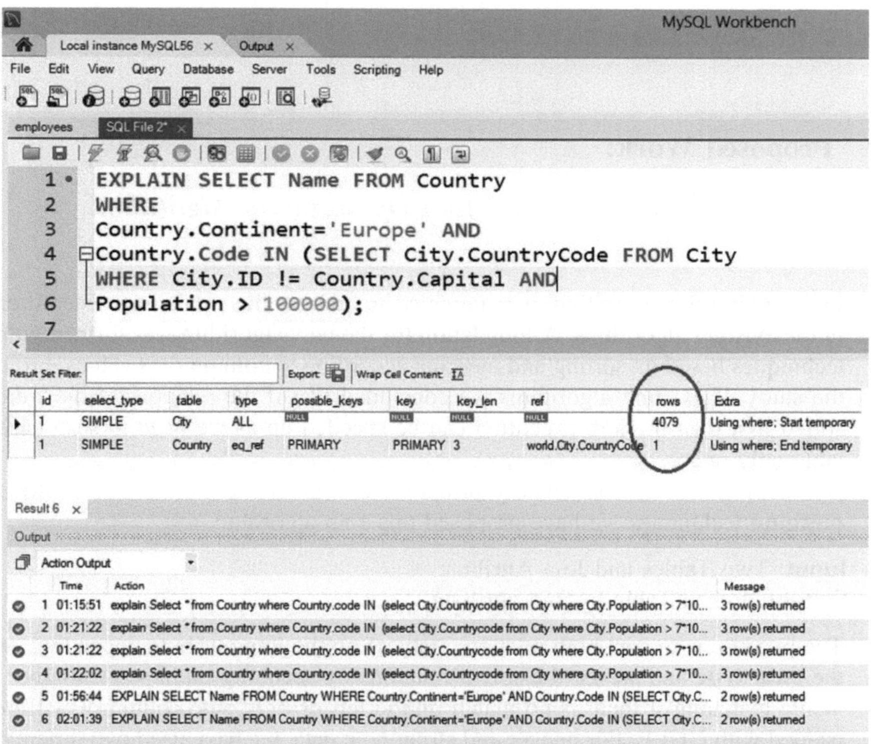

Fig. 5 Subquery in MySql

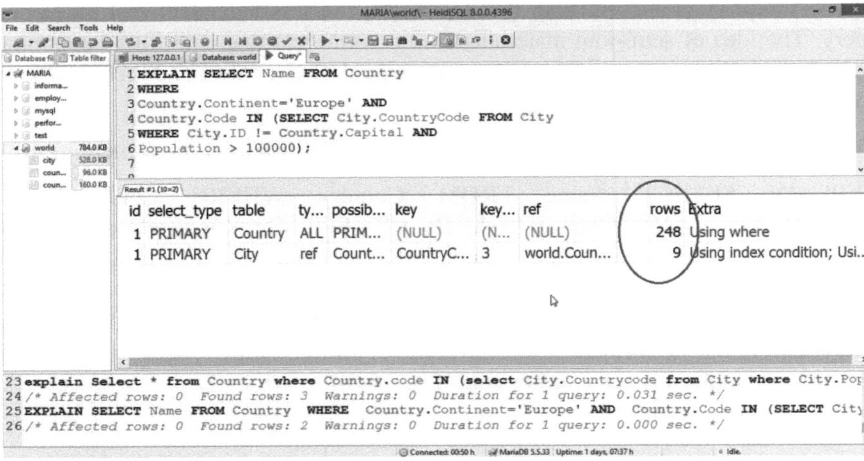

Fig. 6 Subquery in MariaDB

7 Proposed Work:

7.1 Proposed Sorting Based Join Optimization Algorithm

- In designing this algorithm, we attempted to address the weaknesses identified in the existing algorithms. A foundation for the new algorithm was formed with techniques based on sorting and merging according to column data values. From the study of existing algorithms we concluded that if the relations (tables) are sorted first then the join execution can be speeded up as we need to scan each table only once.
- By making use of the DBMS primary key (unique column) and **sorting**, **merging** techniques we have proposed this Join algorithm.

Input: Two Tables and Join Attribute
Output: Joined Table by join attribute

- The purpose is to ***sort*** both relations (tables) on the join attribute depending on the **data type** and ***merge*** the sorted relations by scanning them sequentially.
- While matching if there is no match on the left or right side column it will be padded with NULL. For integer and string type data, we have used two different methods in our function.

7.2 *Benefits of New Algorithm*

- Need to scan each table only once. Hence resources like memory buffer is less used compared to other algorithms.
- Our algorithm can do full outer join, and left or right join unlike other existing algorithms.

8 Implementation Web Based Join Optimization Tool

We intend to propose our algorithm to open source databases. In order to do so, we needed to test our algorithm for its performance. Hence, we created a tool which implements our algorithm with data stored in MySQL.

How the tool works?

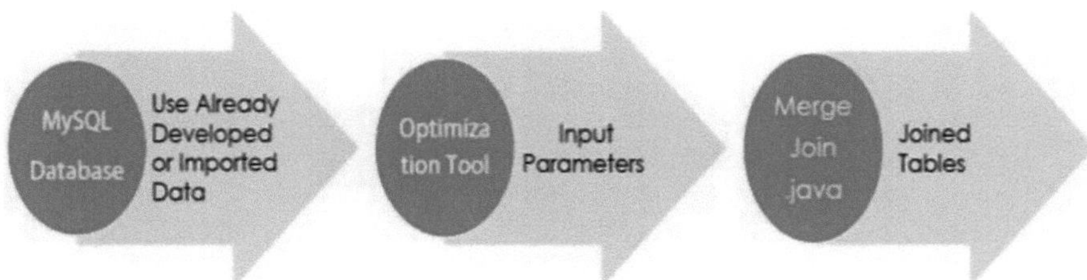

- User can use the **existing databases of MySQL** or create new database which will be stored in MySQL.
- The database tables can be created, data can be inserted and table structure can be defined from tool.
- Once you select the tables that you need to join you can type a query in tool.
- Query will be parsed and related tables are passed as values to our function which has the algorithm implemented within.
- MergeJoin.Java will join the tables and as a result complete Query Execution time along with no. of rows and resultant table will be shown in output (Figures 7, 8, 9, 10 and 11, 12).

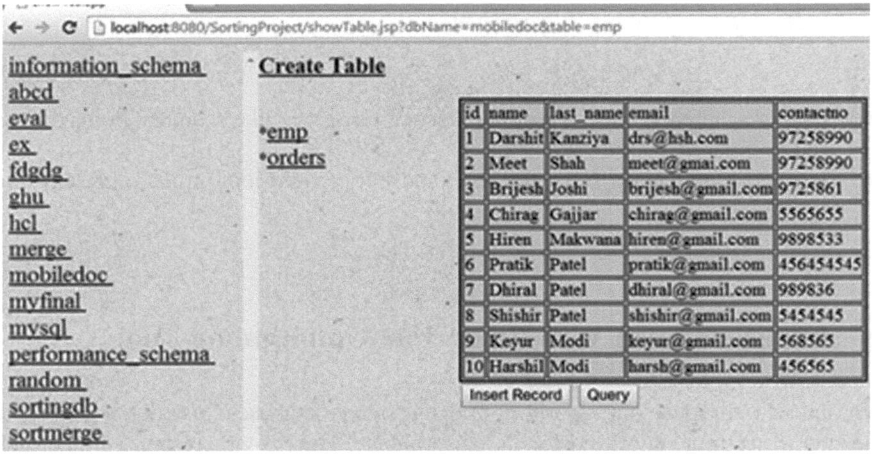

Fig. 7 User can view list of tables and data, once database is selected

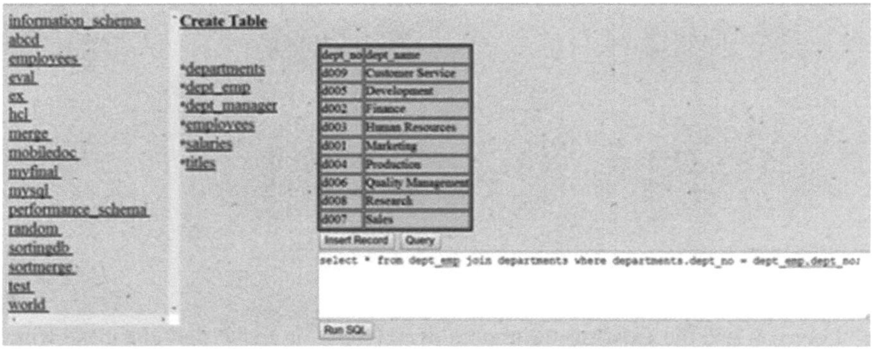

Fig. 8 Query button will allow user to type Join query only. Run SQL button will execute query

9 Result Analysis:

DB used	Time taken
MySQL	0.062 s
MariaDB	0.125 s
Our Optimization Tool	**0.017 s**

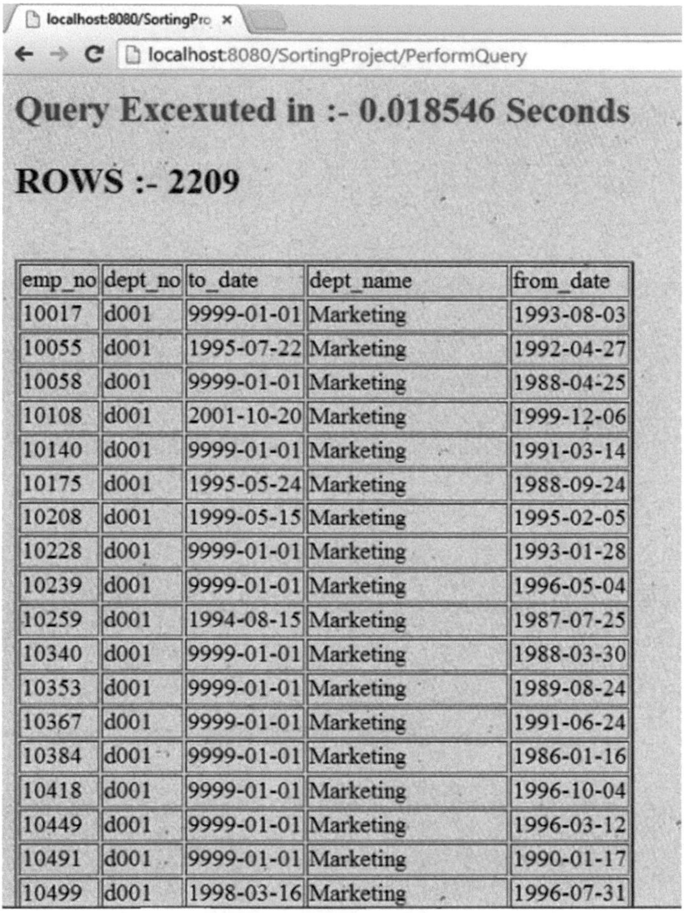

Query Excexuted in :- 0.018546 Seconds

ROWS :- 2209

emp_no	dept_no	to_date	dept_name	from_date
10017	d001	9999-01-01	Marketing	1993-08-03
10055	d001	1995-07-22	Marketing	1992-04-27
10058	d001	9999-01-01	Marketing	1988-04-25
10108	d001	2001-10-20	Marketing	1999-12-06
10140	d001	9999-01-01	Marketing	1991-03-14
10175	d001	1995-05-24	Marketing	1988-09-24
10208	d001	1999-05-15	Marketing	1995-02-05
10228	d001	9999-01-01	Marketing	1993-01-28
10239	d001	9999-01-01	Marketing	1996-05-04
10259	d001	1994-08-15	Marketing	1987-07-25
10340	d001	9999-01-01	Marketing	1988-03-30
10353	d001	9999-01-01	Marketing	1989-08-24
10367	d001	9999-01-01	Marketing	1991-06-24
10384	d001	9999-01-01	Marketing	1986-01-16
10418	d001	9999-01-01	Marketing	1996-10-04
10449	d001	9999-01-01	Marketing	1996-03-12
10491	d001	9999-01-01	Marketing	1990-01-17
10499	d001	1998-03-16	Marketing	1996-07-31

Fig. 9 Joined table output and execution time

From the results, it is clear that MySQL and MariaDB are taking much more time compared to our algorithm based tool.

With different datasets testing was performed and the results are depicted as follows.

Database tables Used with 2000+ records:

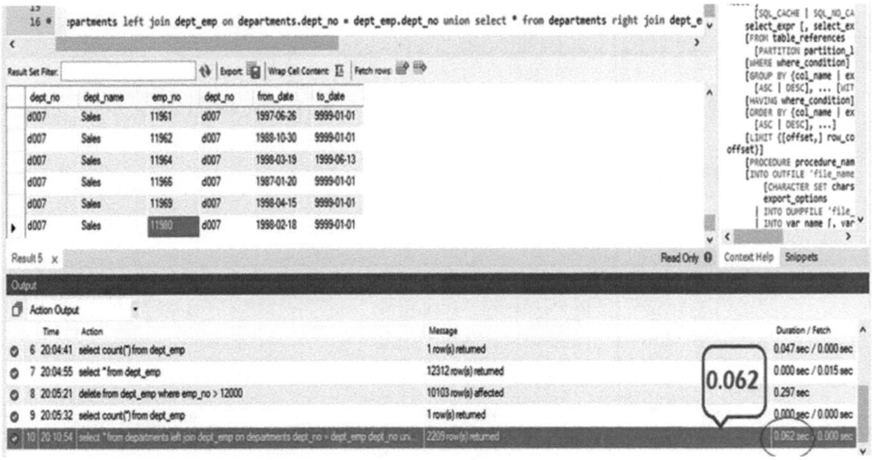

Fig. 10 Same data in MySQL

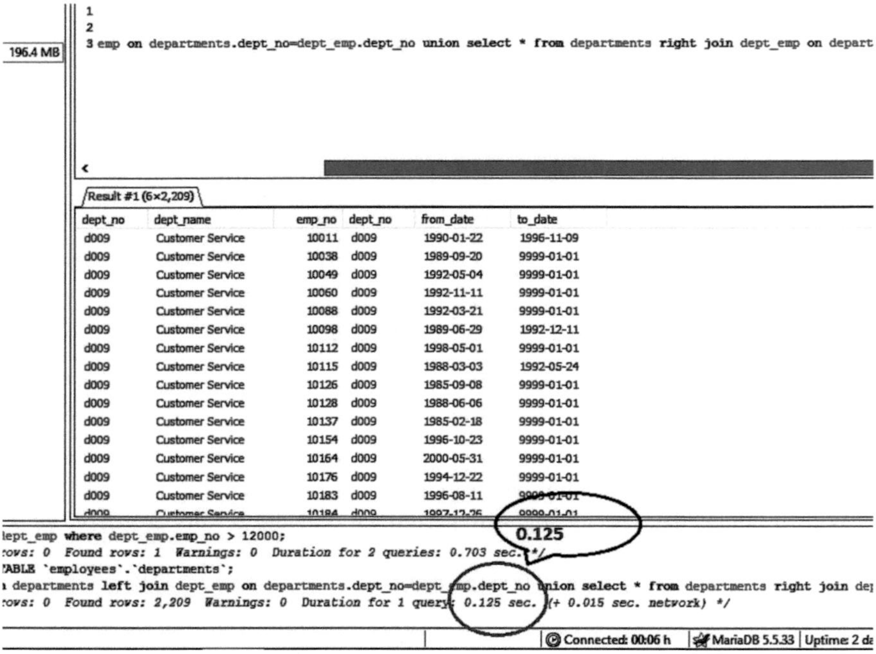

Fig. 11 Same data in MariaDB

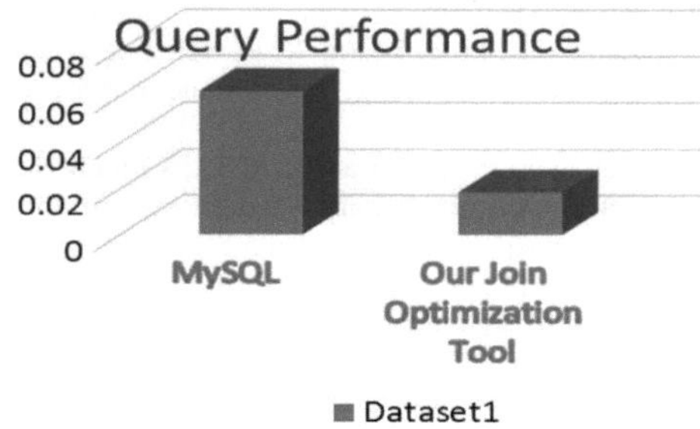

Fig. 12 Comparison of query performance with large datasets

10 Conclusion

MySQL is using nested loop algorithm for join processing which proves to be lesser efficient for larger datasets. MariaDB uses Semi-join Materialization for join processing. We worked on the limitations of the existing algorithms and proposed the new enhanced sorting based join algorithm with techniques like sorting and merging. To test the functionality of the algorithm we implemented web based join optimization tool which is having function implementing our algorithm at its core.

Testing of the tool is done with different datasets from small to large, the result obtained are encouraging which strengthens our belief that our algorithm outperforms existing algorithms. The advanced techniques like hashing, parallel processing etc. may be applied with the algorithm to further enhance the results.

References

1. Connolly, T., Begg, C.: Database Systems: A Practical Approach to Design, Implementation and Management, 3rd edn. Wesley (2011)
2. Johnson, J.L.: Database: Models, Languages, Design. Oxford University Press (1997)
3. Ramakrishnan, R., Gehrke, J.: Database Management Systems, 3rd edn. McGraw Hill (2000)
4. Graefe, G.: Query evaluation techniques for large databases. ACM Comput. Surv. **25**(2) (1993)
5. Ioannidis, Y.: Query Optimization. The Computer Science and Engineering Handbook, pp. 1038–1057 (1997)
6. Giving optimization hints to DB2, IBM, 2003. Available from http://publib.boulder.ibm.com/infocenter/dzichelp/v2r2/index.jsp?topic=/com.ibm.db2.doc.admin/p9li375.html
7. Bruno, N., Chaudhuri, S., Ramamurthy, R.: Power hints for query optimization. In: Proceedings of the International Conference on Data Engineering (ICDE) (2009)
8. Pund, M.A., Jadhao, S.R., Thakare, P.D.: A role of query optimization in relational database. Int. J. Sci. Eng. Res. **2**(1) (2011)
9. Melton, J., Simon, A.: Understanding The New SQL: A Complete Guide. Morgan Kaufman
10. Graefe, G., Dewitt, D.J.: The exodus optimizer generator. In: Proceedings of ACM SIGMOD, San Francisco (1987)
11. http://docs.oracle.com
12. Sudarshan, S., Korth, H.F.: Database System Concepts, 5th edn

13. Sassi, M., Grissa-Touzi, A.: Contribution to the query optimization in the object-oriented databases. Proc. World Acad. Sci. Eng. Technol. **6**, pp. June 2005 1307-6884
14. Ioannidis, Y., Kang, Y.C.: Randomized Algorithms for Optimizing Large Join Queries, March 2013
15. Selinger, P.G., Astrahan, M.M., Chamberlin, D.D., Lorie, R.A., Price T.G.: Access path selection in a relational database system. In: Readings in Database Systems
16. dev.mysql.com/doc/refman/5.5/
17. Kofler, M.: MySQL 5, 3rd edn
18. MySQL Reference Manual, [Online]. Available from http://dev.mysql.com/doc/
19. Ioannidis, Y.E.: Left-deep versus bushy trees: an analysis of strategy spaces and its implications for query optimization. In: ACM SIGMOD International Conference on Management of Data, pp. 168–177 (1991)
20. Chimenti, D., Gamboa, R., Krishnamurthy, R.: Towards an Open Architecture for LDL. In: Proceedings of VLDB, Amsterdam (1989)
21. Heller Stein, J.M.: Predicate migration placement. In: Proceedings of ACM SIGMOD, March 1994
22. Banubakode, A., Haridasa, A.: Adv. Comput. Sci. Technol. **4**(1), 83 (2011)
23. Christodoulakis, S.: Implications of certain assumptions in database performance evaluation. ACM TODS **9**(2) (1984)

Author Index

X.-S. Yang et al. (eds.), *Smart Trends in Systems, Security and Sustainability*, Lecture Notes in Networks and Systems 18,
https://doi.org/10.1007/978-981-10-6916-1

MIX
Papier aus verantwortungsvollen Quellen
Paper from responsible sources
FSC® C105338

If you have any concerns about our products,
you can contact us on
ProductSafety@springernature.com

In case Publisher is established outside the EU,
the EU authorized representative is:
Springer Nature Customer Service Center GmbH
Europaplatz 3, 69115 Heidelberg, Germany

Printed by Libri Plureos GmbH
in Hamburg, Germany